Bioseparation Process Science

BIOSEPARATION PROCESS SCIENCE

Antonio A. García

Matthew R. Bonen

Jaime Ramírez-Vick

Mariam Sadaka

Anil Vuppu

b

**Blackwell
Science**

© 1999 BY BLACKWELL SCIENCE, INC.

Editorial Offices:
Commerce Place, 350 Main Street, Malden, Massachusetts 02148, USA
Osney Mead, Oxford OX2 0EL, England
25 John Street, London WC1N 2BL, England
23 Ainslie Place, Edinburgh EH3 6AJ, Scotland
54 University Street, Carlton, Victoria 3053, Australia
Other Editorial Offices:
Blackwell Wissenschafts-Verlag GmbH, Kurfürstendamm 57, 10707 Berlin, Germany
Blackwell Science KK, MG Kodenmacho Building, 7-10 Kodenmacho Nihombashi, Chuo-ku, Tokyo 104, Japan

Distributors:

USA   Blackwell Science, Inc.
             Commerce Place
             350 Main Street
             Malden, Massachusetts 02148
             (Telephone orders: 800-215-1000 or 781-388-8250; fax orders: 781-388-8270)
Canada   Login Brothers Book Company
             324 Saulteaux Crescent
             Winnipeg, Manitoba, R3J 3T2
             (Telephone orders: 204-224-4068)
Australia   Blackwell Science Pty, Ltd.
             54 University Street
             Carlton, Victoria 3053
             (Telephone orders: 03-9347-0300; fax orders: 03-9349-3016)
Outside North America and Australia
             Blackwell Science, Ltd.
             c/o Marston Book Services, Ltd.
             P.O. Box 269
             Abingdon
             Oxon OX14 4YN
             England
             (Telephone orders: 44-01235-465500; fax orders: 44-01235-465555)

Acquisitions: Nancy Hill-Whilton
Production: Erin Whitehead
Manufacturing: Lisa Flanagan
Typeset by Best-set Typesetter Ltd., Hong Kong
Printed and bound by Edwards Brothers, Inc., Lillington, North Carolina
Printed in the United States of America
99 00 01 02 5 4 3 2 1

The Blackwell Science logo is a trade mark of Blackwell Science Ltd., registered at the United Kingdom Trade Marks Registry

Library of Congress Cataloging-in-Publication Data
Bioseparation process science / [Antonio A. Garcia . . . et al.].
        p. cm.
    Includes bibliographical references and index.
    ISBN 0-86542-568-X
    1. Biomolecules—Separation.   2. Biochemical engineering.
  3. Separation (Technology)   I. Garcia, Antonio A.
  TP248.25.S47B54   1999
  660'2842—DC21
                                                98-41792
                                                CIP

CONTENTS

Preface xv

1. Introduction 1
   1.1   Mass Conservation as an Accounting Method 1
   1.2   Interpreting Differentials and Integrals: World
        Population Statistics 3
   1.3   Accounting for Diffusion, Convection, and Reaction for
        Mass Conservation: The Microscopic Scale 5
   1.4   Summary 7
   1.5   References 7

Part I. Commercial Bioseparations and Product Measurement 9

2. Industrial Bioseparation Processes 11
   2.1   Bioseparation Process Selection 11
        2.1.1 Scale, Concentration, and Price 13
        2.1.2 Product Properties 14
   2.2   Monoclonal Antibodies 16
   2.3   Human Insulin 17
   2.4   Rabies Vaccine 17
   2.5   Penicillin 19
   2.6   Protease 20
   2.7   L-lysine 21
   2.8   Citric Acid 22
   2.9   Summary 23
   2.10 Problems 24
   2.11 References 28

3. Concentration Determination and Bioactivity Assays 29
   3.1   Amino Acids 29
        3.1.1 High-Performance Liquid Chromatography 31
        3.1.2 Reverse-Phase High-Performance Liquid
            Chromatography 32
        3.1.3 Capillary Electrophoresis 32
        3.1.4 Micellar Electrokinetic Chromatography 33
        3.1.5 Electrodialysis 34
        3.1.6 Gas Chromatography 35
   3.2   Peptides and Proteins 35
        3.2.1 Analytical Chromatography 35
        3.2.2 Analytical Electrophoresis 37
        3.2.3 Immunoassays 40

3.3 Nucleic and Polynucleic Acids 41
     3.3.1 Ion-Exchange Chromatography 41
     3.3.2 Reverse-Phase High-Performance Liquid
           Chromatography 42
     3.3.3 Ion-Pair Chromatography 43
     3.3.4 Slalom Chromatography 43
     3.3.5 Gel Electrophoresis 44
     3.3.6 Pulsed-Field Gel Electrophoresis 45
     3.3.7 Capillary Isotachophoresis 45
     3.3.8 Capillary Zone Electrophoresis 45
3.4 Carbohydrates 46
     3.4.1 Monosaccharides 46
     3.4.2 Oligosaccharides 49
     3.4.3 Glycoproteins 51
3.5 Lipids 51
     3.5.1 Fatty Acids 51
     3.5.2 Fats and Oils 56
3.6 Steroids and Antibiotics 56
3.7 Vitamins 57
3.8 Summary 58
3.9 Problems 59
3.10 References 61

Part II. Application of Chemical, Physical, and Biological Properties
         to Bioseparations 65

4. Thermodynamic and Transport Properties 67
4.1 Chemical Equilibria 67
4.2 Solubility 69
     4.2.1 Protein and Amino Acid Solubility 70
4.3 Diffusivity 73
     4.3.1 Uncharged Low-Molecular-Weight Biochemicals ... 73
     4.3.2 Proteins 73
4.4 Isoelectric Points and Charge Dependence on pH 74
     4.4.1 Carboxylic Acids 74
     4.4.2 Amino Acids 76
     4.4.3 Proteins 82
4.5 Hydrophobicity–Hydrophilicity Scales 84
4.6 Acid-Base Scales 84
     4.6.1 Gutmann Donor-Acceptor Theory 84
     4.6.2 Drago E&C Equation 86

      4.6.3 Solvatochromatic Comparison Method 86

      4.6.4 Hard and Soft Acid and Base Theory 86

      4.6.5 Comparison and Correlation of Different Scales 88

  4.7 Metal Ion Binding Constants 88

      4.7.1 Nucleic Acids 89

      4.7.2 Amino Acids 89

  4.8 Summary 90

  4.9 Problems 91

  4.10 References 93

5. Biocolloidal Interactions and Forces 95

  5.1 Short-Range Interactions 95

  5.2 Long-Range Interactions 96

      5.2.1 Van der Waals Forces 96

      5.2.2 Electrostatic Interactions and DLVO Theory 100

      5.2.3 Hydrophobic Effects 102

      5.2.4 Magnetic Interactions 103

  5.3 Summary 104

  5.4 Problems 105

  5.5 References 106

6. Bioaffinity 108

  6.1 Molecular Recognition Processes 108

  6.2 Receptor-Ligand Interactions 110

      6.2.1 Ionic Bonds 110

      6.2.2 Hydrogen Bonds 110

      6.2.3 Hydrophobic Interactions 111

      6.2.4 Van der Waals Forces 111

  6.3 Theoretical Aspects of Receptor-Ligand Affinity 111

      6.3.1 Thermodynamic Approach 112

      6.3.2 Equilibrium Approach 112

  6.4 Specific Interactions 115

      6.4.1 Antibody-Antigen Interactions 116

      6.4.2 DNA-Protein Interactions 117

      6.4.3 Cell Receptor-Ligand Interactions 119

      6.4.4 Enzyme-Substrate Interactions 120

      6.4.5 Biotin-Avidin/Streptavidin Interactions 121

      6.4.6 Lectin-Carbohydrate Interactions 122

  6.5 Summary 122

  6.6 Problems 123

  6.7 References 124

Part III. Bioseparation Methods 125

7. Crystallization and Precipitation 127
  7.1 Saturation and Supersaturation 127
  7.2 Nucleation Phenomena 128
  7.3 Growth of Crystals 130
  7.4 Batch Crystallization 131
      7.4.1 Solution Balance 132
      7.4.2 Solid-Phase Balance 132
      7.4.3 Crystal Size Distribution 134
      7.4.4 Organic Solvent and Salt Precipitation 136
      7.4.5 Growth Rate Dispersion 137
  7.5 Continuous Crystallization 140
  7.6 Yield 141
      7.6.1 Removal of Solvent and Diluent 141
  7.7 Summary 142
  7.8 Problems 142
  7.9 References 145

8. Membrane Filtration 146
  8.1 Membrane Materials 146
  8.2 Driving Forces in Membrane Separations 147
  8.3 General Theory of Microfiltration 147
      8.3.1 Incompressible Cakes 148
      8.3.2 Compressible Cakes 149
  8.4 Microfiltration 149
      8.4.1 Staging in Microfiltration 151
  8.5 Ultrafiltration 152
      8.5.1 Ultrafiltration Process Application 154
      8.5.2 Ultrafiltration Membrane Application
           and Modification 156
  8.6 Reverse Osmosis 156
  8.7 Flux Equations 158
  8.8 Electrodialysis 158
  8.9 Emulsion Liquid Membranes 159
  8.10 Summary 160
  8.11 Problems 160
  8.12 References 163

9. Centrifugation 164
  9.1 Governing Principles 164
  9.2 Advantages and Disadvantages of Centrifugation 166

9.3	Selection of Centrifuges	166
9.4	Types of Centrifuges	168
	9.4.1 Tubular Bowl Centrifuges	169
	9.4.2 Disc-Type Centrifuges	170
	9.4.3 Batch-Basket Centrifuges	172
9.5	Industrial-Scale Centrifugation	173
9.6	Summary	176
9.7	Problems	176
9.8	References	177

10. Chromatography 178

10.1	Detection Methods	178
10.2	Summary of the Types of Chromatography	181
10.3	Stationary Phases	183
10.4	Six Ways to Analyze Chromatographic Processes	187
	10.4.1 Gaussian Solution	187
	10.4.2 Staged Models	190
	10.4.3 Newtonian Continuum Mechanics and Linear Equilibria	196
	10.4.4 Constant Pattern and Saturation Equilibria	200
	10.4.5 Van Deemter Equation	205
	10.4.6 Gel Partitioning Model	205
10.5	Gel-Permeation Chromatography	206
10.6	Ion-Exchange Chromatography	208
10.7	Affinity Chromatography	213
10.8	Hydrophobic Interaction and Reverse-Phase Chromatography	216
10.9	Perfusion Chromatography	216
10.10	Other Chromatographic Methods	217
	10.10.1 Gradient Methods	217
	10.10.2 Displacement Chromatography	218
	10.10.3 Radial-Flow Chromatography	218
	10.10.4 Membrane Chromatography	218
10.11	Scale-Up Strategies and Considerations	219
	10.11.1 Scale-Up Method 1: No Change in Stationary-Phase Particle Size	220
	10.11.2 Scale-Up Method 2: Increasing Stationary-Phase Particle Size	221
	10.11.3 Scale-Up Method 3: Gel Permeation and On-Off Cycling Approach	222

10.12 Summary 223
10.13 Problems 224
10.14 References 228

11. Extraction 230
11.1 Chemical Thermodynamics of Partitioning 230
11.2 Organic-Aqueous Extraction 231
    11.2.1 Extractant/Diluent Systems 233
    11.2.2 Removing Biochemicals from the Organic Phase 237
11.3 Two-Phase Aqueous Extraction 238
    11.3.1 Partitioning Due to Size 239
    11.3.2 The Effect of Protein Charge on Partitioning 240
    11.3.3 Other Effects 241
11.4 Reverse Micelles 243
11.5 Supercritical Fluids 244
11.6 Large-Scale Vessels for Extraction 246
    11.6.1 Mixer-Settlers 246
    11.6.2 Extraction Columns 247
    11.6.3 Centrifugal Contactors 250
    11.6.4 Comparison 251
11.7 Configurations for Stage-Wise Contacting 252
    11.7.1 Cocurrent Contacting 252
    11.7.2 Crosscurrent Contacting 253
    11.7.3 Countercurrent Contacting 254
    11.7.4 A Comparison of Contacting Modes 255
    11.7.5 Graphical Solution 263
    11.7.6 Fractional Extraction 265
    11.7.7 Continuous Countercurrent Extraction 269
11.8 Summary 270
11.9 Problems 271
11.10 References 276

12. Electrophoresis 277
12.1 A Brief Introduction to Some Popular
    Electrophoretic Methods 277
    12.1.1 Gel Electrophoresis 278
    12.1.2 Capillary Electrophoresis 281
    12.1.3 Isoelectric Focusing 282
    12.1.4 Isotachophoresis 283
    12.1.5 Moving Boundary 283
12.2 Basic Concepts of Electrophoresis 283

12.2.1 Electro-osmosis and the Relaxation Effect as
Retardation Forces 286
12.2.2 Situations That Can Hamper Electrophoretic
Separation 286
12.3 Zone Electrophoresis 287
12.3.1 Band Dispersion 288
12.4 Isoelectric Focusing 290
12.5 Isotachophoresis 291
12.6 Two-Dimensional Electrophoresis 292
12.7 Summary 294
12.8 Problems 294
12.9 References 298

13. Magnetic Bioseparations 299
13.1 Magnetic Properties of Materials 299
13.2 Magnetic Particle Classification 305
13.3 Theoretical Considerations 306
13.4 Magnetic Particle Separations 308
13.4.1 High-Gradient Magnetic Separations 309
13.4.2 Affinity Chromatography 310
13.4.3 Aqueous Two-Phase Separations 311
13.5 Applications 312
13.5.1 Cell Separation 312
13.5.2 Immunoassays 313
13.6 Summary 313
13.7 Problems 313
13.8 References 314

14. Solvent Removal and Drying 315
14.1 Methods of Solvent Removal 315
14.2 Theory 316
14.2.1 Vapor-Liquid Systems 317
14.2.2 Liquid-Liquid Systems 321
14.2.3 Liquid-Solid Systems 323
14.3 Rayleigh Distillation 325
14.4 Equipment 327
14.4.1 Evaporation 327
14.4.2 Drying 331
14.5 Summary 334
14.6 Problems 334
14.7 References 336

15. Cell Disruption 337

   15.1   Cells and Cell Membranes 337

   15.2   Cell Disruption Techniques 339

           15.2.1 Mechanical Cell Disruption 340

           15.2.2 Chemical Cell Disruption 348

   15.3   Summary 350

   15.4   Problems 351

   15.5   References 352

Part IV. Bioprocess Synthesis 355

16. Integration of Individual Separation Steps 357

   16.1   Bioseparation Process Heuristics 357

           16.1.1 Reduce Volume Early in the Process Sequence 358

           16.1.2 Save the Most Expensive Step for Last 358

           16.1.3 Follow the KISS Principle 360

           16.1.4 Resolve Components Well as Early as Possible 361

           16.1.5 Minimize Inhibition Mechanisms in the Bioreactor 362

   16.2   Issues in Concurrent Bioseparation and Bioreactor

         Process Development 362

           16.2.1 Take the Lab-Scale Process and Scale It Directly

                 with No Changes 362

           16.2.2 Design a Bioseparation Process Based on the Closest

                 Existing Commercial Product 363

           16.2.3 Pilot-Scale Experimentation with "Spiked"

                 Bioreactor Fluid 363

   16.3   Expert Systems in Process Synthesis 364

   16.4   Integration of Bioreaction and Bioseparation Steps 364

   16.5   Making the Bioreactor Step Bioseparation-Friendly 366

   16.6   Considerations in Final Product Formulation and

         Environmental Impact 367

   16.7   Summary 368

   16.8   Problems 369

   16.9   References 371

17. Production Formulation 372

   17.1   Formulation Characteristics 372

   17.2   Excipients 373

           17.2.1 Thickeners and Binders 373

           17.2.2 Surface-Active Agents 374

           17.2.3 Colors and Flavors 374

           17.2.4 Preservatives 374

17.3   Dosage Forms 375
17.4   Encapsulation 375
17.5   Freeze Drying 377
       17.5.1 Theory 378
       17.5.2 Technique 381
17.6   Summary 383
17.7   Problems 383
17.8   References 383

18. Bioprocess Economics 385
18.1   Resources Available for Cost Estimation 385
       18.1.1 Capital Cost Estimation 386
       18.1.2 Operating Cost Estimation 388
18.2   Economic Decision-Making Models 389
       18.2.1 Internal Rate of Return 391
       18.2.2 Payback Period, Including Interest 391
       18.2.3 Net Present Value 391
       18.2.4 Return on Investment 392
       18.2.5 Choosing Among Projects and Alternative
              Investments 394
18.3   Sensitivity Analyses 394
18.4   Summary 398
18.5   Problems 398
18.6   References 399

Appendix A. The Laplace Transform 400
**Appendix B. Numerical Inversion, van der Laan's Theorem, and Huchel
              and Helmholtz-Smoluchowski Equations** 405
Index 409

PREFACE

In this textbook, the word *bioseparations* is used in the context of biochemical engineering. In this context, the term refers to separation and purification methods for such biological products as biochemicals, proteins, polynucleic acids, and cells. The purpose of this text is to provide students and practitioners of engineering and science with a framework for decision making in the design of bioseparation processes.

Chapter 1 introduces some of the basic concepts central to understanding the remainder of the book. Part I of the text then moves on to discuss some illustrative biotechnology industrial processes and provides the reader with practical information on standard analytical methods. The two chapters that constitute Part I are important bookends covering the full industrial and analytical scale of bioseparations. Chapter 2 covers a broad range of important industrial bioseparation processes, stressing an overview of the arrangement of individual steps in the purification of the final product. Chapter 3 provides a useful discussion of analytical methods, a topic too often ignored in biochemical engineering texts. Process evaluation, however, cannot be carried out without analytical support and analytical methods, which are often emulated and redesigned for commercial-scale production.

Part II touches on several challenging subject areas, describing physical, chemical, and biological interactions with an eye toward exploiting these effects for bioseparations. Physical and chemical interactions are normally covered in some depth in engineering textbooks, and biological interactions are the focus of biochemistry texts. Within this section of the book, we present these different views simultaneously so as to give the student and the practitioner a more complete set of tools with which to design purification and separation processes.

Part III deals with commonly employed "unit" operations (that is, process steps). Its organization parallels the format used in many biochemical engineering textbooks, though we have made an effort to keep the number of symbols and parameters to a minimum so as to focus on the phenomena themselves. In addition, the use of differential calculus is primarily confined to standard first- and second-semester calculus topics, with the exception of the use of Laplace transforms (an orientation to Laplace transforms is provided in the appendices). Our goal in reducing the number of symbols and the level of math complexity is to lower the barriers erected by the use of excessive mathematical jargon and specialized techniques in the multidisciplinary environment of industrial biotechnology. Readers interested in more mathematical content are encouraged to review the specialized textbooks and references cited throughout this book.

Part IV, the final section of the text, covers several key topics necessary to begin the creative process of synthesizing a biological separation process flow diagram. One of the most important tools for designing such processes is computer software, which provides a built-in wealth of expert information. Part IV acknowledges the efficiency and effectiveness of using these tools by *not* subjecting the reader to pencil-and-paper methods that serve only to reinvent the capabilities of currently available software. Once the underlying principles and individual operation analyses of Parts II and III are covered, the practicing bioseparation process designer can learn to use the software design tools presented in Part IV, gaining an understanding of the underlying economics, process integration challenges, and final product formulation issues affecting bioseparations in the real world.

ACKNOWLEDGMENTS

The authors would like to thank the many people who contributed to this book. First, we would like to acknowledge the ideas and technical contributions made by Dong-Hoon Kim while he was a doctoral student at Arizona State University. We also thank the many students who took the Bioseparations course at ASU and provided helpful comments and ideas, especially Otute Akiti, Maria Jaya, Sanjay Agarwal, Himanshu Ranpura, Surasit Chungpaiboonatana, Jing Wu, Daewon Yang, Taehoon Kim, and Gavin Price.

Additionally, we are indebted to Maya Crosby, the energetic editor who solicited the initial book concept and guided the early stages of the project, and to Nancy Hill-Whilton, who has been a most patient and skillful senior editor during the final stages of the book. Our thanks also to Jill Connor who arranged for the reviews and provided continual, enthusiastic support during the many stages needed to complete the book.

Antonio García is very grateful to have a supportive family who has been understanding during the preparation of this book. He especially thanks his wife Beatriz for her care and support. Also, he is thankful for the love and enjoyment of life that his two daughters Rebecca and Jessica share with him every day.

Jaime Ramírez-Vick thanks his wife Nissy for her support and understanding during the writing of this book.

Matthew Bonen wishes to thank Catherine Sheldon for her invaluable support and the occasional loan of her exceptional writing skills.

1

Introduction

Throughout this textbook, the reader will encounter the application of calculus, which is usually associated with the principle of conservation of mass. Traditionally, the use of calculus in textbooks signals that the book is directed toward readers who are studying engineering or who have an engineering background. The approach taken in this book, however, will be to apply useful mathematics to the analysis of bioseparation processes in such a way as to permit access by a wider scientific audience. Both this introductory chapter and Appendix A are designed to help bridge the so-called engineering–mathematics gap—that is, to make the coverage of these topics suitable for a reader who has completed one semester of calculus and physics. Moreover, we have kept the number of defined terms to a minimum whenever possible, and avoided the use of intricate mathematical methods so as to present a uniform and straightforward use of analysis tools.

This chapter deals with three analysis tools applicable to bioseparation processes:
- The principle of mass conservation as an accounting method
- The use of differentials and integrals to help answer important questions
- Chemical species accounting for processes involving diffusion, convection, or reaction

Appendix A introduces the use of Laplace transforms as a convenient method for solving the differential equations with which we model separation processes. The reader can skip this chapter if desired, returning to it as questions arise on the use of mathematics in our subsequent discussions of chromatography, solvent extraction, and crystallization.

1.1 MASS CONSERVATION AS AN ACCOUNTING METHOD

Science students will be familiar with the oft-quoted law of classical physics that states that "mass is neither created nor destroyed." This statement of mass conservation holds unless a nuclear reaction occurs, in which case Einstein's equation relating mass to energy must then be employed. For the purpose of bioseparations, this law holds true in every situation. Thus, in bioseparation process science, the law of mass conservation provides a basis for writing mathematical expressions to help predict the outcome for design variables based on the nature of the process.

The application of the law of mass conservation is entirely analogous to financial accounting or any other type of accounting. An anecdote will illustrate how to perform such mass accounting. Imagine that a professor brings a closed cardboard box and a basket of balls into class. She then requests that a student come to the front of the room to help in adding and removing balls to and from the box. The professor opens the box and allows the student to start by adding eight balls. Then she has him remove two balls. The rest of the class is charged with keeping track of the number of balls being added to and removed from the box.

After several rounds of adding or removing balls from the box, the professor asks the class how many balls are in the box. Nearly everyone in the room answers the professor in unison. The professor, however, tells them that they are all wrong. After checking their tallies, the class members are even more firmly convinced that they have the right number. Yet the professor still claims that they are wrong. Finally, the class has become thoroughly incensed and accuses the professor of not being able to count. "Of course, you have the wrong answer," she exclaims. One student asks, "How can that be? We carefully tracked the number of balls that were added and subtracted, and then double-checked our answer." "But," says the professor, "you did not know that there were four balls in the box when I brought it into class."

This anecdote illustrates the importance of carefully and explicitly constructing an accounting procedure for mass. The proper way of dealing with accounting for the number of balls in the box at any time is to write a procedure, such as the following:

$$\begin{bmatrix} \text{Number of} \\ \text{balls in box} \\ \text{at any time} \end{bmatrix} = \begin{bmatrix} \text{number of} \\ \text{balls in box} \\ \text{initially } (t = 0) \end{bmatrix} + \begin{bmatrix} \text{number of balls} \\ \text{added since the} \\ \text{beginning } (t > 0) \end{bmatrix} - \begin{bmatrix} \text{number of balls} \\ \text{removed since the} \\ \text{beginning } (t > 0) \end{bmatrix} \tag{1.1}$$

A mathematical equation can be created from Equation 1.1 when $b(t)$ is defined as the number of balls in the box at any time, $b(0)$ is the number of balls in the box initially, b_{in} is the number of balls put into the box, and b_{out} is the number of balls removed from the box. After substituting these expressions into Equation 1.1, we have the following equation:

$$b(t) = b(0) + b_{in} - b_{out} \tag{1.2}$$

In most cases, it is more useful to track rates than absolute numbers. For example, a person may want to know how fast his net worth is changing over time. This rate is especially important if the person wants to see whether bankruptcy is imminent or whether he is saving enough for retirement. In that case, one equation could satisfy this need:

$$\begin{bmatrix} \text{Rate of accumulation} \\ \text{of dollars (\$/month)} \end{bmatrix} = \begin{bmatrix} \text{salary rate after} \\ \text{taxes (\$/month)} \end{bmatrix} + \begin{bmatrix} \text{net rate of cash flow due} \\ \text{to investments (\$/month)} \end{bmatrix}$$
$$- \begin{bmatrix} \text{expense rate} \\ \text{(\$/month)} \end{bmatrix} \tag{1.3}$$

For our simpler box problem, we can also change the accounting method to a rate:

$$\begin{bmatrix} \text{Rate of accumulation} \\ \text{of balls (\#/minute)} \end{bmatrix} = \begin{bmatrix} \text{rate of balls into} \\ \text{box (\#/minute)} \end{bmatrix} - \begin{bmatrix} \text{rate of balls out of} \\ \text{box (\#/minute)} \end{bmatrix} \tag{1.4}$$

This expression can be written in terms of a differential equation,

$$\frac{db}{dt} = \dot{b}_{\text{in}} - \dot{b}_{\text{out}} \tag{1.5}$$

where the dots above the letter b refer to the rate. Note that we can integrate Equation 1.5 to obtain a solution when we know the number of balls that are in the box initially. In this case, we have the following equation:

$$\int_{b(0)}^{b(t_{\text{end}})} db = \int_{t=0}^{t=t_{\text{end}}} (\dot{b}_{\text{in}} - \dot{b}_{\text{out}}) dt \tag{1.6}$$

Equation 1.6 can be greatly simplified if the rates at which balls are put into or taken out of the box are averaged over the time during which the class demonstration is performed:

$$b(t_{\text{end}}) - b(t = 0) = (\dot{b}_{\text{in}} - \dot{b}_{\text{out}}) t_{\text{end}} \tag{1.7}$$

It may seem that this section began with the simple idea of counting balls going into and out of a box and ended in a rather complex equation. Equation 1.7, however, has general utility. The value in creating such generalized equations lies in our ability to then use them for design purposes. Although the design of boxes to hold balls might seem trivial, this knowledge would help the professor plan how many balls to take out and put in based on the number of balls she had initially. If one constraint is that the box can hold only 123 balls, Equation 1.7 can help her determine the acceptable average rates for adding and removing balls every minute. Also, this equation can help ensure that some balls are left in the box during the course of the demonstration.

In the next section, we discuss a more practical use of differential and integral calculus for analyzing data.

1.2 INTERPRETING DIFFERENTIALS AND INTEGRALS: WORLD POPULATION STATISTICS

The world population is a statistic that involves everyone. Figure 1.1 provides the most recent world population data (1), as well as historical data. Two important useful analyses can be conducted with these data using the definitions of derivatives and integrals. The analysis using derivatives draws attention to how the rate of population growth changes with time, while the analysis using integrals illustrates the cumulative impact of the world population on resources since 1 A.D.

To determine the rate of increase for the number of people per time, or dP_{pop}/dt where P_{pop} represents the world population, we can determine the slope of the curve at a particular date by using graphical methods. An easier solution, however, is to fit the data to an equation. Thomas Malthus, an English clergyman and political economist, predicted that the world's population would grow more rapidly than the food supply. Malthus' law, therefore, states that the rate of population growth is proportional to population:

$$\frac{dP_{\text{pop}}}{dt} = aP_{\text{pop}} \tag{1.8}$$

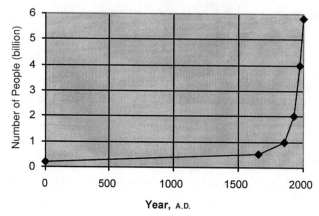

Figure 1.1 World population over time.

An exponential equation can be used to fit this type of behavior:

$$P_{pop}(t) = b \exp(at) + P_{pop}(t = 1 \text{ A.D.}) \tag{1.9}$$

Plotting the population data as a semilog plot shows that the slope changes radically between 1650 and 1850 A.D. (Figure 1.2). Fitting a linear equation for the population between 1800 and 1997 A.D. gives the following:

$$P_{pop}(1997 > t > 1800) = \exp(0.01121t) \tag{1.10}$$

Using Equation 1.10 (or Figure 1.1), we can easily see how the rate of population growth has changed over the years. From the year 1 A.D. until roughly 1800 A.D., the rate of population growth did not increase nearly as dramatically as it did from 1850 to 1997 A.D. Thus differential calculus supplies a compact method of analyzing how this rate has changed and suggests that Malthus' law can be applied; it is also clear, however, that the proportionality constant has changed over the time frame for which data are available. The important implication is that population growth has so far remained unchecked. If the current trend continues, the world population will reach 9.6 billion in 2050—a 60% increase from the 1997 population.

Another useful method for analyzing population data relies on the use of integrals. We can integrate the data in Figure 1.2 by determining the area under the curve. A simple estimation can be performed by integrating Equation 1.11 to give Equation 1.12:

$$\int_{t=1}^{t=1997} P_{pop} dt = \int_{t=1}^{t=1997} (\exp(0.01121t) + 200,000,000) dt \tag{1.11}$$

$$\int_{t=1}^{t=1997} P_{pop} dt = \left[\frac{\exp(0.01121t)}{0.01121} + 200,000,000\, t \right]_{t=1}^{t=1997} \tag{1.12}$$

The result of this integration is 8.7×10^{11} people-years. This value can be used to roughly estimate the overall amounts of food, water, and other necessities that have been provided to people throughout history and the total production of wastes gen-

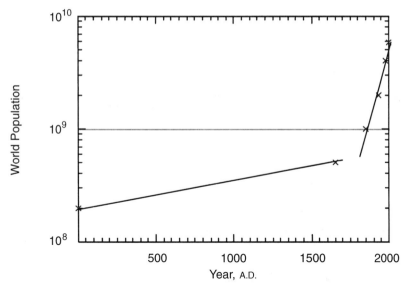

Figure 1.2 World population on a semilog plot showing two different slopes.

erated, assuming average per-person values. In essence, integrating the population data enables us to project the total ecological burden of the human population on the Earth if Malthus' law continues to apply.

1.3 ACCOUNTING FOR DIFFUSION, CONVECTION, AND REACTION FOR MASS CONSERVATION: THE MICROSCOPIC SCALE

When accounting for the transfer of chemical species in bioseparations, we can analyze several basic phenomena by using a mathematical approach similar to that demonstrated earlier in this chapter. At the microscopic scale, a particular bio-chemical compound can be tracked in terms of how it moves through the surface of a "control volume." A *control volume* is a three-dimensional, infinitesimally small piece of space; here it is defined in terms of Cartesian coordinates (Figure 1.3).

The accounting of mass can be carried out by employing the physical laws and chemical transformations that govern motion and dictate reactions. Three basic mechanisms will be discussed in more detail in this textbook: diffusion, convection, and chemical reaction. We measure the rates of these phenomena by using a different manifestation of the law of conservation of mass:

$$
\begin{bmatrix} \text{Accumulation rate} \\ \text{of species A in} \\ \text{control volume} \end{bmatrix} = \begin{bmatrix} \text{net diffusion rate} \\ \text{of species A into} \\ \text{control volume} \end{bmatrix} + \begin{bmatrix} \text{net rate of forced} \\ \text{convection of species} \\ \text{A into control volume} \end{bmatrix}
$$

$$
- \begin{bmatrix} \text{reaction rate of} \\ \text{species A within} \\ \text{control volume} \end{bmatrix}
$$

$$(1.13)$$

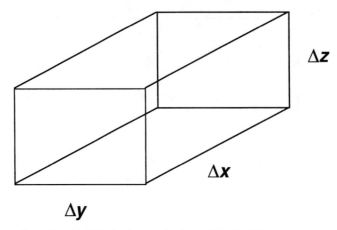

Figure 1.3 Three-dimensional infinitesimal control volume, $\Delta V = \Delta x \Delta y \Delta z$.

To convert Equation 1.13 into a mathematical equation, we must account for all directions by which the biochemical species A can enter and leave the control volume and identify the rate at which species A reacts. Most often, a useful bioseparation device limits the directions in which species can move and so can be described by relatively simple reaction kinetics. Biochemical species can diffuse through a control volume because of the random motion of molecules; in contrast, the convection of species occurs when a fluid flows through the control volume. A chemical reaction may also occur within the control volume. Separate forms of Equation 1.13 are required to account for each of these three mechanisms.

At this point, the most important concepts are diffusion and convection (both of which are one-dimensional manifestations). In addition, the case of a first-order reaction for species A can also be considered. Each of these phenomena can be expressed in terms of Cartesian coordinates.

The most commonly used method of expressing diffusion and convection in mathematical terms is to specify the flux associated with these motion events. The concept of a flux is important because it separates the phenomena that cause mass motion from the shape of the control volume's surface through which the mass enters or exits. Flux is measured in units of moles of species A per unit surface area and per unit time. On the other hand, because the reaction occurs within the control volume, the rate of reaction is usually expressed in terms of moles per unit volume and per unit time. Thus the three important phenomena that can be used to describe the migration of species A through the control volume are given by the following expressions:

Flux due to diffusion (Fick's law): $\quad -D\dfrac{\partial C_A}{\partial x}$ $\qquad\qquad$ (1.14)

Flux due to convection: $\qquad\qquad -UC_A$ $\qquad\qquad\qquad$ (1.15)

Volumetric rate of reaction: $\qquad\quad -kC_A$ $\qquad\qquad\qquad$ (1.16)

We cannot simply substitute these three expressions into Equation 1.13 unless we also specify the volume and surface area of the control volume. For our one-dimensional problem, we merely need to account for the fact that species A is being forced through the control volume or can diffuse through the control volume in the *x*-direction. Thus, in our simplified case, the following expression can be developed:

$$\Delta x \Delta y \Delta z \frac{\partial C_A}{\partial t} = \Delta y \Delta z \left(UC_A \Big|_{x+\Delta x} - UC_A \Big|_x \right) - \Delta y \Delta z \left(D \frac{\partial C_A}{\partial x} \Big|_{x+\Delta x} - D \frac{\partial C_A}{\partial x} \Big|_x \right)$$
$$- \Delta x \Delta y \Delta z k C_A \tag{1.17}$$

Dividing by $\Delta x \Delta y \Delta z$ and taking the limit as each dimension of the control volume approaches zero gives the final result:

$$\frac{\partial C_A}{\partial t} = U \frac{\partial C_A}{\partial x} - D \frac{\partial^2 C_A}{\partial x^2} - k C_A \tag{1.18}$$

When fluxes in the *y*- and *z*-directions must also be considered, Equation 1.18 must be altered to include partial derivatives in the *y*- and *z*-directions. In addition, many more complications result if the implicit assumptions that the diffusivity (*D*) and fluid velocity (*U*) of the material are constants is not valid. We will postpone the introduction of any more mathematics or problem specification details. In later chapters of this book, we will undertake microscopic and macroscopic analyses of biochemical species accounting in specific types of bioseparation processes. For a review or in-depth analysis of the applied mathematics of chemical species accounting, the reader is referred to a wide variety of text and reference books on the subject of mass transfer.

1.4 SUMMARY

In bioseparations, the principles of mass conservation and chemical species accounting are used to analyze and design industrially relevant processes. On the macroscopic level, proper accounting of mass entering or leaving a system during a process must first define the starting point, or initial condition (as was illustrated here by the example of adding and removing balls from a box). The utility of differential and integral calculus can also be demonstrated using world population data. In the example in this chapter, we analyzed world population data by first developing a model (Malthus' law) to explain the data and then specifying the parameters using population statistics.

In the design of bioseparation processes, it is often necessary to predict the behavior of biochemical species that are undergoing microscopic or molecular-scale processes. To achieve this goal, we can describe the flux associated with both diffusion and convection, along with chemical reaction kinetics. Accounting for biochemical species can be performed at the microscopic level using differential calculus. The dual nature of biochemical species accounting at the microscopic and macroscopic level will be illustrated in more detail in later chapters of this book.

1.5 REFERENCE

1. The world almanac and book of facts. New York: Press Publishing, 1998.

PART I

**COMMERCIAL BIOSEPARATIONS
AND PRODUCT MEASUREMENT**

2

Industrial Bioseparation Processes

Bioprocessing is performed in the food, bulk chemical, specialty chemical, and agrochemical industries as well as the biotechnology industry. This broad range of industries is noteworthy for the enormous variation in both the scales of operation employed and the price of the products generated.

For the purposes of this book, *bioseparation processes* are defined as the manufacturing steps undertaken to purify products derived from animals, tissues, cells, or parts of cells. This definition encompasses processes that yield such divergent end products as dairy products and monoclonal antibodies, or citric acid and polio vaccines. The processes that are involved in making these seemingly disparate products can be treated as a single group because the end products, many of which are labile, are created in dilute aqueous solutions that can contain myriad other components. Because of these shared characteristics, a framework can be developed to study biological product purification in a systematic way—hence the term *bioseparation process science.*

A systematic study of bioseparation processes can reveal their underlying principles. Using these principles, a purification strategy applicable to many categories of biological products can then be developed. Hypotheses constructed during the study of bioseparations produce the greatest benefits in the real-world manufacturing environment when they are devised, tested, and refined quantitatively to ensure that they can reliably predict product yield, quality, and processing costs.

The intent of this book is to illustrate the underlying process principles and provide predictive tools for decision making in the manufacturing environment. Paralleling the scientific tradition, which calls for observing nature as the first step in gaining knowledge, this chapter is an "observation" of some commercial bioseparation processes. Each section of this chapter deals with a product that typifies a broader class of biological products of current commercial importance.

2.1 BIOSEPARATION PROCESS SELECTION

Customers make purchasing decisions based on a product's cost, quality, and availability. For a manufacturing facility to be competitive, it must deliver high-quality, low-cost products to the marketplace quickly. A product manufactured through biotechnology must be able to compete on both a cost and quality footing with the same product manufactured by other means. For new high-demand products that can be produced only through biotechnology, however, the corporate decision-making procedure in bioprocess selection may include such considerations as rapid commercial launch of the product before competitors enter the marketplace, ability to meet the projected demand for the product, and ability to manufacture a suitable-quality product. In the health industry, regulatory approvals also introduce new pressures into the standard definition of product quality by dictating that the form, efficacy, and purity of the final product formulation be validated before market approval is granted; the reliability and reproducibility of the process must be verified as well. More generally, compliance with environmental, health, and safety standards in pilot and commercial-scale operations alike is important to the biotechnology industry as a whole. When all of these factors are considered, the needs of the bioprocessing industry can be seen to be both complex and continually changing.

The sample flow diagrams for bioprocessing schemes in this chapter vary in detail. They are not meant to be the final authority on product purification in the bioseparation process side of manufacturing. Instead, they represent rational designs for making products. In general, commercially relevant bioseparation processes include several distinct steps:

1. Separation of solids from the product-laden aqueous solution
2. Product recovery from the aqueous solution
3. Product purification and isolation
4. Final product formulation

When a cell or cell population is the desired product, the separation of solids and liquids may be the entire goal of the bioseparation processing steps. On the one hand, recovering the product without prior cell removal is an attractive option because it reduces both capital costs and the number of processing steps needed. On the other hand, solid/liquid separation is often decoupled from product recovery in commercial processes. If the product is excreted from cells that are cultured in a bioreactor during the manufacturing process, then removing the cells in the first bioseparation step may prove helpful. Such cell removal minimizes enzymatic product degradation and eliminates many of the components that can interfere with later purification steps and introduce occupational and product safety hazards. When the product consists of larger biomolecules, the cells must be ruptured so as to isolate the desired product from the other cell components. Achieving this goal necessitates implementing a cell disruption step before the bioseparations are performed. The solids can then be separated using such equipment as centrifuges and membranes.

Even after removal of cells or cell debris, the product may continue to coexist with numerous contaminating components in the aqueous solution. If a high level of purity is required, the final product formulation is hampered by contaminating components, or the end product is degraded by the presence of such contaminants, product recovery steps are used to extract as much of the product as possible from the solution. Steps to pull out the product generally include membrane separation, extraction, centrifugation, and adsorption.

Purification or isolation of the product usually requires the use of more expensive equipment and materials per unit volume processed. At this stage of the process, it is important that the volume has been reduced during product recovery. To manufacture high-purity products, it may be necessary to use elaborate process schemes and esoteric materials; these processes must be reliable, produce a high yield, and deliver the target level of purity. Another important aspect of the purification step is the detection, using analytical means, of the level and type of components that may persist through to the final formulation step. Typically, this step employs chromatography, although crystallization can play a dual role in purification and final product formulation.

Most biological products are delivered in solid form. Hence, crystallization and precipitation are common steps in product formulation. Alternatively, products in liquid form may be formulated to give them an adequate shelf life and to allow them to be readily used as intended, as in the case of injectable therapeutics. For such applications, the purification step will ideally incorporate the appropriate buffer needed for product formulation. If not, an extra step may be needed to yield the desired formulation.

2.1.1 Scale, concentration, and price

A survey of a range of biotechnology products available illustrates that price and concentration in the bioreactor fluid are generally inversely related; that is, decreasing concentration is associated with increasing price of the end product. A Sherwood plot, shown in Figure 2.1, is used to display this information (1). Nevertheless, a correlation of this type is far from infallible, because many factors may affect price; for example, the number of companies making the same or similar products will affect the price any one company obtains for its product.

Using only the concentration produced in the bioreactor as an indicator of price does not give a complete picture, however. The process scale, or production rate, must also be taken into account. Multiplying bioreactor size or liquid production

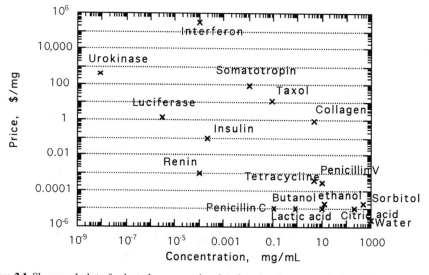

Figure 2.1 Sherwood plot of selected compounds related to the pharmaceutical and biotechnology industries.

Table 2.1 Comparison of Bioreactor Concentration to Batch Size and Production Rate for Selected Biotechnology Products

Product	Concentration (g/L)	Batch Size (m^3)	Production Rate (m^3/day)
Ethanol	70–120	>200	1000
Penicillin	10–50	>200	50–200
Polio vaccine	1–4	<50	1–5
Protease	2–5	220	

rates by the concentration can yield a measure of production rates. The economic feasibility of any manufacturing facility depends upon the end product's price and production rate. Thus smaller-scale production is feasible for higher-value products. Table 2.1 compares concentration levels with batch size and production rates for several products.

2.1.2 Product properties

In Sections 2.2–2.8, we compare commercial bioseparation processes. We can make such comparisons because the individual operations capitalize on—or are driven by—common properties of biological components in solution. Many properties can influence process design, such as whether the product is located intracellularly or extracellularly. Process design must also account for properties such as sensitivities that the product itself or cellular components used to produce it have to extremes of temperature, pH, and ionic strength, or to the presence of enzymes and chemicals. The physical conditions in the bioreactor must be monitored for occurrences such as viscosity increases caused by an undue concentration of cells or unexpected release of polynucleic acids. Unwanted components, such as sticky biopolymers, bioreactor additives, buffers, and residual cell feed components, must be eliminated from the final product solution. In addition, the removal of cells from the solution can be hindered when the target cells and the solution have similar densities. The desired final product form will also exert a strong influence on processing.

A variety of product and component properties can be exploited for separation and purification: size, electrostatic charge, hydrophobicity, aqueous solution solubility, and unique chemical groups or the unique arrangement of chemical functional groups.

The wide range of sizes of the components present in bioreactor fluids enables manufacturers of some products to perform size-based separations. (Figure 2.2 shows size ranges for selected biological components.) Size-based separations may, for example, be carried out by centrifugation or filtration. Centrifuges can separate cells from the bioreactor fluid and, at higher speeds and smaller batch sizes, separate viruses and biopolymers from aqueous solutions. Filtration, a more widely applicable technology, may be performed via microfiltration, ultrafiltration, and reverse osmosis. Forces, such as those induced by charge can be combined with filters and used to selectively remove certain components in electrodialysis. Another

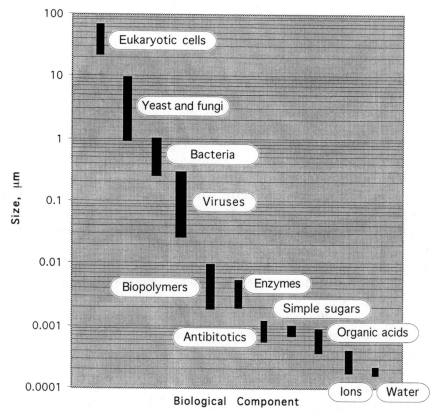

Figure 2.2 A comparison of sizes of biological components.

popular technology combination is size exclusion chromatography, in which porous particles filling a column accomplish size separation.

Proteins, nucleic acids, organic acids, and cell surfaces are charged components that can be separated using ion-exchange and ion-exclusion chromatography, electrophoresis, and ion-exchange membranes. The solubility of these components can also be manipulated by adjusting pH, which enables the use of flocculation, precipitation, adsorption, and crystallization.

Because biotechnology products are manufactured nearly exclusively in aqueous solution, the tendency of water to form hydrogen bonds results in many parts of proteins, lipids, polysaccharides, and organic acid molecules becoming hydrophobic when they lack either electronegative atoms that can hydrogen-bond or ions. This hydrophobicity is manipulated or exploited in hydrophobic interaction chromatography, reverse-phase chromatography, adsorption, extraction, and precipitation.

The more selective separation techniques use chemical groups or biological materials that can interact with specific regions on a biological product, selectively leaving other molecules and components behind in solution. A wide range of sophistication and batch sizes is possible with this approach—from the complexation of acids and bases in extraction and adsorption to bioaffinity chromatography (in which biological ligands such as monoclonal antibodies are attached to the column packing).

2.2 MONOCLONAL ANTIBODIES

High-value therapeutic products are generally manufactured in small batches at production rates that typically do not exceed $1\,m^3$/day. One such type of therapeutic is monoclonal antibodies, which are also important components in clinical and research laboratory assays (2, 3). Antibodies, which are also referred to as immunoglobulins (Ig), can be classified into distinct groups based on their properties and functions. Immunoglobulin G (IgG) is an antibody with a molecular weight of approximately 160,000 daltons; IgM has a molecular weight approximately five times that of IgG. While IgG can be secreted, IgM usually remains on the cell membrane surface.

Monoclonal antibodies can be produced by cells that represent the fusion products of an antibody-producing spleen cell and a cancerous (and thus immortal) myeloma cell. The fusion products, known as hybrid myeloma (hybridoma) cells, can be indefinitely cultured in a nutrient suspension that includes blood serum components. Scaled-up production batches consist of approximately 35 L of solution.

The low production volume and the need for a high level of purity in the end product generally require that the manufacturer of a monoclonal antibody undertake several different purification steps (Figure 2.3). After the bioreactor has

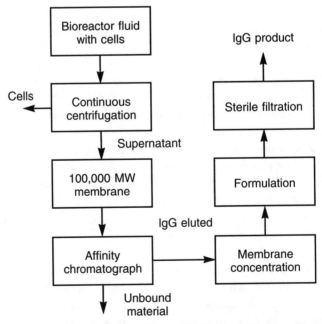

Figure 2.3 Recovery and purification of monoclonal antibodies. (Adapted from: Birch JR, Lennox ES. Monoclonal antibodies: principles and applications. New York: John Wiley and Sons, 1995:238–248; and Pyle DL. Separations for biotechnology 2. New York: Elsevier Science, 1990:472–480.)

reached a specified cell density, the cells are separated from the aqueous solution containing the secreted IgG by a continuous centrifuge. Biochemicals and proteins smaller than approximately 100,000 daltons are removed via an ultrafiltration membrane. The retained IgG-containing solution is then selectively bound to an affinity chromatography column packed with immobilized protein A or protein G; the protein acts as a biospecific ligand for IgG isolation. At lower than physiologic pH levels, IgG is eluted from the column and further concentrated using an ultrafiltration membrane. The membrane concentrates the solution while permitting the addition of a buffer that raises the IgG solution back to a physiologic pH level during the formulation step. Sterile filtration ensures that the product has the desired quality and prolongs its shelf life by guarding against bacterial contamination and growth in the storage solution.

2.3 HUMAN INSULIN

In antibody production, the manufacturer must undertake the challenging step of culturing of mammalian cells to obtain the desired therapeutic protein. In contrast, human insulin can be made by genetically engineering *E. coli* bacteria. Bacterial cells multiply at an extraordinarily rapid rate (doubling approximately every 20 minutes) and can withstand relatively harsh reactor conditions. Nevertheless, the typical product concentrations are quite small—sometimes less than 1 g/L. Moreover, bacterial cells produce pro-insulin, which must be chemically cleaved to yield human insulin (4, 5).

Because human insulin is an injectable therapeutic, it is essential that the end product have high purity; it should also completely lack any bacterial cell products that would prove harmful to humans.

Human insulin is produced in inclusion bodies within the bacterial cell. These inclusion bodies comprise pockets of insulin-rich material sequestered from the rest of the cell cytoplasm. Cell disruption is needed to expose the inclusion body and free the human insulin, thereby allowing the insulin to enter the aqueous solution (Figure 2.4). Cyanogen bromide is used to cleave insulin into what eventually become the A- and B-chains. Dialysis is then used to change the solution's pH, introduce new buffer components, or remove the cleavage reagents.

After extracting the peptide and performing a sulfonation to provide the final product form, the A-chains are separated from the B-chains. These purification steps combine precipitation and several chromatographic steps. For the supernatant A-chain, precipitation of undesirables is performed at pH 5, followed by chromatography using a weak anion-exchange column packing; the final step is reverse-phase high-performance liquid chromatography (HPLC). The dialysate containing the B-chain goes through a precipitation step, which is also followed by anion exchange and size-exclusion chromatography.

2.4 RABIES VACCINE

Virus vaccines are produced in animals, monolayer cultures, and suspended cultures. Harvesting takes place either after the virus is excreted or after the virus lyses the cells.

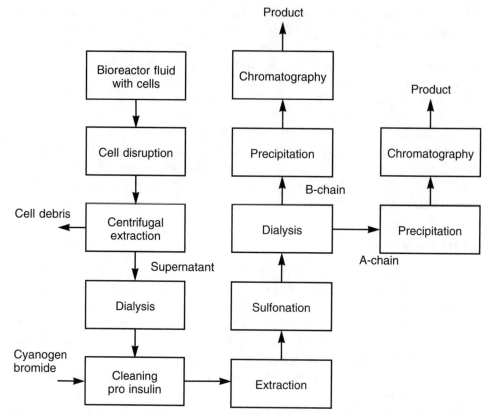

Figure 2.4 Recovery and purification of human insulin. (Adapted from: Walsh G, Headon DR. Protein biotechnology. New York: John Wiley and Sons, 1994:211–220; and Sadana A. Bioseparation of proteins, unfolding/folding validations. San Diego: Academic Press, 1998:43.)

The rabies vaccine can be produced in suspended microcarrier cultures using a simian cell line; the resulting batches may be as large as several hundred liters (4, 6). In the manufacturing process (Figure 2.5), the mammalian cells are immobilized onto a microcarrier bead. The virus does not break the cells open upon infection and replication, so beads and cells infected with rabies can be recycled. Separation of the microcarrier beads and the 0.3 µm virus can be accomplished by ultrafiltration using several membrane modules; the molecular weight cutoffs of these modules range between 10,000 and 100,000 daltons.

Viruses used in vaccines can be either killed or left viable, depending on safety and efficacy concerns. For the rabies vaccine, chemical inactivation is required. This inactivation is performed in a two-step process. First, β-propiolactone is applied at low temperature. Second, hydrolysis takes place along with a gradual warming of the mixture. The entire inactivation process takes 24 hours.

Because of the size of the virus particles, sterile filtration with a 0.2 µm filter is not possible. Instead, this important role is filled by zonal centrifugation using a sucrose gradient; during this centrifugation, the nonantigenic material is removed from the product to minimize any undesirable side effects.

Human albumin is added in the product formulation step as a stabilizing agent.

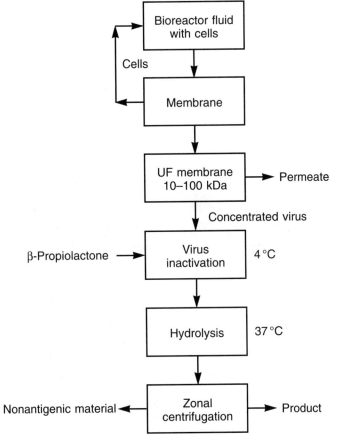

Figure 2.5 Downstream processes for production of rabies vaccine. (Adapted from: Walsh G, Headon DR. Protein biotechnology. New York: John Wiley and Sons, 1994:184–191; and Ellis RW. Vaccines: new approaches to immunological problems. Stoneham: Butterworth-Heinemann, 1992:406–413.)

2.5 PENICILLIN

Figure 2.6 shows the process for recovering and purifying penicillin. The key step in penicillin production is the organic solvent extraction process (5). To obtain active, high-purity penicillin, the extraction steps must employ a special type of centrifugal action extractor, known as a Podbielniak extractor. This equipment shortens the processing time, which in turn minimizes penicillin degradation in the organic solvent.

Because molds are employed in the manufacture of penicillin, the colony formation of molds (mycelia) is amenable to separation using a rotary vacuum filter. First, calcium chloride and a polyelectrolyte are added to the mycelial suspension to form large particles known as flocs. The filtrate containing penicillin is then rapidly processed, via a series of extraction steps, into an organic solvent, usually amyl acetate. The penicillin is transferred from this organic phase into a pH-neutral

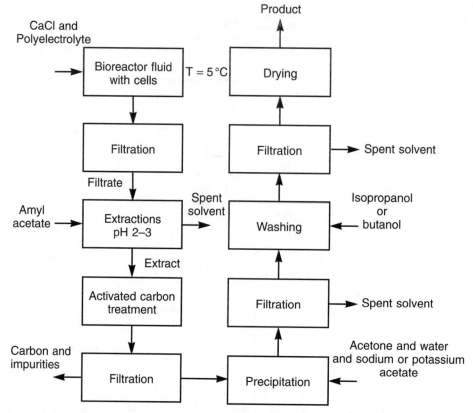

Figure 2.6 Recovery and purification of penicillin. (Adapted from: Atkinson B, Mavituna F. Biochemical engineering and biotechnology handbook. Surrey, England: Macmillan, 1983:990.)

aqueous solution. These steps increase the pencillin's concentration approximately 100 times in the final extractant phase.

To remove impurities, activated carbon is used during or immediately after the extraction steps. Filtration removes this activated carbon and prepares the penicillin for precipitation as a sodium or potassium salt. Precipitation is induced by addition of acetone, followed by washing with an alcohol to remove any lingering impurities. Because at least three different solvents are used, the economics of penicillin production are clearly influenced by proper recovery and recycling of the spent solvents.

2.6 PROTEASE

Protease, an enzyme that can degrade proteins, is used in products such as laundry detergents. Because its applications do not involve food products or injectable theraputics, it need not be produced in a highly purified form. For use in detergents, protease can be produced in either liquid or solid form.

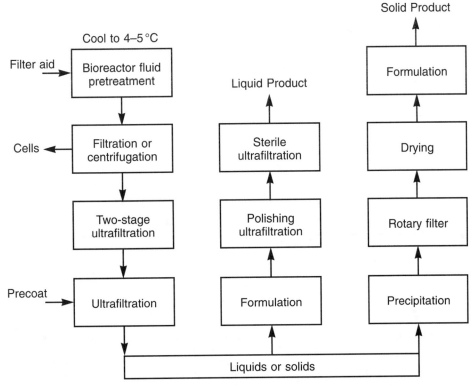

Figure 2.7 Downstream processing for production of liquid or solid protease. (Adapted from: Shuler ML, Kargi F. Bioprocess engineering: basic concepts. Englewood Cliffs: Prentice-Hall, 1992.)

Because the emphasis is on achieving high concentration and high volumetric productivity rather than purity, most of the bioseparation processes for protease involve filtration (Figure 2.7). A cooling step follows the bioreactor phase and precipitation, which serves to minimize bacterial growth and maintain enzyme activity during processing. The filtration steps can either remove low-molecular-weight impurities or concentrate the product. In the solid-product flow scheme, precipitation is generally performed below 5 °C using sodium sulfate, ammonium sulfate, ethanol, or acetone (7).

2.7 L-LYSINE

When the same basic product can be used for both animal and human consumption, different separation processes may be employed to achieve the different purity levels required for these two applications. The amino acid l-lysine, for example, has a net charge of +2 at low pH. This property is exploited in the recovery phase of l-lysine production (1). Using hydrochloric acid, the pH of the l-lysine solution is lowered. The solution is then applied to a large adsorption column containing a strong acid cation exchanger. L-lysine binds strongly under these conditions, and many impurities pass through the column. Elution is accomplished by the addition of an ammonium chloride solution of moderately high concentration.

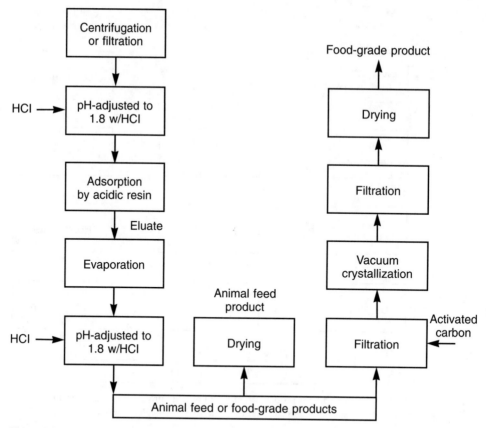

Figure 2.8 Animal and food-grade l-lysine recovery and purification. (Adapted from: Blanch HW, Clark DS. Biochemical engineering. New York: Marcel Dekker, 1996:623–624.)

As the final product is l-lysine-HCl, acidification follows elution from the column (Figure 2.8).

When it is intended for use as an animal feed, the l-lysine product can be spray-dried. To obtain a product that meets human food codes, however, the manufacturer must undertake further purification using activated carbon and crystallization.

2.8 CITRIC ACID

As an acidulant in soft drinks, citric acid is perhaps the most frequently used biotechnology product. The microorganism used to produce citric acid, *Aspergillus niger*, is an acidophile that thrives at low pH levels. Use of *A. niger* therefore simplifies the primary recovery step, extraction; a long-chain amine dissolved in an organic solvent is used to specifically complex with citric acid in its un-ionized form (8).

Figure 2.9 shows the steps involved in the production of citric acid. The amine-citric acid complex is soluble in the organic phase, so the extraction of this complex leaves behind many impurities in the aqueous solution. Back extraction using hot

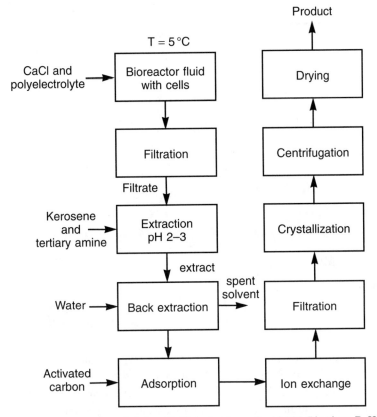

Figure 2.9 Downstream processing of citric acid. (Adapted from: Banid AM, Blumberg R, Hadjo K. U.S. Patent No. 4,275,234. 1981.)

water prepares the citric acid solution for further purification using activated carbon and ion exchange. The final product is crystallized in an evaporative crystallizer and then dried. Depending on the conditions in these units, the end product will be either citric acid-l-hydrate or anhydrous citric acid.

2.9 SUMMARY

Surveying the commercial-scale processes used to recover and purify biotechnology products identifies some key features sought by bioprocess designers: overall low cost; large volume reduction; robustness in tolerating minor fluctuations in process conditions; short residence times to minimize product degradation and to increase productivity; high yield; and high reliability and reproducibility. These considerations play an important role in the selection of process operations, which include cell disruption, flocculation, filtration, centrifugation, extraction, adsorption, chromatography, drying, evaporation, precipitation, and crystallization. These steps are undertaken to transform biotechnology products present in a relatively low concentration in aqueous solution into purified solid products. Table 2.2 indicates

Table 2.2 Typical Product Concentrations During Recovery and Purification

Step	Concentration (g/L)	Weight Percent
Cell culture fluid	0.1–5	0.1–1
Filtration or centrifugation	0.1–5	0.1–2
Primary isolation	5–10	1–10
Purification	50–200	50–80
Precipitation or crystallization	Not applicable	90–nearly 100

the range of concentrations that are typically present during the different stages of recovery and purification.

2.10 PROBLEMS

2.1 Look up the properties of the amino acid tryptophan in reference books such as *Merck Index*, *CRC Handbook*, *Lang's Handbook*, or similar chemical reference texts or databases on the Internet (for example, on the World Wide Web). Also, using Internet databases or the *Food Chemical Codex* in the reference section of your library, look up the purity requirements for tryptophan in foods.

(a) After noting the properties of tryptophan and reviewing the l-lysine process flow diagram (Figure 2.8), develop a bioseparation process flow diagram for manufacturing animal feed and human food-grade tryptophan.

(b) Compare your process with commercial processes described in patents, on the Internet, and in engineering texts or technical journals in your library. Are they similar? Are there any additional steps described in the commercial processes not described in your process? If so, why?

2.2 Designers often rely on heuristics to develop rules (usually called "rules of thumb") to guide them in developing new processes. One common separation process rule of thumb is "Isolate the most abundant component first." In the case of bioprocessing, the most abundant component is, of course, water.

(a) Why can this heuristic be useful in separation process design?

(b) How many of the processes discussed in this chapter use this heuristic?

(c) Why couldn't all commercial processes use this heuristic?

2.3 Develop a generic bioseparation process flow diagram for producing freeze-dried yeast and bacterial cells for use as human food, animal feed, or a bioinsecticide. In your diagram, account for the requirement that the yeast food product must be washed with water, the cells are killed by heating to 90 °C, and the product must be concentrated before drying. Find a reference source (see Problem 2.1) that provides commercial processes for comparison with your generic cell bioseparation process.

2.4 Based on the lactic acid bioseparation process given in Figure 2.10, answer the following questions.

(a) Why does this process differ markedly from the citric acid food-grade process described in this chapter?

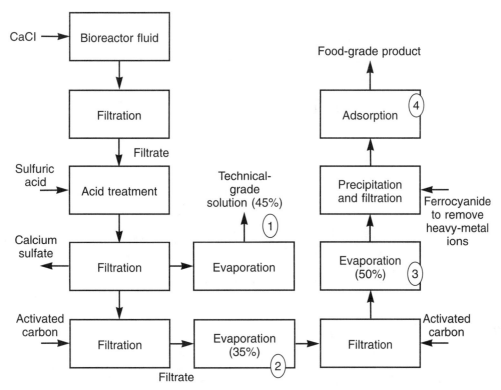

Figure 2.10 Lactic acid bioseparation process.

(b) Methyl lactate, which can be produced by esterifying lactic acid with methanol, is useful in the production of plastics. Esters and methanol can be easily removed from aqueous solution via distillation. Which of the marked points in the process shown in Figure 2.10 indicates where the process stream can be diverted to make methyl lactate?

2.5 Accompanying the heuristic discussed in Problem 2.2 are the following rules of thumb:

(a) Keep the number of steps to a minimum.

(b) Isolate the component that is easiest to remove first.

(c) Bring in a high-resolution step early into the process flow.

(d) Leave the most difficult isolation step for last.

Are these rules contradictory? Do any of these rules, including the heuristic given in Problem 2.2, supersede the others in the commercial separation processes described in this chapter? If so, why?

2.6 Imagine that various types of proteins could be made in aqueous solution by adding the appropriate kinds and amounts of amino acids. The aqueous solution also contains a surface that replaces all of the cellular functions, and this surface can be a part of a membrane system or chromatography column packing. This surface catalyzes the formation of the correct sequence of peptide bonds and allows the protein to develop the correct structure and function needed for a particular product application. Discuss how the protein bioseparation processes would change if the cells could be replaced by such a system. Assume that the solution would need

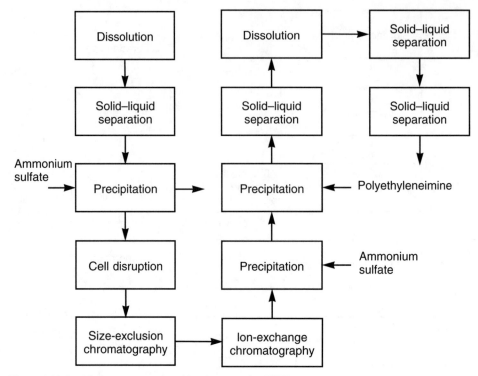

Figure 2.11 Jumbled recovery and purification steps for HGH.

the appropriate buffers along with amino acids and other small molecules, such as metal, inorganic, and organic ions.

2.7 The flow diagram in Figure 2.11 is a jumbled version of a commercial process for recovery and purification of the intracellular product human growth hormone (HGH). The first two precipitation steps should follow product release from the ruptured cells, and the last precipitation step should follow the ion-exchange chromatography step. Reassemble the flow diagram in the proper order to make a highly purified solution of human growth hormone.

2.8 Direct recovery of products during the bioreactor step as well as direct purification from cells or cell debris has been explored for several products. What are the advantages of using such a process for direct product recovery? What are the potential disadvantages? Which, if any, of the following products may be more amenable to this approach?

    (a) Pharmaceuticals
    (b) Human food
    (c) Animal feed
    (d) Industrial chemicals

2.9 Figure 2.12 illustrates the commercial-scale process for recovering and purifying human leukocyte-produced interferon. Explain each of the steps in this process. Why is it relatively simple, even though it yields a human therapeutic product?

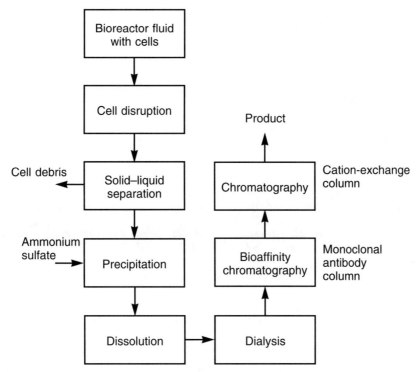

Figure 2.12 Recovery and purification steps for human leukocyte interferon.

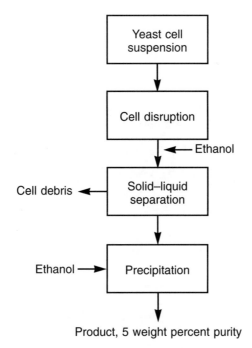

Figure 2.13 Catalase recovery steps.

2.10 The process flow sheet in Figure 2.13 describes the recovery of the enzyme catalase. Develop steps that would improve the purity of the final product from the 5 weight percent obtainable using the given process.

2.11 REFERENCES

1.   Blanch HW, Clark DS. Biochemical engineering. New York: Marcel Dekker, 1996.
2.   Birch JR, Lennox ES. Monoclonal antibodies: principles and applications. New York: John Wiley and Sons, 1995.
3.   Pyle DL. Separations for biotechnology 2. New York: Elsevier Science, 1990.
4.   Walsh G, Headon DR. Protein biotechnology. New York: John Wiley and Sons, 1994.
5.   Sadana A. Bioseparation of proteins, unfolding/folding validations. San Diego: Academic Press, 1998.
6.   Ellis RW. Vaccines: new approaches to immunological problems. Stoneham: Butterworth-Heinemann, 1992.
7.   Shuler ML, Kargi F. Bioprocess engineering: basic concepts. Englewood Cliffs: Prentice-Hall, 1992.
8.   Banid AM, Blumberg R, Hadjo K. U.S. Patent No. 4,275,234. 1981.

3

Concentration Determination and Bioactivity Assays

Bioseparation process selection, development, and operation require the use of analytical methods that provide information on the concentration of the solute of interest, as well as some measure of the concentrations of contaminating species. For some products, concentration measurement is routinely determined through bioactivity assays, because bioactivity is more directly relevant to the price that can be charged for the product and even the product's ultimate marketability. The product's chemical composition may also need to be carefully monitored if the manufacturer must comply with certain regulations, such as those governing the production of pharmaceuticals.

This chapter reviews current techniques in widespread use for the analysis of biochemicals and biopolymers. One of the most important aspects of separation process development is gaining access to reliable, accurate, and rapid analytical methods so as to assess whether the bioseparation process will be viable when it is expanded to production scale. Familiarity with analytical separation techniques and terminologies is also useful for comparing the proposed process with larger-scale separation processes. During scale-up from the laboratory bench to production, product concentrations are—as a rule—moderately high; consequently, very sensitive analytical procedures are not needed to track yield and productivity. This consideration can greatly simplify analytical chromatography procedures. It is important that the bioseparation process developer take this issue into account early on, enabling resources that might have been spent on analysis to be allocated to process development instead.

3.1 AMINO ACIDS

Table 3.1 lists 20 common amino acids. Many analytical methods have been developed that provide rapid and accurate determination of amino acid concentrations. Analytical chromatography, in which amino acid concentrations are determined using ultraviolet (UV) absorbance, is a popular method. Unfortunately, many amino acids do not strongly absorb in the UV range.

Figure 3.1 shows the absorbance spectra for eight common amino acids. Because tryptophan and tyrosine each contain an unsaturated ring, they absorb reasonably well at wavelengths greater than 240 nm. This property can be exploited to distinguish them from many other biochemicals and buffer compounds that absorb in the range of 200 to 230 nm. Only when amino acid concentrations are relatively high can UV detection be effective as a general tool.

Table 3.1 Amino Acids and Their Abbreviations

Amino Acid		Amino Acid	
Alanine	Ala	Leucine	Leu
Arginine	Arg	Lysine	Lys
Asparagine	Asn	Methionine	Met
Aspartic acid	Asp	Phenylalanine	Phe
Cysteine	Cys	Proline	Pro
Glutamic acid	Glu	Serine	Ser
Glutamine	Gln	Threonine	Thr
Glycine	Gly	Tryptophan	Trp
Histidine	His	Tyrosine	Tyr
Isoleucine	Ile	Valine	Val

Source: Adapted from Zubay GL. Biochemistry, 4th ed. Dubuque: William C. Brown, 1998:62.

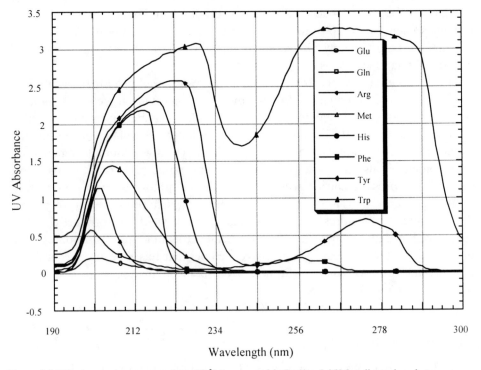

Figure 3.1 UV absorbance spectra of 1×10^{-3} M amino acids. Buffer: 0.05 M sodium phosphate monobasic/dibasic, pH 7.0. Path length: 1 cm.

Because amino acids do not absorb UV light strongly, it may be helpful to chemically modify the target amino acid so as to improve detection sensitivity and minimize background noise. This modification is accomplished by attaching strongly absorbing or fluorescing chemical labels, such as those listed in Table 3.2, to the amino acids. The following sections describe some popular analytical techniques for measuring amino acid concentrations in solution.

Table 3.2 Compounds Used to Label Amino Acids for Detection

Label	Use	Chemical Structure
Dansyl chloride (5-dimethylaminonaphthalene-1-sulfonyl chloride)	Fluorescent label	
Ninhydrin (2,2-Dihydroxy-1,3-inadedione)	Determination of amino acids, 1° and 2° amines	
FITC (fluorescein isothiocyanate)	Fluorescent label	
DNP (dinitrophenol)	Visible and UV indicator	

3.1.1 High-performance liquid chromatography

When high concentrations of amino acids are present, such as during cell culture production of amino acids, high-performance liquid chromatography (HPLC) methods that do not require amino acid labeling can be used to rapidly determine their concentration. Depending on the chemistry of the amino acid side chain, ion-exchange or ion-exclusion chromatography can prove quite helpful in the analysis. For example, HPLC at low pH using a strong anion-exchange support packing in the sulfate or hydroxide form can separate amino acids with basic side chains from solution. Lysine, which has a net charge of +2 at low pH, is rapidly eluted from such a column in a technique known as ion-exclusion chromatography. This approach is characterized by short residence times and rapid analytic results. Strongly ionized species of like charge are eluted first, followed by weakly ionized species of like charge, followed by uncharged species.

Similar methods can be applied to amino acids with acidic side chains, such as glutamic and aspartic acid. Ion-exchange techniques can also be employed for analyzing amino acid products via gradient methods. This method can retain amino acids on the column for a longer time, which serves to discard by-products that may make ion exclusion unfeasible.

3.1.2 Reverse-phase high-performance liquid chromatography

For nearly 30 years, amino acid determinations have been mostly accomplished through the use of various automated ion-exchange postcolumn derivatization (mostly ninhydrin) methods. The more simple technique of reverse-phase (RP) HPLC, however, is now gaining recognition as a powerful method in biological research. The usefulness of this technique in rapidly analyzing low concentrations of amino acids has already been demonstrated in numerous recent publications.

In RP-HPLC, a compound is partitioned between a hydrophobic stationary and a polar aqueous mobile phase. This method is usually performed on columns packed with silica gel to which a C_8 to C_{18} hydrocarbon has been covalently attached. In such systems, the strength of an eluting solvent increases as its polarity decreases. Compounds of similar structure can therefore be expected to elute in order of decreasing polarity. Adequate separation and subsequent quantification of tissue-free amino acids through RP-HPLC is possible. Thus this technique gives reduced analysis time and enhanced sensitivity compared with other analytic methods.

One example using Dabsyl chloride-labeled amino acids resulted in the following retention order:

aspartic acid < tyrosine < asparagine < glutamine < serine < threonine
< glycine < alanine < arginine < proline < valine < methionine < isoleucine
< leucine < tryptophan < phenylalanine < lysine < histidine

This separation was performed using an octadecylsilane-bonded, 15-cm column, starting with 25 mM KH_2PO_4 at pH 7 and using gradient programming with a 70:30 mixture of acetonitrile and methanol (1).

The labeling of amino acids before chromatography begins is known as precolumn derivatization. The use of different precolumn derivatization methods in RP-HPLC has been studied with labeling compounds such as 9-fluorenylmethyl-chloroformate (FMOC-Cl), ortho-phthaldialdehyde (OPA), 1-dimethylaminonaphthalene-5-sulphonylchloride (Dansyl chloride), and phenyliso-thiocyanate (PITC) (2). Using this technique, it is possible to measure free amino acids in physiological fluids. When RP-HPLC is employed, attention must be paid to control factors and limitations inherent in this method and to correct interpretation of the results.

3.1.3 Capillary electrophoresis

Capillary electrophoresis (CE) is a powerful method for separating out compounds such as peptides and amino acids. The high sensitivity (femtomolar level), small volumes (nanoliter), and rapid analysis time (minutes) of CE make it a convenient approach for testing the purity of biological samples separated by conventional HPLC. On the other hand, the small sample volumes used in CE can create problems when this method is used in preparative work. Nevertheless, its high resolution renders it an extremely powerful tool for carrying out the peptide purifications necessary in structural analysis.

Jorgenson and Lukacs (3, 4, 5) were the first to perform capillary zone separations. Capillary zone electrophoresis (CZE) typically employs a fused-silica capillary with an inner diameter of less than 100 µm; a voltage of 10–30 kV is applied over the column, which can measure from 50 to 100 cm in length. The use of microcapillary columns efficiently dissipates the heat generated by the electrophoretic process and handles high plate counts. While 20–50 mM buffers of both basic and acidic pH can be used, high pH buffers are preferred. Such buffers eliminate capillary wall charge and prevent separated species from adhering to the capillary wall.

CZE is a relatively simple technique. Components in the mixture are separated based on their molecular sizes and electric charge ratios. Variable parameters that can improve separation include pH and buffer composition. CZE has been tested for both preparative purposes and direct analysis in the field of amino acid sequencing and composition (6). Experiments have shown that this technique can also produce useful results in further structural analysis, such as direct amino acid determination of such compounds as carboxypeptidase-liberated residues.

High-speed zone electrophoresis has been demonstrated in a fused-silica capillary (7). Using elevated electric fields and short capillary lengths, this method can separate a mixture of FITC-labeled amino acids in less than 2 seconds. Formation of the analyte zone at the head of the capillary is controlled by laser-induced photolysis of a tagging reagent. This gating procedure allows rapid and automated introduction of the sample into the capillary. Joule heating of the buffer, however, limits the speed and efficiency of the separation.

3.1.4 Micellar electrokinetic chromatography

Micellar electrokinetic capillary chromatography (MEKC) has extended the enormous power of CZE to the separation of uncharged solutes. The high efficiency of CZE often suffices to separate charged compounds with very similar mobilities. The introduction of micelles into the electrophoretic medium, however, provides both ionic and hydrophobic sites of interaction simultaneously. Consequently, MEKC proves superior to CZE in separating mixtures containing both charged and uncharged solutes.

The migration behavior of ionizable compounds in MEKC is far more complicated than that of uncharged compounds. As a result, multiple parameters—such as micelle concentration, modifier concentration, and pH—may need to be taken into account to achieve an adequate separation of complex mixtures.

Several techniques are available to optimize capillary electrophoretic separations. For example, the overlapping resolution mapping (ORM) scheme, which was originally developed to optimize HPLC separations, is simple yet effective when applied to capillary electrophoretic separations. Using a systematic optimization scheme based on ORM, the separation of 16 DNP-amino acid derivatives has been carried out (8). This separation was achieved by the simultaneous optimization of buffer pH, surfactant concentration, and modifier concentration in a rapid and simple procedure. Thus a systematic method for optimization might employ a three-dimensional ORM approach to select the conditions that provide the best resolution between the least-separated pair within a desired analysis time.

3.1.5 Electrodialysis

Amino acid analysis is generally carried out using methods based on liquid or gas chromatography. Because these techniques are extremely sensitive and occasionally time-consuming, they are not suitable for continuous monitoring of amino acid concentrations, as might be required in actual production conditions.

A technique for continuously measuring amino acid concentration is flow injection analysis (FIA), using either an enzyme-based detector or a biosensor. One method for continuous determination of the concentration of an amino acid in a mixture involves a combination of electrodialysis with a flow injection system (9). Figure 3.2 depicts a typical FIA electrodialysis apparatus. First, the separation step is carried out with the electrodialysis unit. The concentration of a particular amino acid in a mixture can then be determined by using a nonspecific l-amino acid oxidase column and electrochemical detection of enzymatically produced hydrogen peroxide.

Another flow injection method relies on the pyrolithic conversion of amino acids to nitrogen-containing compounds. These compounds undergo thorough peroxodisulfate digestion using a heated capillary tube containing a platinum wire in a closed system. Aliquots of the digest are subsequently mixed with a malachite green chromogenic reagent solution in a symmetrical system. The system makes a spectrophotometric determination of nitrate at 650 nm. When this technique was applied to analyze proline, cysteine, alanine, and tyrosine concentrations in aqueous mixtures, it was found to be reproducible and accurate (10).

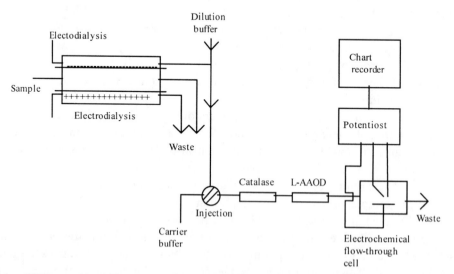

Figure 3.2 The combined FIA electrodialysis system. (Adapted from: Cooper JC, Danzer J. Enhanced selectivity in flow-injection analysis for l-amino acids using electrodialysis with amino acid oxidation. Analytica Chimica Acta 1993;282:369.)

3.1.6 Gas chromatography

In the last two decades, extensive advances have been made in the separation of amino acids by gas chromatography (GC). The high polarity of the amino acid molecules makes it necessary to prepare volatile derivatives. This derivatization is commonly achieved in two steps. The first step yields a carboxylic acid alkyl ester. The second step perfluoroacylates the amino groups and remaining acylatable groups. The separation of the trifluoroacetyl (TFA) alkyl esters of the amino acids is performed in packed columns. This method of separation has been reported in the literature for TFA amino acid n-butyl esters.

For capillary columns, another type of derivative, N(O)-HFB amino acid esters, has been used. A variety of HFB amino acid alkyl esters have been obtained by applying various esterifying alcohols.

The HFB derivatives of amino acids may demonstrate higher volatility than the corresponding TFA analogues in a capillary column. Because the HFB group has electron-capturing properties, electron capture detection (ECD) with a glass capillary column has been introduced to gas chromatographic analysis.

Studies have been performed on the analysis of amino acids using fused-silica, open tubular capillary columns (FSOT), along with ECD or flame ionization detection (FID) (11). When HFB amino acid isobutyl esters were employed, the ECD response proved several hundred times more sensitive than the FID response. This method has been successfully applied to analyze most of the trace amino acids found in fossil brachiopods and black shales.

It is also possible to analyze amino acids using thermal methods, such as thermogravimetry (TG) and differential-scanning calorimetry (DSC). For example, α-amino acids have been analyzed with simultaneous TG-DSC measurements and kinetic calculations by dynamic TG techniques (12). With such techniques, compounds with similar structures can be grouped by their thermograms, which have the same shape (this method does not give enough information on their kinetic variation, however). The thermodynamic and kinetic data hold different significance, except in series where the rings of the side chains influence the thermal decomposition of the compounds.

3.2 PEPTIDES AND PROTEINS

Protein separation and analysis is most often carried out by electrophoresis or analytical liquid chromatography (LC). With HPLC, detection is routinely accomplished through UV absorption at 280 or 254 nm. Currently used electrophoretic techniques include gel electrophoresis, sodium dodecyl sulfate polyacrylamide gel electrophoresis (SDS-PAGE), isoelectrofocusing, and immunoelectrophoresis. Common methods of detection in electrophoresis include silver or coomassie blue staining and radiography.

3.2.1 Analytical chromatography

The more common techniques used for protein analyses are gel permeation (GPC), ion exchange, hydrophobic interaction, and reverse-phase chromatography. Except for GPC, which is not well suited for the fractionation of peptides, these techniques are often employed for peptide analysis as well.

GPC is a desirable method for detecting proteins, whose molecular weights vary widely. This method is a relatively simple analysis to run, as only one mobile phase is used in an isocratic system and peak assignment is based on a calibration curve (this curve is typically supplied by the manufacturer). In the calibration curve shown in Figure 3.3, the general useful range of analysis lies between 3000 and 70,000 daltons, an area over which the curve is reasonably linear. The sharp increase in the curve's slope above 100,000 daltons indicates that the gel pores keep out globular proteins with molecular weights higher than this cutoff level.

In ion exchange, several strong cation and anion chromatography packings have been developed specifically for protein separation. The chromatofocusing ion-exchange method uses a packing that combines strong and weak basic amines. The elution of proteins in this system relies on the proteins' isoelectric point, in a highly sensitive manner that can resolve proteins with pI differences as small as 0.05 units. To minimize nonspecific, irreversible protein binding, these chromatographic packings use hydrophilic polymer matrices. The presence of very small, nonporous particles (generally 10 μm) provides a high surface area with rapid liquid-to-solid transfer rates. In general, strong cation-exchange packings have sulfonic acid groups and can separate proteins that differ in their net charge when pH varies between 2 and 6. The most positively charged protein is eluted last. At the other end of the spectrum, packings that have strong anion-exchange groups, such as quaternary amines, separate proteins between pH 8 and 12, with the most negatively charged protein being eluted last. Both anion and cation exchange take advantage of an

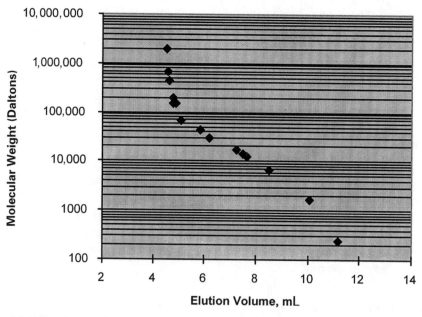

Figure 3.3 Calibration curve for SigmaChrom GFC-100 Column, 280 nm (30 cm × 7.5 mm ID, 12–15 μm particles). Mobile phase: 50 mM Tris-HCl/100 mM Kcl, pH 7.5, 0.5 mL/min. Sample: 20 μL, 1–10 mg/mL globular proteins.

NaCl gradient to elute the proteins. In addition, chromatofocusing uses a special buffer change that develops a pH gradient within the column, causing the proteins to elute in a sharp band at their respective isolectric points. The pH range used is quite wide, but generally varies between 4 and 10.

Hydrophobic interaction chromatography (HIC) and RP-LC both separate proteins and peptides based on their amino acid residues' side-chain hydrophobicity. The HIC packing contains a hydrophilic matrix with attached unsaturated organic molecules, such as phenyl groups; in contrast, the RP-LC surface is covered with alkanes of C_8 or C_{18} length. Thus HIC employs aqueous mobile phases and a decreasing salt concentration gradient to elute the protein, while RP-LC involves a mixture of polar organic solvents.

In addition to its use in analytical determination, RP-LC is applied to detect variations in peptide and protein structure as part of the quality-control process during biopharmaceutical production. A common RP-LC method deploys acetonitrile as an organic phase modifier to control retention time. Protein retention is quite sensitive to acetonitrile concentration when used with peptides and small molecules. This feature is exploited to prepare gradient methods whose slopes represent a 0.5% to 1% increase in organic modifier concentration per minute. The ending modifier concentration is no more than 95%, which causes water to be retained in the organic phase.

The second important factor in RP-LC is the use of an ion-pairing agent, such as TFA, which aids in solubilizing proteins and peptides. The analytical chromatographer must make additional adjustments, including a manipulation of the solution's pH and temperature. These adjustments are guided by experimental design, with a pH range of 2 to 6.5 and a temperature range from room temperature to 50°C being explored to improve resolution or analysis speed.

Many analytical laboratories have been reducing their solvent usage because of the high cost of solvent recycling and disposal. An environmentally preferred alternative is to use columns with smaller inner diameters; such columns can reduce flow rates from the typical 1 mL/min to 200 µL/min and even 5 nL/min. These narrow-diameter columns have several important side benefits:

- Smaller samples can be analyzed (as little as five nanomoles of protein).
- Detection sensitivity increases because the samples move through the column as smaller plugs of fluid.
- Highly sensitive and powerful analysis tools, such as mass spectrometers, can be incorporated in the analysis train.

The trade-off is that capillary and microbore columns of 1-mm inner diameter may require specialized pumps and plumbing. A reasonable compromise would be to use columns with an inner diameter of 2 to 2.5 mm, which should reduce flow rates to between one-tenth and one-third of that in standard columns.

3.2.2 Analytical electrophoresis

If the separation of enzymes, such as protease, must be accomplished by non-destructive means so as to permit later assessment of biological activity, gels must be prepared under nondenaturing conditions. One technique that satisfies these criteria is polyacrylamide gel electrophoresis (PAGE). The speed of electrophoretic migration of a protein depends not only on its charge, but also on its molecular size and shape. Thus it is possible to separate two enzymes with identical net charges but different sizes by varying the pore size of the support as a function of the acrylamide concentration (which can vary between 3% and 30% on a weight-per-volume basis) and the amount of cross-linker used. Gels with less than 2.5% concentration of acrylamide permit proteins with a molecular weight close to 1,000,000 daltons to be separated; gels with a 30% concentration do the same for polypeptides with molecular weights close to 2000 daltons. Separation by charge is achieved by operating between pH 3 and 11, which produces the maximum charge differences between neighboring protein species. If the goal is to analyze very complex media or isolate enzymes in very small amounts, the resolution may be increased through two-dimensional operation.

For analytical purposes, proteins are often separated after denaturing. Using gels with an increasing acrylamide concentration permits the separation of a larger range of molecular weights with increased resolution. Electrophoresis can be performed with a single buffer (as in continuous electrophoresis) or with several buffers (a method known as discontinuous electrophoresis). The latter technique permits proteins to be concentrated in extremely small zones prior to their migration.

3.2.2.1 Isoelectric focusing

Isoelectric focusing offers two distinct advantages over the standard methods of electrophoresis. First, the former technique displays extraordinary sensitivity; in fact, it can resolve proteins that differ in pI by only 0.001 to 0.01 units. Second, it is time-independent, so only negligible deterioration of the zone definition occurs over the course of the procedure. The resolution is improved by using high field strengths of 100 to 300 V/cm in 0.2- to 1-mm gel layers. Affinity electrophoresis and iso-tachophoresis can also be used.

3.2.2.2 Protease detection in gels

Protease can be detected in gels using nonspecific methods. Because such techniques imply an irreversible alteration of the enzyme, their only use involves detection, not analysis. These methods include the thymol-sulfuric acid method, the periodic acid-Schiff base method, and the fluorescein isothiocyanate-labeled lectin method (13). Nevertheless, it is often desirable to detect proteases (which can also be glycoproteins) more specifically.

Specific detection allows for the preservation of enzymatic activity, enabling determination of protease bioactivity at a later time. It necessitates eliminating electrophoretic reagents that are likely to interfere in the determination, such as ampholine and SDS, and can involve the renaturation of the enzyme. Several techniques are used, such as removal of SDS by incubation in buffered 25% isopropanol or renaturing of the enzyme by a combination of non-ionic detergent, glycerol, and substrate.

To retrieve the enzyme, the gel is cut into pieces and an elution buffer used. The enzyme can then be assayed by conventional methods. The measured enzymatic activity depends on the degree of denaturation produced by the particular electrophoretic technique and the amount of recovery from the gel. Spectrophotometric assays, for example, are often used to detect proteases. They rely on fluorochrome or chromogenic indicators conjugated to the substrate. For subtilisin detection, N-CBZ-Gly-Gly-Leu-*p*-nitroanilide is always chosen as a chromogenic substrate. Proteolytic activity is then determined by measuring the amount of indicator released in the supernatant. Alternatively, indicators such as fluorescamine or *o*-phthaldehyde may be added to the supernatant to detect digestion products. Another option involves the use of a radiolabeled substrate. Radioassays rely on the release of radiolabeled products into the supernatant, where concentrations can then be determined by gamma or scintillation counting.

Assays may measure enzymatic activity in situ after electrophoretic detection ends. In simultaneous capture, for example, the substrate is converted by an enzyme into a product that couples immediately with a reagent to form an insoluble, colored product. In a technique known as post-incubation coupling, a substrate is incubated with the enzyme to form a product that can be detected by a color change following the addition of a reagent. Autochromic methods, which make direct observations of enzymatic activity by noting changes in the optical properties of either substrate or reaction product, are also used to detect enzymes such as protease without removing those components from the gel. Sandwich-type incubation is possible, in which a matrix containing an auxillary indicator enzyme or high-molecular weight substrate is added to the gel. Finally, staining for enzymatic activity can be accomplished through copolymerization of the substrate in the gel, a technique applied to high-molecular-weight substrates such as gelatin or casein. To prevent the enzyme from acting on the substrate during electrophoresis under enzymatic conditions, an inhibitor or chelating agent may be included, a less than optimal pH may be used, or another means may be employed to halt these reactions.

Example 3.1

To illustrate the use of gel electrophoretic methods, we will discuss the detection of an alkaline protease using PAGE (14). The electrophoresis was performed using a 7.5% (weight/volume or w/v) acrylamide slab gel (8×10 cm, 1 mm in thickness) with a running buffer of 25 mM Tris/192 mM glycine at pH 8.3. The gel was stained for the presence of protein with coomassie brilliant blue R-250; it was then destained with 7.0% (volume/volume or v/v) acetic acid. Protein bands that possessed protease activity were visualized by using an agar sheet that contained 1.0% (w/v) casein, 50 mM borate/NaOH buffer at pH 10.0, and 2.0% (w/v) agar. The slab was laid on the top of the agar sheet and was left at 40 °C for several hours.

After flooding with 5.0% (w/v) trichloroacetic acid, the bands of protein that had protease activity appeared as clear zones in the overlay agar sheet. The molecular mass of this protease was estimated by SDS-PAGE in the conventional way, with 12.5% (w/v) slab gels using the same running buffer as before, with the addition of 0.1% (w/v) SDS.

3.2.3 Immunoassays

Antibody recognition has proved to be a powerful tool in chromatographic and electrophoretic analytical methods. Immunoassays rely on the direct visualization of antibody binding to a protein antigen via immunoassay technology. Two of the more popular immunoassays currently in use are radioimmunoassays (RIAs) and enzyme-linked immunosorbent assays (ELISAs). ELISA is a solid-phase method, while RIA is conducted in solution. Both techniques use microtiter wells and auto-mated sample application, washing, and detection procedures.

Although several types of ELISA protocols exist, all of them share a common set of steps. Antigen and an antibody specific for that antigen will bind in an immunorecognition reaction, which can have association constants ranging from 10^4 to $10^{10}\,M^{-1}$ or greater. This condition is descibed kinetically by Equations 3.1 and 3.2, where Ab is the antibody, Ag is the antigen, and AgAb is the antibody–antigen complex.

$$Ab + Ag \leftrightarrow AgAb \tag{3.1}$$

$$K = \frac{[AgAb]}{[Ag][Ab]} \tag{3.2}$$

Antibody binds to antigen through the formation of noncovalent bonds between the amino acid residues at the binding site of the antibody and the antigen surface (15). After the complex forms, washing removes the free antigen and antibody from the microwell. To determine activity, an agent is added that allows for automated detection, usually based on a color change in the microwell solution.

Two forms of ELISA exist: competitive and noncompetitive forms. In non-competitive ELISA (also known as "sandwich ELISA"), the antibody is initially attached to the solid surface, which is usually a microwell material consisting of polystyrene or polyacrylamide. The next three steps are familiar: addition of an antigen, incubation, and washing to remove free antigen. An excess of enzyme-labeled antibody is then added. To detect the antibody, an enzyme substrate is used; the quantity of the colored product formed is proportional to the amount of antigen present in the microwell. In this technique, the name "noncompetitive" refers to the fact that the measurement being taken directly reflects the number of occupied sites.

In competitive ELISA, antibody is again attached to the microwell wall. After washing the microtiter well, the complementary antigen is added, followed by the application of an enzyme-labeled antigen that reacts with the antibody. Subsequent washing removes free antigen, after which an enzyme substrate is added for detec-tion. In this technique, the amount of product formed is inversely proportional to the amount of the unlabeled antigen bound by the antibody. This assay is classified as a "competitive" ELISA because the actual measurement taken indicates the number of unoccupied sites. To interpret the results, a calibration curve of known standards is required.

Many variations of these assays exist, but the basic theme for all ELISAs remains the same. In general, noncompetitive assays are more sensitive than competitive assays, as the latter involve limited amounts of reagent. Noncompetitive assays, which use an excess of antibody, are also less sensitive to environmental conditions and contaminants that may exist in the solution (15).

In RIA, competitive binding between radioisotope-labeled and unlabeled antigen for a set amount of antibody is used to quantify the amount of antigen in a sample. If a known amount of labeled antigen and antibody is used, an unknown quantity of unlabeled antigen can be determined (as long as a calibration curve has been calculated prior to the procedure). After both unlabeled and labeled antigens have reacted with the antibody, the amount of bound antigen counts, B, and the amount of bound counts if no unlabeled antigen is used, B_0, are identified. The radioactivity of complexed, labeled antigen is inversely proportional to the concentration of free, unlabeled antigen. A linear calibration curve is made possible by plotting $\ln\left[\dfrac{(B/B_0)}{1-(B/B_0)}\right]$ versus the natural log of the concentration of free, unlabeled antigen.

3.3 NUCLEIC AND POLYNUCLEIC ACIDS

In the past two decades, much research has focused on developing analytical techniques suitable for nucleic acids. The research has been especially fruitful in the specific area of nucleic acid analysis. This section reviews current techniques used in the analytical determination of nucleic acids and their derivatives. We will first present the chromatographic methods, followed by a discussion of the gel and capillary electrophoretic methods.

3.3.1 Ion-exchange chromatography

In ion-exchange chromatography, the stationary phase has electric charges on its surface, created when anions or cations are placed on resin or gel surface. The mobile phase contains ions and ionic sample molecules, which must compete for a place on the surface of the stationary phase. In cation-exchange chromatography, the resin incorporates anions (Figure 3.4); in anion-exchange chromatography, cations are placed on the surface of the resin. Charges are neutralized by counterions in the mobile phase.

The first simultaneous separation of bases, nucleosides, and nucleotides was performed in 1978 using Aminex-14, a totally porous anion-exchange resin (16). Microparticulate, chemically bonded, totally porous packings were developed in 1973. To create a strong anion-exchange packing, quaternary amines are generally bonded to the silica through silanol groups, while strong cation-exchange packings rely on sulfonic acid groups. This material offers the advantages of increased sample capacity, greater efficiency, and more rapid mass transfer, which results in shorter retention times. A microparticulate, chemically bonded, strong anion-exchange column has been used to separate 21 nucleotides with excellent results (17).

Many separation techniques using ion-exchange chromatography have been developed for bases, nucleosides, and nucleotides. Nevertheless, ion exchange has

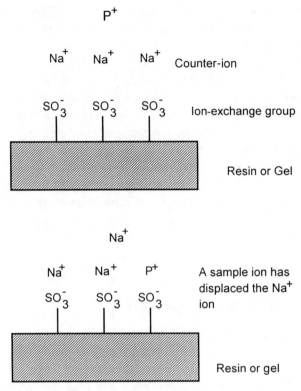

Figure 3.4 An example of a cation exchanger in the sodium form. (Adapted from: Meyer VR. Practical high-performance liquid chromatography. New York: John Wiley and Sons, 1988.)

some major disadvantages, including long column equilibration times, lack of reproducibility, and weak intercolumn performance. In addition, this method cannot separate the three groups of nucleic acids simultaneously with high resolution (18).

3.3.2 Reverse-phase high-performance liquid chromatography

RP-HPLC offers many advantages relative to ion-exchange chromatography—namely, high resolution of compounds with diverse structures and properties, column stability, reproducibility, and rapid analysis. Because the silica backbone of reverse-phase columns decomposes at pH levels lower than 2 and higher than 8, however, the eluents used cannot exceed these limits and the solutes under consideration must remain stable in this pH range.

Additionally, the solutes should not be ionized, because charged species exhibit a greater affinity for the more polar mobile phase than for the hydrophobic stationary phase; consequently, charged species will not be retained with RP-HPLC. At pH 5, for example, nucleosides and bases are mostly neutral and thus easily separated by RP-HPLC. Using a phosphate buffer at pH 5.7 and gradient elution, an excellent separation of these compounds in both standard mixtures and biological fluids can be achieved (19, 20). Isocratic separations are more desirable, however, because they require less instrumentation, do not produce equilibration time between runs, and involve simple quantification methods. Using an isocratic RP-HPLC system, a mixture of 20 ribonucleotides, deoxyribonucloides, and bases has been separated (21). Similarly, a mixture of 18 major and minor deoxyribonucleosides and ribonucleosides was separated in 40 minutes through the use of RP-HPLC (22).

3.3.3 Ion-pair chromatography

Ion-pair chromatography (IPC) facilitates the separation of ionized and non-ionized compounds. It involves the coating of a stationary support, such as silica gel or cellulose, with a reagent capable of forming ion pairs with the sample (23). To elute the ion pair, a comparatively nonpolar mobile phase is applied. In its current iteration, IPC involves the addition of a pairing agent to the mobile phase to increase the retention of the ionic compound; this compound forms a neutral ion pair with the pairing agent.

As nucleotides exhibit an overall negative charge at pH values greater than 3, IPC is considered a viable tool in nucleotide analysis. Because of their acidity, quaternary amines are generally employed as the counter-ions in this procedure. Various adenine nucleotides have been separated using a C_{18} column, an isocratic program with a mobile phase consisting of an acetonitrile-water solution, and tetrabutylammonium phosphate as the pairing agent at a pH of 5.8 (24). The first separation of monoribonucleotide isomers using IPC was accomplished under isocratic conditions using tetramethylammonium hydroxide as a pairing agent (25). In addition, Zwitterionic detergents have been used to separate nucleotides (26).

IPC has value in separating bases. The major purine and pyrimidine bases can be separated by RP-HPLC, but several other bases—cytosine, 5-methylcytosine, guanine, thymine, and uracil—cannot be completely separated under normal chromatographic conditions (27). Fortunately, the separation can be improved by the addition of a pairing agent. Because the target components are bases, the pairing agent used in this case is a negatively charged heptane sulfonate. The desired resolution can be achieved using a C_{18} column, a mobile phase of phosphate buffer, and the sulfonated pairing agent at pH 5.6.

3.3.4 Slalom chromatography

The principle of slalom chromatography was developed by Hirabayashi and Kasai (28). Large, double-stranded DNA molecules can be separated by size by deploying an ordinary HPLC system and a size-exclusion column. Interestingly, the order of elution in this approach is the opposite of that expected for size-exclusion chromatography, with larger fragments being eluted later than smaller ones. The separation process strongly depends on the flow rate and size of the packing particles, rather than the pore size or chemical nature of the packings (29, 30). This phenomenon cannot be explained by any previously known mechanism. Kasai has suggested that the mechanism involved might reflect the hydrodynamics of the system, rather than equilibrium concerns (31).

A column that is packed with hard, spherical beads has narrow and tortuous open spaces. When DNA molecules are injected into the column, they unfold and extend, a motion promoted by the laminar flow generated as the solvent passes through the narrow channels. These extended molecules must flex quickly in the rapid flow of mobile phase to pass through the tiny openings. It is quite possible that the longer the DNA molecule, the more difficulty it encounters in entering the openings. Thus size-dependent separation occurs in the opposite order to size exclusion. Experiments done by Hirabayashi and Kasai (28) indicate that this explanation will suffice as a first approximation. Independently made observations have supported this hypothesis (32).

Important characteristics of slalom chromatography include the lack of interaction between the DNA fragments and the column packing, and the observation that only the particle size of the packing is important. Thus smaller particles resolve smaller DNA fragments, and larger particles resolve larger DNA fragments. In addition, the pore size and chemical nature of the particles used are not significant factors. Thus the separation depends largely on the flow rate, and to a lesser extent on the temperature.

3.3.5 Gel electrophoresis

Electrophoresis of nucleic acids may be carried out in agarose, polyacrylamide, or agarose-acrylamide composite gels. In the case of duplex DNA fragments, a neutral pH will allow the DNA to carry an overall negative charge. Fragments loaded into a sample well at the negative cathode end of the gel will then move toward the positive anode. The electrophoretic mobility of DNA fragments larger than oligonucleotides depends on the fragment size, rather than base composition or sequence.

Agarose gels can be used to analyze double-stranded DNA fragments from 70 base pairs, using 3% (w/v) agarose gel, to 800,000 base pairs, using 0.1% agarose gel (33). For any set of conditions, there is an upper fragment size at which resolution deteriorates sharply. Likewise, gels become difficult to use below 0.5% agarose. For these reasons, fragments larger than 50,000 base pairs should be analyzed using an alternative method, such as pulsed-field electrophoresis. Polyacrylamide gels are viable options for fragments between 6 base pairs, using 20% acrylamide (34), and 1000 base pairs, using 3% acrylamide. The smallest detectable amount of non-radioactive DNA has been found to be less than 1 ng.

Denaturing gel systems of several types are available for the analysis of single-stranded DNA fragments. As with the double-stranded DNA, the fragments migrate with a size-dependent mobility, though the mobility of single strands of DNA moving in nondenaturing gels does depend somewhat on secondary structure. To obtain better resolution, the system may include gels that offer a gradient of increasing denaturing power. That is, the mobility of a fragment drops sharply at the point in the gel where complementary strands begin separating. The final fragment position is therefore sequence-dependent. In combination with size separation, this process makes two-dimensional separation of DNA fragments possible.

3.3.6 Pulsed-field gel electrophoresis

Conventional agarose gel electrophoresis, as described in Section 3.3.5, can separate only DNA fragments of less than 50 kilobases. In contrast, pulsed-field gradient gel electrophoresis (PFGE) can separate DNA molecules as long as 2000 kilobases (35). In this electrophoretic method, DNA travels through a concentrated agarose gel under the influence of two nonuniform electric fields, which occupy nearly perpendicular positions. The fields are then pulsed (i.e., switched on and off in an alternating fashion). In conditions of high gel concentrations and high voltage gradients, DNA molecules are assumed to stretch along the direction of the field so as to penetrate the pores of the gel and make a net forward movement. The molecules must therefore reorient themselves in the gel every time the field is pulsed if they are to travel in a direction approximately at right angles to the axis along which they are stretched. The longer the DNA molecule, the longer the time taken to find the new orientation, and the more the gel restrains the molecule.

Several modifications of this method have been proposed in an effort to improve the resolution and running characteristics of the gel (36–38). These variations are sometimes referred to by different terms, depending on the type of apparatus and the arrangements of the electrodes. The technique modifications have also extended the fragment size that can be separated, with DNA molecules larger than 6000 kilobases in size being clearly resolved (39).

3.3.7 Capillary isotachophoresis

In capillary isotachoporesis (ITP), separations are carried out in small capillaries using a discontinuous electrolyte system consisting of a leading electrolyte (LE) and a terminating electrolyte (TE). The LE has the most mobile ions, and the TE the least mobile. In the typical procedure, a 1- to 10-μl sample is injected between the LE and TE zones and allowed to separate (40). To prevent diffusion between the separated sample zones, substances such as methylcellulose are added to enhance the solution's viscosity. ITP analysis is performed at constant current. Because the separated zones will experience different electric field strengths, diffusion is restrained, creating sharp zone boundaries in a process known as the zone-sharpening effect.

According to Kohlrausch's law (41), the lengths of the separated homogeneous sample zones reflect the LE concentration. That is, depending on the concentration of the LE electrolyte, the target compounds will be either concentrated or diluted. Consequently, compounds present in only small amounts can be detected more easily in complex mixtures than in most other separation processes.

3.3.8 Capillary zone electrophoresis

The separation of nucleic acids and their fragments represents a challenging application of CZE. The various parameters that can influence the separation of oligonucleotides have been thoroughly investigated (42). When polycytidines serve as the model compounds, varying the pH between 5 and 8 and the ionic strength between 20 and 200 mM apparently has little effect on the separation. If spermine is added to the background electrolyte, however, the migration order becomes inverted as the migration of larger oligonucleotides slows. If the background electrolyte contains SDS plus spermine, then the separation order changes yet again. Reportedly, the best separations have been achieved with an electrolyte containing 60 mM SDS and 3 mM spermine.

If ampholine is added to the background buffer, a wide size range of nucleic acids can be separated under conditions similar to those required for the isotachophoretic separation of proteins (43). Somewhat surprisingly, the chain lengths of the nucleic acids do not serve as the major factor in determining the individual mobilities under these conditions. Polynucleotides having more than 1 kilobase appear as sharp peaks, but this sharpness disappears when the sample is treated with DNase, indicating heterogeneity and incomplete separation of the products. To rectify this development, high-performance capillary electrophoretic (HPCE) separations have been attempted. This technique uses gel and open-tube capillary columns with fluorescein-labeled, single-stranded oligonucleotides for DNA sequencing (44).

Capillary polyacrylamide gel electrophoresis has been used to pull out deoxycytidine-terminated DNA fragments (45). A postcolumn, laser-induced fluorescence detector minimizes the background noise caused when the gel-filled capillary scatters light. To date, a detection limit of 10^{-20} moles of fluorescein-labeled DNA has been achieved.

Polyacrylamide gels with smaller pore sizes than agarose are well suited for separating samples of 10 to 35 kilobases. Small polynucleotides have been successfully isolated on such gels (44, 46). To gain a deeper insight into the overall performance and reproducibility of such systems, polyacrylamide gel columns were tested with homopolymeric standard samples. Researchers reported plots of migration times versus molecular size over a range of 30 to 160 base pairs (47). In the plots, heteropolymeric samples separated in polyacrylamide-filled capillaries exhibited a linear relationship between migration time and the number of bases in the sample. The same relationhip held true for three different oligonucleotide samples. Consequently, gel-filled capillary separations appear feasible for molecular mass determinations.

3.4 CARBOHYDRATES

3.4.1 Monosaccharides

Monosaccharides possess a great variety of structures with different physical and chemical properties, so no single method is suitable for the quantitative or qualitative analysis of all monosaccharides. The method of choice depends on the desired accuracy and resources available. HPLC and gas-liquid chromatography (GLC) are the primary techniques used for quantitative analyses of monosaccharide mixtures. Single monosaccharides may be identified by other techniques, such as mass spectrometry (MS), infrared spectrometry (IR), proton nuclear magnetic resonance (^{1}H-NMR), and carbon-13 nuclear magnetic resonance (^{13}C-NMR). These classes of compounds can also be quantified with enzymatic assays or colorimetric analysis. Depending on the chemical and physical properties of the compounds, different modes of chromatography may be employed to separate and analyze carbohydrates.

3.4.1.1 High-performance ion-exchange chromatography

Typically, sulfonated polymeric columns containing metal-loaded cation-exchange packings are used to detect monosaccharides; temperatures for these systems hover near 85 °C. The counter-ions most commonly used are Ca^{2+}, Ag^+, H^+, and Pb^{2+}. Cation exchangers in the H^+ form are best suited for separating carbohydrate mixtures, including acids and amino sugars (48). Ca^{2+} is used mainly for corn sugars and syrups at low temperatures, approximately 1.5 °C. Pb^{2+} is used for mixed hexoses and pentoses. Ag^+, on the other hand, has utility when oligosaccharides need to be separated from monosaccharides. It is important to remove Na^+ and K^+ from the samples when using metal-loaded columns, as these ions will interact with the column packing and reduce the efficiency of the separation process.

In addition, amino-bonded silica columns may be used in conjunction with an acetonitrile/water mobile phase. In this process, polar and hydrophobic interactions as well as partitioning between the water-enriched stationary phase and the acetonitrile-rich mobile phase pull out the desired compounds. Anion-exchange chromatography, executed at high pH using strong basic anion-exchange resin, is a reproducible and highly preferred method. Anion exchange can separate compounds that are negatively charged under such basic conditions. Table 3.3 lists typical conditions of ion-exchange chromatography for monosaccharides.

Table 3.3 Examples of Typical Conditions for the HPLC of Monosaccharides

Column Type	Mobile Phase	Separation Mechanism	Commercial Columns
Anion exchange (4° ammonium)	Sodium hydroxide	Anion exchange	Dionex Carbopac MA1
Anion exchange (aminopropylsilane bonded silica, OH⁻ form)	Acetonitrile/water	Hydrogen bonding between hydroxyls and amines	Varian MicroPak AX-5
Cation exchange (sulfonate, Ca^{2+} form)	Water	Ion-moderated partition	Waters Sugar-Pak I
Cation exchange (sulfonate, Ag^{2+} form)	Water	Ion-moderated partition	Bio-Rad Aminex HPX-42A

Source: Adapted from Chaplin MF, Kennedy JF, eds. Carbohydrate analysis: a practical approach. United Kingdom: Oxford University Press, 1994.

Pulsed amperometric detection (PAD) is an extremely sensitive electrochemical detection method in which the oxidation of anionic carbohydrates at a pH of 13 uses a gold electrode balanced at positive potential (48). The electrode can be pulsed between high positive and negative potentials, an action that cleans the electrode surface. Using this method, the total and free carbohydrate content in soluble coffee was analyzed using anion-exchange chromatography (49). The mobile phase consisted of deionized (DI) water and NaOH. The column was maintained at ambient temperature, the flow rate of the eluents persisted at 1 mL/min, and the injected volume comprised 10 to 20 mL of standard and sample solutions. This method yielded a reproducible separation of all major carbohydrates found in soluble coffee, including glucose, fructose, galactose, ribose, sucrose, fucose, mannitol, rhamnose, arabinose, and xylose (49).

3.4.1.2 Gas-liquid chromatography

Gas-liquid chromatography (GLC) is a sensitive technique capable of analyzing nanomolar quantities of carbohydrates. It is generally less prone to interference from other classes of compounds, such as salts and proteins, than other chromatography techniques. GLC separation operates based on differential extractive distillation of the components in a mixture. To use this method, volatile derivatives of the compounds must be prepared. GLC derivatization via trimethylsilylation is generally applicable to monosaccharides (50), with the majority of monosaccharides producing a mixture of isomers upon derivatization.

Wall-coated open tubular (WCOT) columns containing fused silica can resolve a large number of components. These columns offer tenfold better resolution and fivefold faster speed than packed-bed columns.

Internal standards—such as myo-inositol and pentaerythritol—are usually necessary for GLC analysis.

A flame ionization detector (FID) is usually employed as the method of detection. FID responds to all carbohydrate-related molecules in a linear fashion over an extremely wide range. In this technique, hydrogen is mixed with a carrier gas such as helium, and the mixed gases are burned in air from an external source. Analyte molecules that elute from the column are subsequently burned in the flame, releasing electrons. The electrons migrate to the electrodes in the detector, producing an electric current that is amplified and measured (51). FID is inexpensive to purchase and operate.

3.4.1.3 Gas-liquid chromatography–mass spectrometry

Electron impact mass spectrometry (EI-MS), when used in conjunction with GLC, can analyze complex mixtures more rapidly than virtually any other method. It takes advantage of both mass spectra and retention time factors. For GLC purposes, carbohydrates are normally converted to volatile derivatives—also a requirement for EI-MS. EI-MS is based on the principle of positive ionization of the molecules after bombardment with an electron beam. The mass spectra patterns obtained from monosaccharide derivatives can then be compared with spectra from known materials, allowing the chromatographer to identify the molecules. GLC-MS is a very useful technique for analyzing partially methylated alditol acetates, as known standards are not always available (48).

Other techniques, such as IR, can be used to confirm the identity of monosaccharide molecules by comparing their spectra with the spectra of known materials. Fourier transform IR spectroscopy (FTIR) is also used for on-line quantitative analysis of relatively simple sugar solutions, such as soft-drink beverages.

3.4.2 Oligosaccharides

The analysis of oligosaccharides involves qualitative identification, quantitative determination, and analysis of the linkages between different residues. A variety of methods are available for analyzing oligosaccharides, including thin-layer chromatography (TLC), gel permeation chromatography (GPC), affinity chromatography, ion-exchange chromatography, capillary electrophoresis, and infrared spectroscopy.

3.4.2.1 High-performance ion-exchange chromatography

Several unique column chemistries have been developed for carbohydrate separation. One common technique, ion exclusion, is used in carboxylic acid separation as well. Metal affinity chromatography for oligosaccharide analysis was developed by replacing hydrogen ions with either potassium, calcium, magnesium, silver, or lead ions after using strong cation exchangers. Amino groups may also interact with oligosaccharides and have application in separating mono-, di-, and trisaccharides.

Choosing the best column for a given separation requires knowledge of prior results or experimentation. In general, calcium ion columns have utility in separating sugars in high-fructose corn syrup. Lead (Pb^{++}) ion columns can separate xylose, galactose, and mannose, which cannot be resolved using calcium ion columns. Other columns also have specific strengths. Consulting with several vendors prior to settling on a column and method is recommended.

Anion-exchange HPLC can separate oligosaccharides with a polymerization degree of 50 or lower. The analysis is performed at high pH, and pulsed amperometric detection is commonly used for the actual separation process. The results obtained depend upon the molecular size, sugar composition, and the type of linkages between the monosaccharide units. To fractionate oligosaccharides in order of decreasing molecular size, anion-exchange resins utilizing carbonate or bicarbonate forms are popular. Using acetic acid, formic acid, or lithium chloride eluents, it is possible to fractionate acid-containing oligosaccharides such as uronic acid or aldonic acid (48). The analysis of oligosaccharides and monosaccharides in body fluids by anion-exchange chromatography of their borate complexes has been carried out with 2-cyanoacetamide labeling (50).

3.4.2.2 Capillary electrophoresis

Capillary electrophoresis (CE) provides rapid and efficient separations of complex mixtures of ions. Figure 3.5 shows a typical CE system, which consists of two electrolyte buffer reservoirs, a fused-silica capillary, a high-voltage power supply, and a detector linked to a data acquisition system.

The sample is introduced into the capillary either by pumping action or by creating a pressure difference between the inlet and outlet of the capillary. In most cases, sample volumes lie in the nanoliter to picoliter range. Electro-osmotic flow separates the molecules in the sample. Excess positively charged ions flow toward the cathode, enhancing the isolation of charged ions. Possible detectors include optical, conductive, or electrochemical options.

Hexosamine-containing oligosaccharides are often separated and characterized by CE (50). In contrast, neutral oligosaccharides cannot be detected directly by this technique and must undergo derivatization. Neutral oligosaccharides are converted to primary amines by reductive amination. CZE separates oligosaccharide derivatives according to the degree of polymerization, up to a degree of 20 (48). Borate or phosphate buffers are frequently employed, along with an applied voltage in the kilovolt range.

3.4.2.3 High-performance liquid affinity chromatography

The general principle of affinity chromatography is the most specific chromatographic method. Its high specificity derives from the fact that the participating components are biologically matched, both spatially and electrostatically. In this technique, a ligand is bound to the support and the sample is adsorbed from the solution in a reversible process. The matched sample component is retained in the

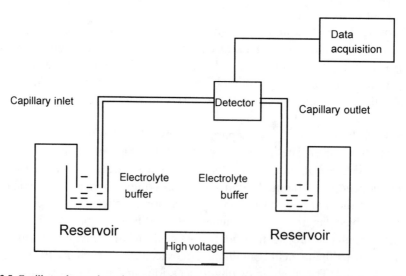

Figure 3.5 Capillary electrophoresis system. (Adapted from: Ward KM, Lehmann CA, Leiken AM, eds. Clinical laboratory instrumentation and automation: principles, applications, and selection. Philadelphia: W.B. Saunders, 1994.)

stationary phase; the other molecules pass through with the mobile phase. The sample is separated from the stationary phase by either adding an eluent with great affinity to the ligand or changing the solution's ionic strength or pH.

Affinity chromatography has been used to analyze glucose-containing oligosaccharides (Glc$_4$) (52). The ligand used was a murine monoclonal antibody, IgG2b, that specifically binds glucose-containing oligosaccharides. The antibody was coupled to a Selectisphere-10-activated tresyl column, which consists of a macroporous silica bed. Upon isocratic elution, the system was able to detect 10 ng of oligosaccharide, with a linear response reaching to 250 ng, using a pulsed amperometric detector. The sample required no derivatization, and the analysis could be completed in 20 minutes. In general, affinity HPLC methods provide better resolution and more rapid analysis than other methods for oligosaccharides. To optimize the separation process, the temperature can be varied (52).

Other methods for analysis of oligosaccharides include MS-GLC (to obtain structural information of oligosaccharides), nuclear magnetic resonance (NMR; a nondestructive technique to obtain structural information), and colorimetric methods (for gross determination of carbohydrate content).

3.4.3 Glycoproteins

Glycoproteins are proteins that bear carbohydrate chains. High-performance affinity chromatography is very useful in analysis of clinical samples that contain such proteins. Lectins—sugar-binding proteins—bear at least two sugar-binding sites, which explains why they precipitate polysaccharides, glycoproteins, and glycolipids (53). These substances, which are used in the study of glycoproteins, are classified according to their monosaccharide specificity. For example, *Canavalia ensiformis* (Con A) and *Lens culinaris* (LCA) are lectins that have monosaccharide specificity to α-D-glucose and α-D-mannose, respectively (48). Lectins can be used as detectors in combination with electrophoresis because of the high specificity toward the constituent monosaccharide.

Various electrophoretic techniques are used to analyze glycoproteins. Phases employed include polyacrylamide gels, cellulose acetate membranes, and filter paper. A two-dimensional technique on polyacrylamide gel gives the best separation and is commonly used when working with complex biological fluids (50).

3.5 LIPIDS

3.5.1 Fatty acids

Several different chromatographic techniques are applied to analyze carboxylic acids. The relevancy of a specific method for routine analytical practice relates to the simplicity and rapidity which which it can make a determination, factoring in the time needed for any sample preparation.

Gas chromatography has been used for the separation of carboxylic acids (54). Because most carboxylic acids are relatively involatile, suitable volatile derivatives must be prepared. The complicated preparation step required ultimately limits the appeal of GC.

An alternative, high-performance liquid chromatography, affords a rapid and simple technique for analyzing certain mixtures of carboxylic acids. Four main methods are commonly used: adsorption chromatography, ion-exchange chromatography, ion-exclusion chromatography, and reverse-phase chromatography. Another applicable technique is capillary zone electrophoresis.

3.5.1.1 Elution chromatography

In elution chromatography, excess solvent sweeps a pulse of solutes through a packed column. The packed column contains solid adsorbents that retard the elution of specific solutes. The pulse enters one end of the column as a narrow, concentrated peak, but it exits the other end dispersed and diluted by the additional solvent. Pulses of different solutes leave the column outlet at different times, allowing each one to be drawn off separately (Figure 3.6). Detection requirements vary with the separation goal. For analysis of a single component, for example, the ideal detector would sense only the material of interest and would not respond to any other component. Although UV photometric detectors are the most convenient options today,

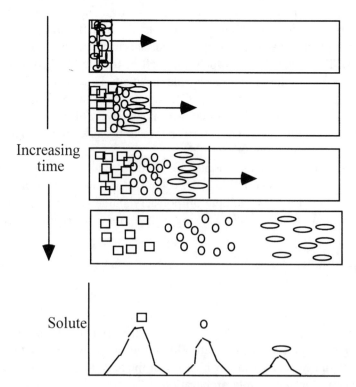

Figure 3.6 Schematic diagram of elution chromatography. (Adapted from: Belter PA, Cussler EL, HU WS. Bioseparations: Downstream processing for biotechnology. New York: John Wiley and Sons, 1988:184.)

Table 3.4 Solid Phases Used for Elution Chromatography

Scale	Type	Stationary Phase	Mechanism	Remarks
Large	Adsorption	Alumina, silica gel	Van der Waals forces	Low selectivity, inexpensive
	Partition	Diatomaceous earth/glycol	Liquid/liquid distribution	Low selectivity, inexpensive
	Ion exchange	Polymers of sulfonated styrene, dextran	Weak ionic or covalent bonding	Good selectivity
Moderate	Gel filtration	Dextran, polyacrylamide	Differences in absorption	Good selectivity
	Affinity	Bound enzymes	Specific chemical reactions	Exceptional selectivity
Small	Thin layer (TLC)	Alumina, hydroxyapatite	Weak van der Waals forces	Standard laboratory
	Paper	Cellulose	Weak van der Waals forces	Analytical tool only

Adapted from: Belter PA, Cussler EL, Hu W-S. Bioseparations: Downstream processing for biotechnology. New York: John Wiley and Sons, 1988:186.

other detectors (such as potentiometric and conductometric detectors) have been used as well.

The packings are similar to those used for adsorption (Table 3.4), but the mechanisms involved in elution chromatography differ from those employed in adsorption. The goal in adsorption chromatography is to capture the solute on the adsorbent and elute it in concentrated form; in elution chromatography, the goal is purification of the product, even though it may remain in dilute solution (55).

3.5.1.2 Adsorption chromatography

In adsorption, molecules in a fluid phase concentrate on a solid surface without any chemical change. This phenomenon occurs because physical surface forces from the solid phase attract and hold certain molecules—namely the substrate—from the fluid surrounding the surface. Silica, alumina, activated carbon, and macroporous non-ionic polymers frequently serve as adsorbents. Low-molecular-weight organic solvents are the most common eluents.

Charcoal adsorbents, for example, have been used to remove carboxylic acids from aqueous solutions. Table 3.5 lists the bed volumes of the feed streams in such a system, as well as the amount of 1.0 N NaOH required for regeneration. Because charcoal does not possess the same fixed ion-exchange groups as ion-exchange resins have, it is slower to regenerate, especially when in contact with more hydrophobic solutes. The extent of adsorption is inversely proportional to the water solubility of the compound, and the adsorption of monofunctional organic carboxylic acids is always greater than that for polyfunctional molecules of the same molecular weight. Substitution of hydrophilic groups (—OH) lowers the quantity of solute adsorbed and increases the speed of regeneration. Conversely, substitution

Table 3.5 Adsorption of Organic Acids on Charcoal, Feed Concentration 0.1%

Solute	Breakthrough Point (Bed Volumes)	Regeneration Requirements (Bed Volumes)
Formic acid	4.0	< 2
Acetic acid	21	< 2
Oxalic acid	22	< 2
Malonic acid	36	< 2
Propionic acid	42	2.5

of hydrophobic groups (—CH) increases the quantity of solute adsorbed and decreases the efficiency of regeneration (56).

Another example involves the adsorption of carboxylic acids on styrene-based XAD-6 (a hydrophobic adsorbent) and methacrylate-based XAD-7 (a hydrophilic adsorbent) (57). For both resins, as the chain length of the acid increases, the amount of adsorption increases.

3.5.1.3 Ion-exchange chromatography

In ion-exchange chromatography, true heteropolar chemical bonds are reversibly formed between ionic components in the mobile and stationary phases. The ion-exchange beds may have either hydrophobic or hydrophilic matrices that contain ionizable groups.

For example, itaconic acid has been extracted from a molasses fermentation liquor by adsorbing it onto a strongly basic anion resin. The itaconic acid was then eluted quantitatively from the resin with 3 N NaOH (58). This acid was also purified by passing the liquor through a cation resin in the potassium form (59). After neutralization of the effluent from this column, the solution was passed through an ion-exchange column membrane system. The dialyzate subsequently moved through a weakly basic resin column, where the itaconic acid was eluted with a concentrated solution of potassium bicarbonate. In this procedure, ion exchange is typically followed by conductometric detection of the organic carboxylic acid.

Another ion-exchange chromatography example involves the quantitative analysis of organic acids in sugar-cane process juice. Separation of these acids was achieved on Aminex HPX-87H cation-exchange columns, and researchers improved resolution by connecting two columns in series and equilibrating them at different temperatures (60). The same method was also successful in separating the optical isomers of aconotic acid.

3.5.1.4 Ion-exclusion chromatography

Ion-exclusion chromatography (IEC) involves the separation of an ionic component from a non-ionic component. The ionic component is excluded from resin beads by ionic repulsion, while the non-ionic component is distributed into the liquid phase inside resin beads. The ionic solute travels only in the interstitial volume.

Consequently, it will reach the end of the column before the non-ionic solute, which must travel a more tortuous path through the ion-exchange beads.

IEC is particularly useful because it can easily distinguish organic acids from one another as well as from inorganic anions of strong acids (61). In this type of chromatography, relatively strong acid eluents are required to isolate many acids simultaneously. Such elution conditions are not completely compatible with conductometric detection, however, because sensitive detection requires low background conductance.

3.5.1.5 Reverse-phase chromatography

As stated earlier in this chapter, reverse-phase liquid chromatography involves the partitioning of a compound between a hydrophobic, stationary phase and a polar, aqueous, mobile phase. The strength of an eluting solvent increases as its polarity decreases. In addition, compounds of similar structure can be expected to elute in order of decreasing polarity. In one application of this technique, shikimic acid was analyzed with a 25-cm C_{18} Microsorb-MV reverse-phase column and a mobile phase consisting of 0.013 M tetrabutylammonium hydroxide, 0.0087 M trizma base, 0.09 M acetonitrile, and approximately 0.0045 M acetic acid in water (62).

3.5.1.6 Capillary zone electrophoresis

In electrophoresis, as noted earlier, electrically charged particles are transported in a direct-current electric field. The media employed in biochemical applications of this technique are usually aqueous solutions, suspensions, or gels. In capillary electrophoresis (CE), a supporting medium is not necessary, and analysis can be carried out in free aqueous solution.

The most commonly used capillary material for CE is fused silica, although glass and Teflon have been employed as well. The silanol groups on the silica capillary walls acquire a negative charge with positive counter-ions in the aqueous electrolyte buffer. These hydrated counter-ions are attracted to the cathode, creating an electro-osmotic flow toward this pole. Hence, if an electric field is applied across the fused-silica capillary, the mobile ions of the solution will migrate with their hydrating water toward the cathode, resulting in a bulk flow of the solution. The presence of this electro-osmotic flow effectively isolates both negatively and positively charged species in the same run (63).

Low-molecular-weight carboxylic acids can be separated and quantified by capillary zone electrophoresis (CZE) by using an on-column conductivity detector. The addition of 0.2–0.5 mM TTAB (tetradecyltrimethylammonium bromide) serves to control the electro-osmotic flow, ensuring that all carboxylate anions pass through the detector. Unlike in other CZE detection methods, retention time and peak area are directly related in conductivity. This relationship confers a unique advantage on this technique, as only a single internal standard can permit accurate determination of absolute concentrations in a mixture; no separate calibration for each component is necessary. Although some carboxylic acids have very similar structures, their mobilities under suitable conditions differ sufficiently to achieve full resolution. This technique has proved useful for separating optically active isomers of fumaric acid and related acids (64). In addition, it has been used to quantify carboxylic acids in dairy products, such as lactate in yogurt (64).

Everaerts, Beckers, and Verheggen (65) have proposed three major methods for separating similar species via isotachophoresis. In this technique, separations occur based on differences in mobility, pKa values, and interaction with different electrolytes (these ideas also apply to CZE). The difference in the species' mobilities is greatly influenced by pH.

3.5.2 Fats and oils

Lipids are commonly analyzed using thin-layer chromatography (TLC). Standard column chromatography is too slow with these substances, and detection problems may arise because all compounds may not be eluted. In TLC, an adsorbent—typically silicic acid, cellulose, or aluminum oxide—is coated onto glass, plastic, or aluminum plates. The mobile phase primarily consists of a polar organic solvent saturated with water.

TLC closely resembles paper chromatography, which provided the first demonstration of the chromatographic separation principle. Solutes (such as lipids) are retained based on their type of interaction, such as hydrogen bonding, ion exchange, or a specific chemical interaction.

One especially attractive feature of TLC is the direct visualization of the solutes, made possible by spraying the plate with a coloring agent that reveals the presence of a solute as a spot. For lipids, ninhydrin and Phospray are typically used as coloring agents for visualization. Spot sensitivities can be as low as $2\,\mu g$/spot. After this identification, HPLC or GC can be used to quantify the solute amount by scraping off the spot and diluting it in the appropriate sample buffer.

The combination of TLC with HPLC has provided an important advance in separation technology. In high-performance thin-layer chromatography (HPTLC), small-diameter, monodispersed adsorbent particles replace the traditional adsorbents. Robotics perform the various steps, such as sample spraying, concentration determination, and data storage using video cameras and digital file capture. HPTLC can surpass HPLC's abilities in terms of both analysis reliability and quantification of results.

3.6 STEROIDS AND ANTIBIOTICS

Reverse-phase chromatography is used to determine the concentration of steroids and antibiotics, both of which comprise important classes of drugs. Pore size is an important factor in these applications. A typical C_{18} stationary phase consists of 3–$5\,\mu m$ bonded silica particles with a pore size close to $10\,nm$. Table 3.6 provides examples of steroid (66) analyses, as well as erythromycin and antibiotic (67) analyses.

Table 3.6 Mobile Phases Used in RP-LC for Separating Antibiotics

Group	Antibiotic	Mobile Phase
β-Lactam	Penicillin	44% acetonitrile, 0.1% TFA (pH 1.8)
	Cephalosporin	20% acetonitrile, 0.1% TFA (pH 1.8)
Macrolide	Roxithromycin	Sodium phosphate, 54% acetonitrile (pH 6–7)
Tetracycline	Tetracycline	Phosphoric acid, acetonitrile, methanol (pH < 2.6)
Carbapenem	Imipenem	Acetonitrile, 0.1% acetic acid, triethylamine (pH 7)

Consider the following example of the effect of hydrophobicity on retention time. One steroid separation used a mobile phase of phosphoric acid and acetonitrile with an increasing concentration of acetonitrile to separate a mixture of steroids. This process produced an elution order of cortisone, estradiol, ethinylestradiol, testosterone, mestranol, and progesterone. The more hydrophilic steroids, with more oxygen atoms and lower molecular weights (as shown in Table 3.7), eluted first.

3.7 VITAMINS

Analysis methods commonly used for vitamins are subsets of analytical liquid chromatography. In general, fat-soluble vitamins are separated by the following methods:
- Chromatographic techniques that exploit differences in hydrophobicity
- Reverse-phase chromatography
- Polarity, using columns with specific chemical functional groups that can interact with polar groups on the vitamin

Water-soluble vitamins are usually analyzed by one of two methods:
- Ion exclusion
- Use of supports with specific chemical groups that exploit polar interactions

For high-resolution analysis of optically active vitamins, chiral columns that can specifically retain one optical isomer are preferred. Many of the finer details in the separation and analysis process depend upon the chemistry of the vitamin under consideration.

Consider the example of biotin separation. Biotin, also known as vitamin H, is a polar compound with substantial hydrophobicity. At a pH of 1 to 4, d-biotin's solubility in water is only 0.3 g/kg at 23 °C. Because it has a carboxylic acid group, however, this vitamin's solubility increases at pH values greater than 5, exceeding 10 g/kg at a pH of 7 and 23 °C.

Biotin can be analyzed by RP-LC using a C_{18} bonded-silica support. To detect biotin with reasonable sensitivity (down to a concentration of 20 mg/L), postcolumn derivatization with DACA (4-dimethylaminocinnamaldehyde) is needed. Figure 3.7 shows how chromatography is applied to a mixture of d-biotin and desthiobiotin

Table 3.7 Chemical Structures of Steroids Separated Using RP-LC

Steroid	Structure
Cortisone (340.46 daltons)	
Estradiol (272.4 daltons)	
Ethinylestradiol (296.4 daltons)	
Testosterone (288.4 daltons)	
Mestranol (310.4 daltons)	
Progesterone (314.5 daltons)	

using a Spherisorb ODS-2 column. Detection occurs in the visible region (546 nm), as both biotin and desthiobiotin form Schiff base conjugates with DACA. This analytical method, while complex, can separate biotin from its metabolic precursors or degradation products.

3.8 SUMMARY

Many solutes produced in biotechnological applications can be analyzed using similar methods. Today, the most commonly used method for quantifying biochemical and biopolymer concentrations is high-performance liquid chromatography. Electrophoretic techniques—especially capillary electrophoresis—are

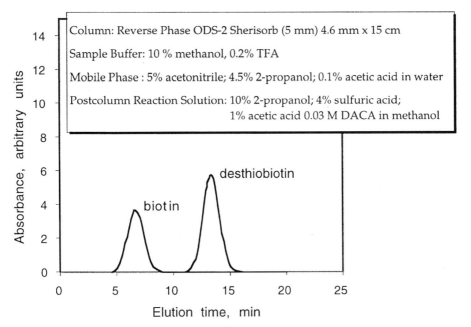

Figure 3.7 Analysis of biotin and desthiobiotin using RP-LC and postcolumn derivatization to enhance detection sensitivity.

becoming more popular for protein, oligonucelotide, and carboxylic acid analyses, however. Among HPLC techniques, ion-exchange chromatography and reverse-phase chromatography can deal with nearly all types of biological molecules. Many useful methods, such as thin-layer chromatography, can also complement HPLC methods.

Bioactivity can be analyzed by electrophoretic, HPLC, and immunoassay methods. In HPLC, retention time and tryptic maps, which give sequence information, provide effective measures of bioactivity. Immunoassays can determine whether changes occur in the region where antibodies bind, presumably signaling whether the antibodies continue to show bioactivity.

3.9 PROBLEMS

3.1 Figure 3.8 shows the chromatograph that resulted from cation exchange of five amino acids using a column packed with sulfonated polystyrene particles. Given the conditions listed below, explain the retention order. It may be useful to hypothesize about which chemical property or properties dictate retention time. Then test your hypothesis until it explains the data. Where would you expect phenylalanine to elute?

    (a) Before glycine

    (b) After histidine

    (c) Between tyrosine and lysine

    (d) Between glycine and methionine

    (e) None of the above

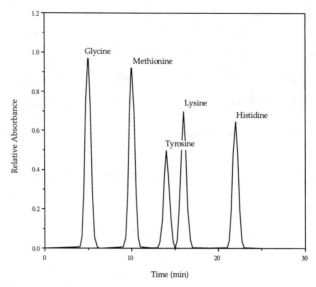

Figure 3.8 HPLC chromatograph of several amino acids.

Why? Note that some peak overlap may occur with any chromatographic method.
3.2 You have been asked to measure the concentration and bioactivity of tissue plasminogen activator (TPA), a therapeutic peptide produced by recombinant cell culture.

(a) How would you determine the concentration of this peptide in a production sample?

(b) Describe at least two ways to determine the concentration of bioactive product, either directly or indirectly. Note that TPA's bioactivity is quite closely linked to its structure and sequence.

3.3 Your company has developed monoclonal antibodies that bind to a new therapeutic protein. The monoclonal antibody will be used to recover the protein during the manufacturing process via affinity chromatography. What would you recommend as an analytical method for determining product concentration? As antibodies bind to only one specific region on the accessible surface of the protein product, would the monoclonal antibody be an effective analyte for measuring bioactivity?

3.4 Prompted by the increasing interest in developing less toxic alternatives to ethylene glycol, biotechnologists are working on a recombinant bacterial cell culture process for making propylene glycol. Propylene glycol has a melting point of $-59\,°C$ and a boiling point of $188.2\,°C$ at atmospheric pressure. At high temperatures, it tends to oxidize, forming propionaldehyde, lactic acid, acetic acid, and pyruvic acid. What would be a useful analytical technique for measuring the concentration of this product in aqueous solution? What would be a useful analytical technique for simultaneously measuring glucose and propylene glycol?

3.5 What are the advantages of analyzing a solution containing albumin and IgG by electrophoresis versus by chromatography? What are the disadvantages?

3.6 Develop an experimental design for studying the effect of temperature and pH on a reverse-phase separation of two peptides by tabulating the conditions that would be studied. What is your rationale for choosing these combinations? How many experiments are needed? Are there ways to simplify this effort?

3.10 REFERENCES

1. Stocchi V, Piccoli G, Magnami M. Reversed-phase HPLC separation of amino acid derivatives for amino acid analysis and microsequencing studies at the picomole level. Analytical Biochemistry 1989;178(1):107–116.
2. Furst P, Pollack L, Graser TA. HPLC analysis of free amino acids in biological material: an appraisal of four pre-column derivatization methods. J Liq Chromatogr 1989;12:2733.
3. Jorgenson J, Lukacs KD. Free-zone electrophoresis in glass capillaries. Clin Chem 1981; 27:1551–1553.
4. Jorgenson J, Lukacs KD. Capillary zone electrophoresis. Science 1983;222:266–272.
5. Jorgenson JW. Electrophoresis. Anal Chem 1986;58:743A–760A.
6. Bergman T, Agerberth B. Direct analysis of peptides and amino acids from capillary electrophoresis. FEBS Letters 1991;283:100.
7. Monnig CA, Dohmeier DM, Jorgenson JW. Sample gating in open tubular and packed capillaries for high-speed liquid chromatography. Anal Chem 1991;63:807.
8. Yik YF, Li SFY. Resolution optimization in micellar electrokinetic chromatography: use of overlapping resolution mapping scheme in the analysis of dinitrophenyl-derivatized amino acids. Chromatographia 1993;35:560.
9. Cooper JC, Danzer J. Enhanced selectivity in flow-injection analysis for l-amino acids using electrodialysis with amino acid oxidation. Analytica Chimica Acta 1993;282:369.
10. Burguera JL, Burguera M. Pyrolysis-flow-injection analysis–spectrophotometric determination of amino acids in aqueous solutions. Analytica Chimica Acta 1991;261:23.
11. Liu DM. Capillary gas chromatographic analysis of amino acids in geological environments with electron capture detection. J High Res Chromatogr 1989:239.
12. Rodante F. The calorimetric study of some alpha-amino acids bearing heteroatoms in their side chains. Thermochimica Acta 1992;1994:197.
13. Manchenko GP. Handbook of detection of enzymes on electrophoretic gels. Boca Raton: CRC Press, 1994.
14. Kobayashi T. Purification and properties of an alkaline protease from alkalophilic bacillus sp. KSM–K16. App Mic Biotech 1995;43:473–481.
15. Deshpande SS. Enzyme immunoassays: from concept to product development. New York: Chapman and Hall, 1996.
16. Floridi A, Palmerini CA, Fini C. Simultaneous analysis of bases, nucleosides and nucleoside mono- and polyphosphates by high-performance liquid chromatography. J Chromatogr 1977;138:203–212.
17. McKeag M, Brown PR. Modification of high pressure liquid chromatography nucleotide analysis. J Chromatogr 1978;152:253–254.
18. Perrone PA, Brown PR. Ion-pair chromatography of nucleic acid derivatives. In: Hearn MTW, ed. Ion-pair chromatography. New York: Marcel Dekker, 1985:259.
19. Hartwick RA, Assenza SP, Brown PR. Identification and quantitation of nucleosides, bases and other UV-absorbing compounds in serum, using reversed-phase high-performance liquid chromatography. I. Chromatographic methodology. J Chromatogr 1979a;186:647–658.
20. Hartwick RA, Krstulovic AM, Brown PR. Identification and quantitation of nucleosides, bases and other UV-absorbing compounds in serum, using reversed-phase high-performance liquid chromatography. II. Evaluation of human sera. J Chromatogr 1979b;186:659–676.
21. Rustum YM. High-pressure liquid chromatography. I. Quantitative separation of purine and pyrimidine nucleosides and bases. Anal Biochem 1978;90:289–299.
22. Assenza SP, Brown PR, Goldberg AP. Isocratic reverse-phase HPLC for the separation of deoxyribonucleosides and ribonucleosides. J Chromatogr 1983;277:305–307.
23. Eksborg S, Schill G. Ion pair partition chromatography of organic ammonium compounds. Anal Chem 1973;90:2092.
24. Juengling E, Kammermeier H. Rapid assay of adenine nucleotides or creatine compounds in extracts of cardiac tissue by paired-ion reverse-phase high-performance liquid chromatography. Anal Biochem 1980;102:358–361.

25. Al-Moslih MI, Dubes GR, Masoud AN. Separation of isomeric and nonisomeric monoribonu-cleotides by isocratic ion pair HPLC. J High Resolut Chrom Commun 1981;4:173.

26. El Rassi Z, Horváth CS. Chromatog 1982;15:75–82.

27. Erhlich M, Erhlich K. Separation of six DNA bases in ion pair–reverse phase high pressure liquid chromatography. J Chromatogr Sci 1979;17:531–533.

28. Hirabayashi J, Kasai K. Nucleic Acid Res Symp Series 1988;20:67.

29. Hirabayashi J, Kasai K. Size-dependent, chromatographic separation of double-stranded DNA which is not based on gel permeation mode. Anal Biochem 1989;178:336–341.

30. Hirabayashi J, Ito N, Noguchi K, Kasai K. Slalom chromatography: size-dependent separation of DNA molecules by a hydrodynamic phenomenon. Biochem 1990;29:9515–9521.

31. Kasai K. Size-dependent chromatographic separation of nucleic acids. J Chromatogr 1993;618: 203–221.

32. Boyes BE, Walker DG, McGeer PL. DNA restriction fragments in a size-exclusion column by a non-ideal mechanism. Anal Biochem 1988;170:127–134.

33. Fangman WL. Separation of very large DNA molecules by gel electrophoresis. Nucleic Acids Res 1978;5:653–665.

34. Jovin TM. Multiphasic zone electrophoresis. 3. Further analysis and new forms of discontinuous buffer systems. Biochem 1973;12:890–898.

35. Schwartz DC, Cantor CR. Separation of yeast chromosome-sized DNAs by pulsed field gradient gel electrophoresis. Cell 1984;37:67–75.

36. Carle GF, Olson MV. Separation of chromosomal DNA molecules from yeast by orthogonal-field–alternation gel electrophoresis. Nucleic Acids Res 1984;12:5647–5664.

37. Anand R. Pulse field gel electrophoresis: a technique for fractionating large DNA molecules. Trends Genetics 1986;2:278–283.

38. Chu G, Vollrath D, Davis RW. Separation of large DNA molecules by contour-clamped homoge-neous electric fields. Science 1986;234:1582–1585.

39. Smith CL, Matsumoto T, Niwa O, Klco S, Fan JB, Yanagida M, Cantor CR. An electrophoretic karyotype for *Schizosaccharomyces pombe* by pulsed field gel electrophoresis. Nucleic Acids Res 1987;15:4481–4489.

40. Bruchelt G, Niethammer D, Schmidt KH. Isotachophoresis of nucleic acid constituents. J Chro-matogr 1993;618:57–77.

41. Kohlrausch F. Ann Physik und Chemie 1897;62:209.

42. Dolnik V, Liu JP, Banks JF, Notovny HV, Bocek P. Capillary zone electrophoresis of oligonucleotides. Factors affecting separation. J Chromatogr 1989;480:321–330.

43. Yamamoto H, Manabe T, Okuyama T. Gel permeation chromatography combined with capillary electrophoresis for microanalysis of proteins. J Chromatogr 1989;480:277–83.

44. Heiger DN, Cohen AS, Karger BL. Separation of DNA restriction fragments by high performance capillary electrophoresis with low and zero crosslinked polyacrylamide using continuous and pulsed electric fields. J Cromatogr 1990;516:33–48.

45. Swerdlow H, Wu SL, Harke H, Dovichi NJ. Capillary gel electrophoresis for DNA sequencing. Laser-induced fluorescence detection with the sheath flow cuvette. J Chromatogr 1990;516:61–7.

46. Cohen AS, Najarian D, Karger BL. Separation and analysis of DNA sequence reaction products by capillary gel electrophoresis. J Chromatogr 1990;516:49–60.

47. Paulus A, Gassmann E, Field MJ. Calibration of polyacrylamide gel columns for the separation of oligonucleotides by capillary electrophoresis. Electrophoresis 1990;11:702–708.

48. Chaplin MF, Kennedy JF, eds. Carbohydrate analysis: a practical approach. United Kingdom: Oxford University Press, 1994.

49. Prodolliet J, Bugner E, Feinberg M. Determination of carbohydrates in soluble coffee by anion-exchange chromatography with pulsed amperometric detection: interlaboratory study. J AOAC Int 1995;378:768–782.

50. Kakehi K, Honda S. Profiling of carbohydrates, glycoproteins and glycolipids. J Chromatogr 1986; 379:27–55.

51. Ward KM, Lehmann CA, Leiken AM, eds. Clinical laboratory instrumentation and automation: principles, applications, and selection. Philadelphia: W. B. Saunders, 1994.

52. Wang TW, Kumlien J, Ohlson S, Lundblad A, Zopf D. Analysis of a glucose-containing tetrasaccha-ride by high performance liquid affinity chromatography. Anal Biochem 1989;182:48–53.

53. Hodgson J. Chromatography of complex carbohydrates. Biotechnology 1991;9:149–150.

54. Harmon MA, Doelle HN. Gas chromatographic separation and determination of microquantities of the esters of the tricarboxylic acid cycle acids and related compounds. J Chromatogr 1969;42: 157.

55. Belter PA, Cussler EL, Hu WS. Bioseparations. New York: John Wiley and Sons, 1988.
56. Dechow FJ. Separation and purification techniques in biotechnology. Park Ridge: Noyes Publications, 1989.
57. Paleos J. Adsorption from aqueous and nonaqueous solutions on hydrophobic and hydrophilic high surface-area copolymers. J Coll Interf Sci. 1969;31:7.
58. Tsao JCY, Huang T, Wang P, et al. J Chinese Chem Soc (Peking) 1964;2:2–3.
59. Kobayski T, Nakamura I, Nakagawa M. U.S. Patent No. 3,873,425. 1975.
60. Blake JD, Clarke ML, Richards GN. Determination of organic acids in sugar cane process juice by high-performance liquid chromatography: improved resolution using dual Aminex HPX-87H cation-exchange columns equilibrated to different temperatures. J Chromatogr 1987;398:265.
61. Okada T. Redox supressor for ion-exclusion chromatography of carboxylic acids with conductometric detection. Anal Chem 1988;60:1666.
62. García AA, Miles DA, Sadaka M. Recovery of Shihimic acid using temperature-swing complexation extraction and displacement back extraction. Iso Puri 1994;2:75.
63. Compton SW, Brownlee RG. Capillary electrophoresis. Biotechniques 1988;6:432.
64. Huang X, et al. Quantitave analysis of low molecular weight carboxylic acids by capillary zone. Anal Chem 1989;61:766.
65. Everaerts JL, Beckers JL, Verheggen PEM. Isotachophoresis. In: J Chromatogr. Library 6. Amsterdam: Elsevier, 1976.
66. Bio-Rad Laboratories. HPLC columns methods and applications. Bio-Rad Laboratories, 1996:86.
67. Cserhatin T, Oros G. Biomedical chromatography. BMC 1996;10:117.

PART II

APPLICATION OF CHEMICAL, PHYSICAL, AND BIOLOGICAL PROPERTIES TO BIOSEPARATIONS

4

Thermodynamic and Transport Properties

In this chapter, we discuss the chemical and physical properties of biomolecules taken into account when quantifying biomolecular behavior during separations. In some cases, these properties can be adequately described by theoretical models. Often, however, we must determine these values experimentally. Because biomolecular separations are carried out by exploiting particular molecular properties, process design relies on the estimation, measurement, or theoretical derivation of values for these properties; the separation rationale can then be extended to other systems based on the conclusions drawn from this analysis.

In biological systems, we have a special interest in aqueous solution phenomena. Water plays a unique role in bioseparations, simultaneously creating opportunities to isolate particular components and imposing process limitations. The ability of water molecules to strongly interact with one another gives rise to an important phenomenon, the hydrophobic effect. Biomolecules and their subunits are frequently classified by how hydrophobic (or, conversely, how hydrophilic) they are. A hydrophobic compound, for example, will minimize its contact with water by moving to an adjacent organic phase or interface. By considering these characteristics, we can predict the conformation of a given biomolecule in aqueous solution.

Chemical equilibrium also plays an important role in aqueous solutions. In solution, hydronium ions are transferred or molecules interact through electron sharing, processes known as acid-base interactions. The ability of water to both hydrogen-bond and act as a Lewis base by sharing electrons on the oxygen atoms greatly influences polar and acid-base interactions between solutes in aqueous solution. It is often assumed that water acts as a "medium" in acid-base interactions when chemical equilibria are written. In fact, water molecules can greatly modify the equilibrium and change the relative acid or base strengths compared with gas-phase interactions. The high dielectric constant of water also enables ion solvation and transfer.

4.1 CHEMICAL EQUILIBRIA

Chemical equilibrium is an important and powerful tool that facilitates in-depth analysis of biomolecules in aqueous solution. It then becomes possible to design separation processes that manipulate these biomolecules. Although this topic is covered in introductory chemical courses, we will review it here, noting how chemical equilibrium applies in cases involving multiple phases. It is also sometimes convenient to recast interactions or phase transfer phenomena in terms of chemical reactions, especially when they are part of a thermodynamic system that includes a chemical reaction.

In our discussion of chemical equilibria in an aqueous solution, B will stand for the biomolecule of interest and A for a reactant. The role of water is normally left unstated when analyzing a process involving a single phase:

$$A + B \rightleftharpoons AB \tag{4.1}$$

$$K = \frac{[AB]}{[A][B]} \tag{4.2}$$

In Equation 4.2, K is the chemical equilibrium constant. The terms in brackets denote the activity of each species in aqueous solution; this activity is defined as the product of the concentration and the activity coefficient, with equilibrium constants being used as the units of concentration. Because activity coefficients can be difficult to predict with any accuracy, and these coefficients are difficult to obtain experimentally for solutes in complex aqueous solutions, we can state them as a ratio:

$$K = \frac{\gamma_{AB}}{\gamma_A \gamma_B} \frac{C_{AB}}{C_A C_B} \tag{4.3}$$

Taking the logarithm of Equation 4.3, we have the following equation:

$$\ln K = \ln\left(\frac{\gamma_{AB}}{\gamma_A \gamma_B}\right) + \ln\left(\frac{C_{AB}}{C_A C_B}\right) \tag{4.4}$$

These terms can be related to the Gibbs free energy change for the reaction:

$$\Delta G = -RT \ln K \tag{4.5}$$

$$\Delta G = -RT \ln\left(\frac{\gamma_{AB}}{\gamma_A \gamma_B}\right) - RT \ln\left(\frac{C_{AB}}{C_A C_B}\right) \tag{4.6}$$

The reaction can then be analyzed by breaking down the theoretical driving forces by determining the linear free-energy relationships (LFERs) (1). At this level, we can start to understand and plan how to best use specific interactions in bioseparations. LFERs divide the free energy change into several parts, making it easier to predict the entire change or measure it through simple experimentation. The concept of LFERs is akin to establishing a thermodynamic path, but the theoretical value of the free energy change for a particular step is simply approximated.

For example, if we are attempting to show how electrostatic and hydrophobic interactions promote the reaction discussed in Equation 4.1, the LFERs would reflect the contributions of each of these phenomena:

$$\Delta G = \Delta G_{\text{electrostatic}} + \Delta G_{\text{hydrophobic}} \tag{4.7}$$

$$\ln K = \ln K_{\text{electrostatic}} + \ln K_{\text{hydrophobic}} \tag{4.8}$$

We will now discuss some useful LFERs that can be applied to equilibrium bioseparations. When a separation process involves a complex between a biomolecule and reagent that is distributed between two phases, the equilibrium reaction can be written as before; we must, however, denote the phase in which the reaction occurs. In addition, an equilibrium distribution constant, K_D, is defined in the same manner as the chemical equilibrium constant. If an organic phase containing species A is in contact with an aqueous phase, the following equations can be written:

$$B_{aqueous} \rightleftharpoons B_{organic} \tag{4.9}$$

$$K_D = \frac{B_{organic}}{B_{aqueous}} \tag{4.10}$$

$$B_{aqueous} \rightleftharpoons AB_{organic} \tag{4.11}$$

$$K = \frac{[AB]_{organic}}{K_D[B_{aqueous}][A_{organic}]} \tag{4.12}$$

If we measure the concentration of B in the aqueous phase and the concentration of A in the organic phase, the equilibrium may then be written in those terms.

4.2 SOLUBILITY

To measure solubility, we must examine the equilibrium between solid and liquid phases. At equilibrium, the chemical potential of each component in the liquid phase is equal to its chemical potential in the solid phase. This relationship is described by the following equation:

$$\mu_{solute}^{liquid\ phase} = \mu_{solute}^{solid\ phase} \tag{4.13}$$

The chemical potential can be further defined as consisting of the standard-state chemical potential and the activity of the solute in a particular phase. To describe a multiphase system, we will use a convention where the subscript 2 refers to the solute, I denotes the liquid phase, and II denotes the solid phase:

$$\mu_2^{I,0} + RT \ln(a_2^I) = \mu_2^{II,0} + RT \ln(a_2^{II}) \tag{4.14}$$

The activity may also be expressed as the product of an activity coefficient and the mole fraction of the solute when we develop the equilibrium equation:

$$\mu_2^{I,0} + RT \ln(x_2^I \gamma_2^I) = \mu_2^{II,0} + RT \ln(x_2^{II} \gamma_2^{II}) \tag{4.15}$$

Activity coefficients of solutes in a liquid mixture can be expressed by empirical means through Margules or van Laar equations; more rigorous models, such as UNIQUAC, may be used as well. These expressions account for the nonideal behavior of solution components. In general, the activity coefficients depend on the solute's mole fraction along with the properties of, or concentrations in, the liquid phase.

For the solid phase, we can assume that the mole fraction of water is negligible (an assumption that is adequate as a first approximation). If we then assume that the standard-state fugacities are equal to the pure-solute fugacities, the activity coefficient of the solute is unity. Thus Equation 4.15 becomes

$$\ln(x_2^I) = \frac{\mu_2^{II,0} - \mu_2^{I,0}}{RT} - \ln \gamma_2^I \qquad (4.16)$$

As Equation 4.16 suggests, it is generally useful to explore the relationship between the natural log of the solubility and the solution conditions that affect the activity of the solute. The difference in the standard-state chemical potentials indicates a direct relationship between temperature and solubility; note, however, that the activity coefficient also depends on temperature. Using the Margules or Scatchard-Hildebrand equations for the activity coefficient (2), we see that $\ln(\gamma_2) = f(1/T)$. Thus temperature effects on solubility may be approximated with the logarithm of the solubility, as inversely proportional to temperature.

Useful extensions to the fundamental theoretical expression in Equation 4.16 have been derived for amino acids, proteins, and nucleic acids (3–5).

4.2.1 Protein and amino acid solubility

Table 4.1 summarizes equations that have been derived to predict protein solubilities. These solubilities are important factors in separations that work by depending on selective precipitation of solutes. Proteins and amino acids can be precipitated by cooling or concentrating the solution through solvent evaporation. Proteins may also be precipitated by adding other substances to the solution so as to alter the protein solubility. Common compounds that can be added to precipitate proteins include polar organic solvents, non-ionic polymers, polyelectrolytes, metallic ions, inorganic salts, and mineral acids and bases. Mineral acids and bases, for example, may alter the solution's pH.

One of the most simple chemical precipitation techniques involves the use of inorganic salts. These salts follow a trend known as the Hofmeister series, shown in Table 4.2. In general, the Hofmeister series shows that the more highly charged ions are more effective when used to precipitate proteins.

Cooling the solution is a popular technique in bioseparations. It both lowers solubility and helps prevent denaturation of an active product. Figures 4.1–4.3

Table 4.1 Protein Precipitation Equations and Terms

Precipitating Agent	Equation	Definition of Terms
Nonionic polymers	$\log S = \beta - K_m$	m = polymer molality $K = \dfrac{v}{2.303}\left[\dfrac{r_s + r_r}{r_r}\right]$;
		$\beta = \dfrac{\mu_i - \mu_i^o}{RT}$
Organic solvents	$\log S = k/D^2$	$k = f$(solution dielectric constant) D = solution dielectric constant
Salts	$\log S = \alpha - K_s I$	$\alpha = f$(pH, T, solute) $K_s = f$(salt, solute) I = ionic strength

Source: Adapted from Glatz CE. Precipitation. In: Ansejo J, ed. Separation processes in biotechnology. New York: Marcel Dekker, 1990:329–356.

Table 4.2 Hofmeister Series in Order of Most to Least Effective Precipitating Agent

Cations	Anions
Th^{4+}	citrate^{3-}
Al^{3+}	tartrate^{2-}
H^+	F^-
Ba^{2+}	IO_3^-
Sr^{2+}	phosphate^{2-}
Ca^{2+}	sulfate^{2-}
Mg^{2+}	acetate$^-$
Cs^+	borate$^-$
Rb^+	Cl^-
NH_4^+	ClO_3^-
K^+	Br^-
Na^+	NO_3^-
Li^+	ClO_4^-
	I^-
	CNS^-

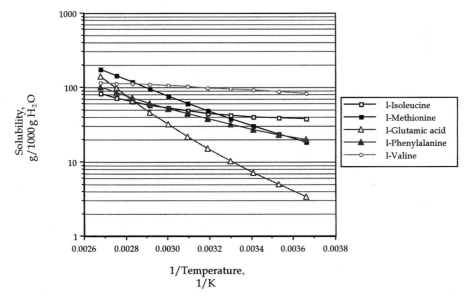

Figure 4.1 Effect of temperature on the solubility of isoleucine, methionine, glutamic acid, phenylalanine, and valine in water. (Adapted from Greenstein JP, Winitz M. Chemistry of the amino acids. Malabar: Robert E. Kreiger, 1986:564.)

illustrate amino acid solubilities as a function of temperature. Most of the amino acid data follow the form of Equation 4.16. Notable exceptions include l-leucine, l-proline, and l-valine, all of which deviate from this theoretical model because their aqueous activity coefficients have a complex dependence on temperature.

At their isoelectric points (pIs), ionogenic solutes such as proteins and amino acids display minimum solubilities. Often precipitation can be accomplished by simply adjusting the pH of a concentrated solution to the isoelectric point of the solute of interest. Section 4.4 discusses the isoelectric points of amino acids and proteins in more detail.

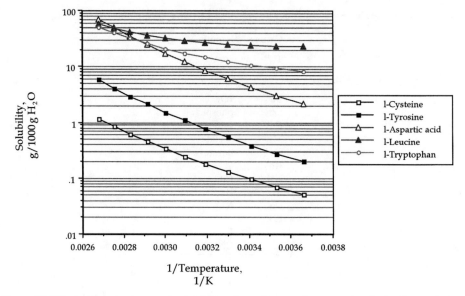

Figure 4.2 Effect of temperature on the solubility of cysteine, tyrosine, aspartic acid, leucine, and tryptophan in water. (Adapted from Greenstein JP, Winitz M. Chemistry of the amino acids. Malabar: Robert E. Kreiger, 1986:564.)

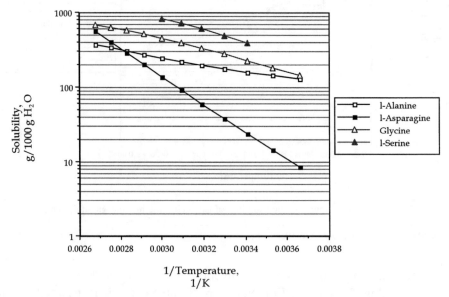

Figure 4.3 Effect of temperature on the solubility of alanine, asparagine, glycine, serine, and histidine. (Adapted from Greenstein JP, Winitz M. Chemistry of the amino acids. Malabar: Robert E. Kreiger, 1986:564.)

4.3 DIFFUSIVITY

Many separation papers and textbooks describe Fickian diffusivities. These diffusivities are determined by assuming a binary solution and passive transport of the biomolecules down a chemical potential gradient. Fick's law is defined as

$$J_A = -D_{AB} \frac{\partial C_A}{\partial x} \tag{4.17}$$

where J_A is the flux of species A. The following sections describe how diffusivities of species may be estimated when data are unavailable at a particular temperature.

4.3.1 Uncharged low-molecular-weight biochemicals

An accurate, yet simple semi-empirical correlation is provided through Sitaraman's modification of the Wilke-Chang equation. Equation 4.18 applies to uncharged biochemical species, such as alcohols, carboxylic acids below their lowest pK_a, and low-molecular-weight sugars:

$$D_{AB} = 16.79 \times 10^{-14} \left(\frac{M_B^{1/2} \Delta H_B^{1/3} T}{\mu_B V_A^{1/2} \Delta H_A^{1/3}} \right) \tag{4.18}$$

Sitaraman's equation uses the enthalpy terms ΔH_A and ΔH_B, where A and B represent the solute and solvent, respectively. These terms account for chemical interactions caused by inelastic collisions between a polar solute and water. The molecular weight of the solute is M_B; T is the temperature in Kelvins; μ_B is the viscosity of water; and V_A is the molar volume of the solute at its normal boiling point. This equation reportedly produces less than a 12% error for solutes in water.

4.3.2 Proteins

Macromolecules usually follow the Stokes-Einstein equation for diffusivity. This equation relates diffusivity to the molecular radius (r), which can be estimated from the protein molecular weight:

$$D = \frac{kT}{6\pi\eta r} \tag{4.19}$$

A more accurate approach uses hydrodynamic radii to calculate the diffusivity, rather than assuming that proteins are spheres (6). The denominator in Equation 4.19 is replaced with a frictional coefficient, f, used in the frictional ratio which is the Perrin factor, f/f_0, where f_0 is the frictional coefficient for a sphere:

$$f_0 = 6\pi\eta \left(\frac{3M\tilde{v}}{4\pi} \right)^{1/3} \tag{4.20}$$

In Equation 4.20, M is the protein-specific mass; $\tilde{v}$ is the protein-specific volume, generally $0.75\,g/cm^3$. The viscosity of water (η) at $20\,°C$ is 0.01 Poise. Perrin factors, which are calculated for a variety of shapes, are most commonly based on elliptical axial profiles. They employ a ratio of the long axis to the short axis. Through this technique, diffusivity measurements can be used to estimate the general shape of various biopolymers and viruses. In general, Perrin factors range from 1 to 2. The exceptions involve extremely long, rodlike biopolymers, such as long-chain DNA and polysaccharides.

Example 4.1

Estimate the diffusivity of IgG_1 at room temperature. Immunoglobin type G_1 has a molecular weight of 150 kDa.

Solution

Based on the fact that IgG_1 has a reasonably compact Y-shape, we can assume that the Perrin factor will be nearly 1. In determining f_0, the specific mass is calculated by taking the molecular weight and dividing by Avogadro's number to get the mass of one molecule. Using the specific volume of $0.75 \, g/cm^3$, f_0 can be calculated as follows:

$$f_0 = 6\pi(0.01P)\left(\frac{3(150,000 \, g/mol)(0.75 \, g/cm^3)}{4\pi}\right)^{1/3} = 6.7 \times 10^{-8} \tag{4.21}$$

The diffusivity is then calculated as follows:

$$D = \frac{(1.38 \times 10^{-16} \, ergs/mol \cdot K)(298K)}{7.1 \times 10^{-8}} = 6.7 \times 10^{-7} \, cm^2/s \tag{4.22}$$

4.4 ISOELECTRIC POINTS AND CHARGE DEPENDENCE ON pH

4.4.1 Carboxylic acids

A weak acid and water react as follows:

$$HA + H_2O = H_3O^+ + A^- \tag{4.23}$$

In this equation, HA is the general formula for an acid. The equilibrium expression of Equation 4.23 is then given by Equation 4.24:

$$K = \frac{[H_3O^+][A^-]}{[HA][H_2O]} \tag{4.24}$$

K is the equilibrium constant. Because water is the solvent, the change in the water concentration can be considered negligible. Therefore, we can assume $[H_2O]$ to be a constant for dilute solutions. A new constant can be obtained by multiplying K and $[H_2O]$:

$$K_a = K[H_2O] \tag{4.25}$$

K_a is generally called the acid dissociation constant. Then, Equation 4.23 becomes

$$K_a = \frac{[H_3O^+][A^-]}{[HA]} \tag{4.26}$$

Taking the logarithm of both sides of Equation 4.26 gives the following equation:

$$\log K_a = \log[H_3O^+] + \log \frac{[A^-]}{[HA]} \tag{4.27}$$

Equation 4.28 can then be derived from the mathematical definitions of pH and pK_a (pH = $-\log[H_3O^+]$, $pK_a = -\log K_a$):

$$pH = pK_a + \log \frac{[A^-]}{[HA]} \tag{4.28}$$

Equation 4.28 is commonly known as the Henderson-Hasselbalch equation. It is applicable to dilute solutions of acids whose deviation from ideal behavior is minimal.

Using the previously calculated values of K_a for carboxylic acids (Table 4.3), we can obtain the ratio of ionized to un-ionized acid molecules by determining the pH of a solution. For dicarboxylic and tricarboxylic acids, the following equations can be used to determine the concentrations of the various species. For dicarboxylic acids:

$$\log \frac{[HA^-]}{[H_2A]} = pH - pK_{a_1} \tag{4.29}$$

$$\log \frac{[A^{2-}]}{[H_2A]} = 2pH - pK_{a_1} - pK_{a_2} \tag{4.30}$$

Table 4.3 Dissociation Constants of Selected Carboxylic Acids at 25 °C

Carboxylic Acid	pK_{a_1}	pK_{a_2}	pK_{a_3}
Acetic acid	4.76		
Ascorbic acid	4.17	11.57	
n-Butyric acid	4.82		
Chorismic acid	3		
Citric acid	3.13	4.76	6.40
Formic acid	3.74		
Fumaric acid	3.02	4.38	
Glyoxylic acid	3.18[30]		
Lactic acid	3.73		
Malic acid	3.46	5.10	
Oxalic acid	1.25	3.67	
Oxaloacetic acid	2.22	3.89	13.03 (enol OH)
Pyruvic acid	2.39		
Quinic acid	3.66		
Shikimic acid	4.15		
Succinic acid	4.21	5.72	
Tartaric acid	3.03	4.37	

Similarly for tricarboxylic acids:

$$\log \frac{[H_2A^-]}{[H_3A]} = pH - pK_{a_1} \tag{4.31}$$

$$\log \frac{[HA^{2-}]}{[H_3A]} = 2pH - pK_{a_1} - pK_{a_2} \tag{4.32}$$

$$\log \frac{[A^{3-}]}{[H_3A]} = 3pH - pK_{a_1} - pK_{a_2} - pK_{a_3} \tag{4.33}$$

These equations indicate which species is predominant at a particular pH. Often this information is useful in developing ion-exchange, electrophoresis, and crystallization processes.

Temperature has only a very small effect on the dissociation constant of carboxylic acids. Between 0 and 100 °C, the dissociation constant decreases by only 0.02 per 10 °C. Thus, for physiological conditions, the pK_a of a carboxylic acid can be assumed to be constant. For amines, however, temperature has a significant effect on pK_a and cannot be neglected (7).

4.4.2 Amino acids

The common amino acids can be divided into three groups according to their side chain, or R-group:

* Group 1 amino acids have R-groups that are not ionizable.
* Group 2 amino acids have negatively charged R-groups at pH 7.0.
* Group 3 amino acids have positively charged R-groups at pH 7.0.

The acid dissociation constants for common amino acids are shown in Table 4.4 (5, 8–10).

Figure 4.4 shows the charge states for various amino acids with side chains that cannot ionize. The deprotonation ratio of the α-carboxyl group can be derived from Equation 4.28:

$$\frac{[A^-]}{[HA]} = 10^{(pH-pK_a^c)} \tag{4.34}$$

Then, the fraction of [HA] is calculated as follows:

$$[HA] = \frac{1}{1 + 10^{(pH-pK_a^c)}} \tag{4.35}$$

The positive charge of the α-amino group, based on the fraction of [HA], can be expressed as follows:

$$\text{Charge of } \alpha\text{-amino group} = \frac{1(+1)}{1 + 10^{(pH-pK_a^c)}} \tag{4.36}$$

The deprotonation ratio of the α-amino group can also be derived from Equation 4.36 using the variables of pH and pK_a^a.

Table 4.4 Acid Dissociation Constants of Common Amino Acids

Amino Acid	pI	α-COOH	pK_a α-NH$_3^+$	Side Chain
Glycine	5.97	2.36	9.56	
Alanine	6.02	2.31	9.70	
Valine	5.97	2.26	9.49	
Leucine	5.98	2.27	9.57	
Proline	6.30	1.9	10.41	
Isoleucine	6.02	2.4	9.7	
Methionine	5.06	2.10	9.05	
Phenylalanine	5.48	2.17	9.11	
Tryptophan	5.88	2.34	9.32	
Serine	5.68	2.13	9.05	
Threonine	5.60	2.10	8.96	8.16
Cysteine	5.02	1.96	10.29	10.11
Tyrosine	5.67	2.17	9.04	
Asparagine	5.41	2.15	8.72	
Glutamine	5.70	2.16	9.01	3.70
Aspartic acid	2.98	1.94	9.62	4.20
Glutamic acid	3.22	2.18	9.59	10.68
Lysine	9.74	2.19	9.12	12.48
Arginine	10.76	2.3	9.02	6.02
Histidine	7.59	1.7	9.09	

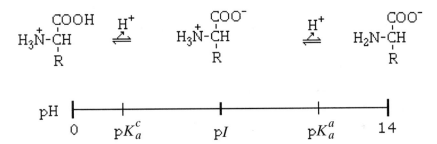

Figure 4.4 The state of charges of amino acids according to pH.

$$\frac{[A^-]}{[HA]} = 10^{(pH - pK_a^c)} \tag{4.37}$$

The fraction [A$^-$] in the deprotonation of the α-amino group is obtained from Equation 4.37:

$$[A^-] = \frac{1}{1 + \dfrac{1}{10^{(pH - pK_a^c)}}} \tag{4.38}$$

The negative charge of the α-carboxyl group (the fraction of [A$^-$]) can be found as follows:

$$\text{Charge of α-carboxyl group} = \frac{1(-1)}{1 + \dfrac{1}{10^{(pH - pK_a^c)}}} \tag{4.39}$$

The net charge of the amino acid can then be obtained by adding Equations 4.36 and 4.39:

$$\text{Net charge of amino acid} = \frac{1(+1)}{1+10^{(pH-pK_a^c)}} + \frac{1(-1)}{1+\dfrac{1}{10^{(pH-pK_a^a)}}} \tag{4.40}$$

Figure 4.5 and Figure 4.6 show the net charge, as calculated from Equation 4.40, of the amino acids without ionogenic side chains. The net charge of the amino acids that can have a negatively charged carboxyl R-group (aspartic acid and glutamic acid) can be determined by employing a similar approach (Figure 4.7).

The charges on the α-carboxyl group and the α-amino group of these amino acids can be calculated in much the same way. The deprotonation ratio of the R-group at a certain pH can be derived from Equation 4.37, assuming an ideal solution:

$$\frac{[A^-]}{[HA]} = 10^{(pH-pK_a^s)} \tag{4.41}$$

The fraction of $[A^-]$ liberated during deprotonation of the side chain is obtained from Equation 4.41:

$$[A^-] = \frac{1}{1+\dfrac{1}{10^{(pH-pK_a^s)}}} \tag{4.42}$$

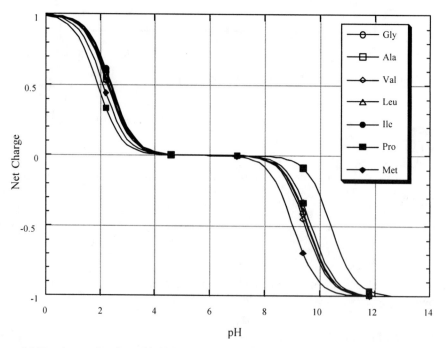

Figure 4.5 Net charge of amino acids (Gly, Ala, Val, Leu, Ile, Pro, and Met) as a Function of pH (calculated from Equation 4.15).

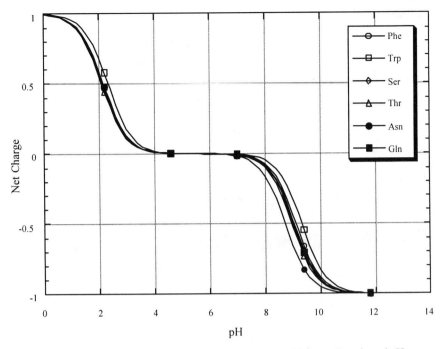

Figure 4.6 Net charge of amino acids (Phe, Trp, Ser, Thr, Asn, and Gln) as a Function of pH (calculated from Equation 4.15).

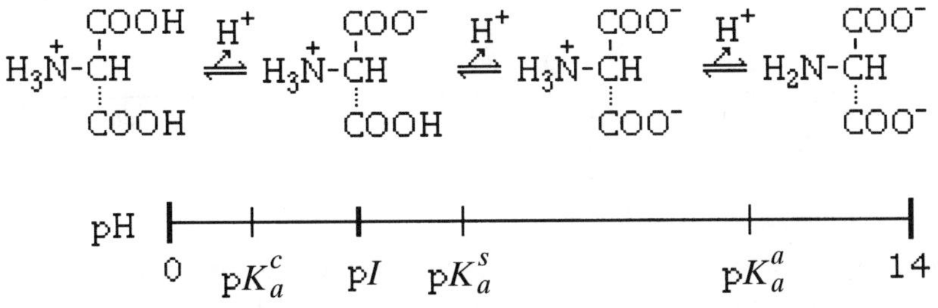

$$\underset{\underset{COOH}{|}}{\overset{\overset{COOH}{|}}{H_3\overset{+}{N}-CH}} \quad \overset{H^+}{\rightleftarrows} \quad \underset{\underset{COOH}{|}}{\overset{\overset{COO^-}{|}}{H_3\overset{+}{N}-CH}} \quad \overset{H^+}{\rightleftarrows} \quad \underset{\underset{COO^-}{|}}{\overset{\overset{COO^-}{|}}{H_3\overset{+}{N}-CH}} \quad \overset{H^+}{\rightleftarrows} \quad \underset{\underset{COO^-}{|}}{\overset{\overset{COO^-}{|}}{H_2N-CH}}$$

pH |———+————+———+———————————+———|
   0   pK_a^c pI pK_a^s pK_a^a 14

Figure 4.7 State of charges of amino acids (carboxyl side chain) according to pH.

The negative charge of the side chain, based on the fraction of $[A^-]$, is calculated as follows:

$$\text{Charge of side chain} = \frac{1(-1)}{1+\dfrac{1}{10^{(pH-pK_a^s)}}} \tag{4.43}$$

Thus the net charge of an amino acid with a negatively charged side chain can be obtained from Equations 4.36, 4.39, and 4.43, assuming minimal deviations from the ideal solution:

$$\text{Net charge} = \frac{1(+1)}{1+10^{(pH-pK_a^c)}} + \frac{1(-1)}{1+\dfrac{1}{10^{(pH-pK_a^a)}}} + \frac{1(-1)}{1+\dfrac{1}{10^{(pH-pK_a^s)}}} \tag{4.44}$$

Table 4.5 Macro Forms and Microspecies of Tyrosine

Formula	Microspecies	Charges		
		Carboxyl	Amino	Phenolic
H_3L^+	(1) $HO\text{-}\langle O \rangle\text{-}CH_2\overset{\overset{+NH_3}{\vert}}{C}HCOOH$	0	+	0
H_2L	(2) $HO\text{-}\langle O \rangle\text{-}CH_2\overset{\overset{+NH_3}{\vert}}{C}HCOO^-$	−	+	0
	(3) $^-O\text{-}\langle O \rangle\text{-}CH_2\overset{\overset{+NH_3}{\vert}}{C}HCOOH$	0	+	−
	(4) $HO\text{-}\langle O \rangle\text{-}CH_2\overset{\overset{NH_2}{\vert}}{C}HCOOH$	0	0	0
HL^-	(5) $^-O\text{-}\langle O \rangle\text{-}CH_2\overset{\overset{+NH_3}{\vert}}{C}HCOO^-$	−	+	−
	(6) $HO\text{-}\langle O \rangle\text{-}CH_2\overset{\overset{NH_2}{\vert}}{C}HCOO^-$	−	0	0
	(7) $^-O\text{-}\langle O \rangle\text{-}CH_2\overset{\overset{NH_2}{\vert}}{C}HCOOH$	0	0	−
L^{2-}	(8) $^-O\text{-}\langle O \rangle\text{-}CH_2\overset{\overset{NH_2}{\vert}}{C}HCOO^-$	−	0	−

Source: Adapted from Martell AE, Motekaitis RJ. Determination and use of stability constants, 2nd ed. New York: VCH Publishers, 1992.

Likewise, the side chains of tyrosine and cysteine can have a negative charge at high pH. Martell and Motekaitis (11) have illustrated ionized tyrosine as macro forms and microspecies (Table 4.5). In tyrosine, with the pK_a values given in Table 4.5 ($pK_a^c = 2.17$, $pK_a^a = 9.04$, $pK_a^s = 10.11$), the majority of the microspecies of the formula H_2L are of form (2), because $pK_a^c \ll pK_a^a < pK_a^s$. Form (6) is the most likely form among the microspecies of the formula HL^- because the amino group of tyrosine has a greater tendency toward protonation than the phenolic group at the same pH ($pK_a^a < pK_a^s$). Cysteine, in contrast, has a thiol group ($pK_a^c = 1.96$, $pK_a^s = 8.16$, $pK_a^a = 10.29$). The net charge of tyrosine and cysteine can be calculated by Equation 4.44, just as in the case of aspartic and glutamic acids.

Figure 4.8 shows the net charge of amino acids (tyrosine, cysteine, glutamic acid, and aspartic acid) as a function of pH (calculated from Equation 4.44). The net charges of amino acids that can have positively charged R-groups (lysine, arginine, and histidine) can be determined in the same manner (Figure 4.9). The charges associated with the α-carboxyl and α-amino groups for these amino acids can be obtained in much the same way as for α–amino acids that have no charged side chains.

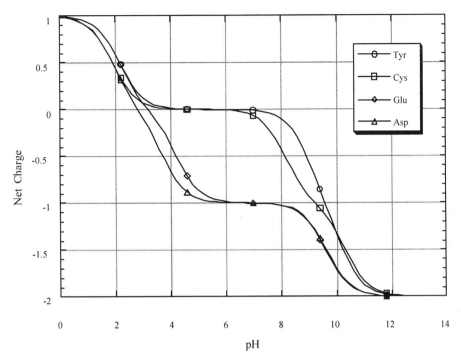

Figure 4.8 Net charge of amino acids (tyrosine, cysteine, glutamic acid, and aspartic acid) as a function of pH (calculated from Equation 4.19).

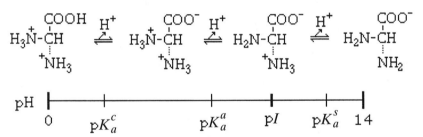

Figure 4.9 State of charges of amino acids with positively chargeable R-groups according to pH.

The ratio $[A^-]/[HA]$ in the deprotonation of the side chain can be obtained from Equation 4.28:

$$\frac{[A^-]}{[HA]} = 10^{(pH - pK_a^s)} \tag{4.45}$$

The fraction of $[HA]$ in the deprotonation of the side chain is derived from Equation 4.45:

$$[HA] = \frac{1}{1 + 10^{(pH - pK_a^s)}} \tag{4.46}$$

The positive charge of the side chain, based on the fraction of $[HA]$ of the side chain, is calculated as follows:

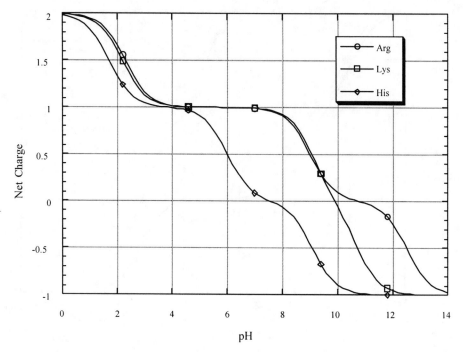

Figure 4.10 Net charge of amino acids (arginine, lysine, histidine) as a function of pH (calculated from Equation 4.23).

$$\text{Charge of side chain} = \frac{1(+1)}{1 + 10^{(\text{pH} - \text{p}K_a^s)}} \tag{4.47}$$

Then, the net charge of amino acids that can have a positively charged side chain is obtained by summing Equations 4.36, 4.39, and 4.47:

$$\text{Net charge} = \frac{1(+1)}{1 + 10^{(\text{pH} - \text{p}K_a^c)}} + \frac{1(+1)}{1 + 10^{(\text{pH} - \text{p}K_a^s)}} + \frac{1(-1)}{1 + \dfrac{1}{10^{(\text{pH} - \text{p}K_a^a)}}} \tag{4.48}$$

Figure 4.10 shows the net charges of arginine, lysine, and histidine as a function of pH, as calculated with Equation 4.48.

Equations 4.40, 4.44, and 4.48 apply only when we can ignore deviations from nonideal behavior, which is usually the case for low amino acid concentrations. These equations can be easily applied to calculate isoelectric points for various amino acids, because the isoelectric point occurs when the amino acid has no net charge. Table 4.6 lists experimentally determined isoelectric points.

4.4.3 Proteins

Their highly complex structure makes it a more straightforward process to experimentally measure isoelectric points for proteins than to calculate them using the Henderson-Hasselbalch equation. Moreover, in a polypeptide chain, the ionization

Table 4.6 Isoelectric Points of Common Amino Acids

Amino Acid	Isoelectric Point
Aspartic acid	2.98
Glutamic acid	3.22
Cysteine	5.02
Asparagine	5.41
Phenylalanine	5.48
Threonine	5.60
Glutamine	5.70
Tyrosine	5.67
Serine	5.68
Methionine	5.06
Tryptophan	5.88
Isoleucine	6.02
Valine	5.97
Glycine	5.97
Alanine	6.02
Leucine	5.98
Proline	6.30
β-Alanine	6.90
Histidine	7.59
Lysine	9.74
Arginine	10.76

Table 4.7 Isoelectric Points of Selected Proteins

Protein	Isoelectric Point
Pepsin	1
Human calmodulin	4.46*
Egg albumin	4.6
Isomerase	4.75
Serum albumin	4.9
Bovine serum albumin	5.6†
Urease	5.0
β-Lactoglobulin	5.2
γ-Globulin	6.6
Hemoglobin	6.8
Myoglobin	7.0
Chymotrisinogen	9.5
Cytochrome C	10.65
Lysozyme	11.0

*World Wide Web calculation (12), using Ca binding at site 1.
† World Wide Web calculation (12), using FT chain terminii at 25-607.

of one amino acid group affects the ionization of functional groups located farther down the chain (10). Note, however, that a protein will have an isoelectric point (pI) similar to that of the predominant amino acid residue. Table 4.7 lists selected protein isoelectric points.

When no data are available, several algorithms can be used to estimate the isoelectric point. In fact, a World Wide Web site is devoted to the calculation of protein isoelectric points based on tabulated data or sequence information (12). The main drawback to using this resource is that the calculations do not incorporate tertiary

Table 4.8 Grouping of Amino Acids by Hydrophobicity

Hydrophobic	Moderately Polar	Very Polar
Alanine	Glycine	Proline
Cysteine	Serine	Asparagine
Valine	Threonine	Glutamic acid
Isoleucine	Histidine	Arginine
Methionine	Tyrosine	Aspartic acid
Phenylalanine		Glutamine
Tryptophan		Lysine

Source: Adapted from Rose GD, Geselowitz AR, Lesser GJ, Lee RH, Zehfus MH. Hydrophobicity of amino acid residues in globular proteins. Science 1985;229:834–838.

structure effects and post-translation modifications. Nevertheless, this and similar sites provide helpful starting points for planning precipitation or electrophoresis experiments.

4.5 HYDROPHOBICITY–HYDROPHILICITY SCALES

Many different ways and scales to measure amino acid hydrophobicity have been proposed. One recent attempt compares the free energy change of transferring an amino acid side chain from water to an organic phase (13). Amino acids are grouped into categories based on the mean area of the amino acid residue that is buried during protein folding (Table 4.8).

Table 4.9 compares three popular quantitative scales. The predictions in the table are important for reverse-phase chromatography and can explain protein behavior in liquid-liquid partitioning; they are also important in hydrophobic chromatography. Using the concept of LFERs, the log of the retention time in reverse-phase chromatography can be shown to correlate with hydrophobicity scale parameters (1). Resources are available on the World Wide Web for the futher study of protein hydrophobicity (14, 15).

4.6 ACID-BASE SCALES

Lewis acids and bases can be characterized by several methods other than the Brønsted-Lowry pK_a scale. Three popular scales that are useful for biological molecule characterization are the Gutmann donor and acceptor numbers, the Drago E&C equation, and the solvatochromic comparison method. In addition, the qualitative hard-soft acid-base (HSAB) theory developed by Pearson (16–18) provides insight into many biomolecular interactions. The following sections briefly describe these methods and compare their applicability using chemicals that represent common biochemical functional groups.

4.6.1 Gutmann donor-acceptor theory

Gutmann's scales measure Lewis acidity and basicity. Donor numbers (DN) are defined as the molar enthalpy of reaction of a highly diluted donor solvent (D) and antimony pentachloride in 1,2-dichlorethane:

Table 4.9 Hydrophobicity Scales for Amino Acids

Amino Acid	$A°-<A>$	$\Delta G°$ (kcal)	$\log(RH_{vap}/RH_{H_2O})$
Alanine	86.6	0.5	1.43
Arginine	162.2		−8.00
Asparagine	103.3		−7.12
Aspartic acid	97.8		−4.91
Cysteine	132.3		−0.91
Glutamine	119.2		−6.90
Glutamic acid	113.9		−4.74
Glycine	62.9	0	1.76
Histidine	155.8	0.5	−7.51
Isoleucine	158.0		1.58
Leucine	164.1	1.8	1.68
Lysine	115.5		−3.21
Methionine	172.9	1.3	−1.09
Phenylalanine	194.1	2.5	−0.56
Proline	92.9		
Serine	85.6	−0.3	−3.72
Threonine	106.5	0.4	−3.59
Tryptophan	224.6	3.4	−4.33
Tyrosine	177.7	2.3	−4.49
Valine	141.0	1.5	1.46

$A°-<A>$ is the mean buried area (30).
$\Delta G°$ (kcal) is the free energy of transfer of an amino acid side chain from water to an organic solvent phase (13).
$\log(RH_{vap}/RH_{H_2O})$ is the free energy change for transfering an amino acid side chain from water to the vapor phase (3).

$$D + SbCl_5 \rightarrow D - SbCl_5 \tag{4.49}$$

$$DN = -\Delta H_{D-SbCl_5} \tag{4.50}$$

The major assumptions in the use and interpretation of donor numbers are as follows:

1. Only 1:1 adducts are formed.
2. DN measures the equilibrium constant for the 1:1 adduct formation, as well as the energy of the D–SbCl₅ bond.
3. The relative strengths derived using SbCl₅ hold for other acceptor acids.

Several researchers have questioned the validity of the third assumption. In fact, Gutmann cautions against the use of DN for predicting interactions between soft donor-acceptors, because SbCl₅ is classified as a hard acceptor (16).

The terminology of "soft" and "hard" acids and bases stems from a delineation put forward by Pearson (17, 18) to help unify the observations of earlier workers. Pearson believed that it was useful to separate acids and bases into two categories. Generally speaking, a *hard acid* is one that prefers to associate with a hard base, and pK_a values can rationalize equilibrium and kinetic data for these species. Likewise, *soft acids* prefer soft bases, although acid-base interactions of the Lewis type are more important in this case. Because the DN scale is based on a hard acid, pK_a values should correlate with DN for solvents that exhibit ionizing acidity/basicity.

Acceptor numbers (AN) are derived in a somewhat analogous fashion. Triethyl phosphine oxide (Et_3PO) is the reference donor. In this case, however, pure solvent (or acceptor) is used. The ^{31}P chemical shift induced by adduct formation serves as the basis of the scale. Arbitrarily, an AN value is assigned to hexane. The other values are then scaled accordingly using the ^{31}P chemical shifts. Interpretations and uses of AN follow the same basic assumptions stated for DN.

4.6.2 Drago E&C equation

Drago and coworkers (19) have proposed the following E&C equation:

$$-\Delta H = E_A E_B + C_A C_B \tag{4.51}$$

This equation can quantitatively predict the enthalpy of adduct formation ($-\Delta H$) for a Lewis acid-base interaction. E_A and E_B are believed to reflect electro-static properties. C_A and C_B reflect the covalent properties of the acceptor and donor, respectively, which form a 1:1 adduct in the gas phase or in a weak solvent.

The parameters were derived empirically from experimentally determined $-\Delta H$ values by fixing four sets of E&C parameters and using a least-squares regression analysis. Over time, the published E&C values have changed as more (and more accurate) enthalpy data have become available; they have also changed to reflect new choices of compounds to include in the system.

4.6.3 Solvatochromic comparison method

A set of solvatochromic parameters (π^*, α, and β), used in developing LFERs, has been compiled (20). The parameter π^* reflects the solvent's dipolarity/polarizabil-ity, as well as its acidity. The β term measures the solvent's ability to donate an elec-tron pair, and is related to its Gutmann donor number. The α scale, an acidity parameter, measures the solvent's ability to donate a proton.

One strength of this method is that the parameters were determined by aver-aging and comparing data obtained using different spectral techniques and systems; data from a variety of other thermodynamic experiments were also incorporated into the calculations. The researchers argue that other acid-base scales lump solvent interactions together, while their system more realistically assesses the impact and importance of the different phenomena involved.

4.6.4 Hard and soft acid and base theory

The general principle underlying the hard and soft acid and base theory (HSAB) is that "hard acids prefer to coordinate with hard bases, and soft acids prefer to coor-dinate with soft bases." A hard acid is small, carries a high positive charge, and lacks valence electrons that are easily distorted or removed because of their strong attrac-tion to the nucleus. In contrast, a soft acid is large, has a low positive charge, and

Table 4.10 Classification of Lewis Acids

Hard	Borderline	Soft
H^+, Li^+, Na^+	Cu^{2+}, Zn^{2+}, Ni^{2+}	Ag^+, Cu^+, Au^+
K^+, Be^{2+}, Mg^{2+}	Fe^{2+}, Co^{2+}, Pb^{2+}	Pt^{2+}, Hg^+, Hg^{2+}
Ca^{2+}, Sr^{2+}, Fe^{3+}	Sn^{2+}, Sb^{3+}	Tl^+, Cd^{2+}, Pt^{4+}, Pd^{2+}

Table 4.11 Classification of Lewis Bases

Hard	Borderline	Soft
H_2O, OH^-, F^-	$C_6H_5NH_2$, C_5H_5N	RSH, RS^-, R_2S
Cl^-, CH_3COO^-, SO_4^{2-}	N_3^-, Br^-, NO_2^-	SCN^-, $S_2O_3^{2-}$, I^-
NO_3^-, R^*OH, RO^-	SO_3^{2-}	R_3P, $(RO)_3P$, C_2H_4
NH_3, RNH_2		C_6H_6

*The symbol R stands for an alkyl group such as CH_3 or C_2H_5.

contains several valence electrons that are easily distorted or removed. The classifications of Lewis acids and bases are shown in Table 4.10 and Table 4.11, respectively.

Several theories have been proposed to explain the preferential hard-hard and soft-soft acid-base interactions. These theories tend to consider different aspects of acid-base behavior. The ionic-covalent theory is the oldest and most obvious explanation of acid-base interactions (21–24). Hard acids, which have high positive charges and small sizes, most tightly hold hard bases, which have high negative charges and small sizes. Thus hard acids bind hard bases primarily by ionic forces. In contrast, soft acids bind soft bases primarily by covalent bonds. Soft acids, which contain several easily distorted valence electrons, form strong covalent bonds with soft bases; in the latter compounds, the valence electrons are also easily distorted.

The π-bonding theory is applied chiefly to metallic complexes. Soft acids incorporate loosely held outer d-orbital electrons. They can form π bonds by donating these electrons to suitable ligands in which empty d-orbitals are available, such as P, As, S, or I (25, 26).

The electron correlation theory suggests that London dispersion forces between atoms in the same molecule may stabilize the molecule (27, 28). These forces depend on the product of the polarizabilities of the interacting groups, and so can be significant when both groups are highly polarizable. The additional stability bestowed by London forces takes the form of a complex created by the linkage of a polarizable acid (soft acid) and a polarizable base (soft base).

4.6.4.1 Interactions between soft metals and sulfur groups

Selective soft acid–soft base interactions serve as the underlying concept for immobilized soft-metal affinity chromatography (ISMAC). Sulfur groups in biological molecules are soft donor groups and prefer to form complexes with soft acceptors.

Table 4.12 Stability Constants of Metal Ion-Base Complexes

		Thiourea (Log K)	Histidine (Log K)
Soft metal ions	Ag^+	7.11 ± 0.07*	—
	Cd^{2+}	1.3 ± 0.1*	5.74*
Borderline metal ions	Cu^{2+}	0.8†	10.16 ± 0.06*
	Zn^{2+}	0.5*	6.51 ± 0.06*

*Stability constant in aqueous solution (ionic strength; 0.1) at 25 °C.
†Stability constant in aqueous solution (ionic strength; 1.0) at 25 °C.
Source: Adapted from Martell AE, Smith RM. Critical stability constants. Vol. 5. New York: Plenum Press, 1982; and Martell AE, Smith RM. Critical stability constants. Vol. 6. New York: Plenum Press, 1989.

Soft metal ions include such metals as Ag(I), Cu(I), Au(I), Cd(II), Pd(II), Pt(II), and Pt(IV).

Table 4.12 lists the stability constants for complexes between metal ions and bases. The sulfur group of thiourea (a soft base) complexes with Ag^+ ions (soft acids) much more strongly than with Cu^{2+} ions (borderline acids). Complexes of histidine and borderline metal ions, such as Cu^{2+} and Zn^{2+}, possess higher stability constants than histidine and soft metal ion complexes. Such examples show the selective affinity between borderline metal ions and borderline bases, and between soft metal ions and soft bases.

The selectivity of IMAC can likely be changed from nitrogen to sulfur chemical functional groups by using an immobilized soft metal ion. In addition, the preferences of soft metal ions to complex with soft base groups can be manipulated to selectively separate biochemicals that contain accessible soft base (most notably, sulfur) groups. As mentioned in the Chapter 3, both batch and column experiments support this modification.

4.6.5 Comparison and correlation of different scales

Because of inherent similarities in the way that all of these acid-base scales were developed, it is not surprising that parameters from different systems agree with one another. Moreover, as previously stated, the Gutmann DN parameters are based on the measurement of a hard acceptor's interaction with a donor. This origin would suggest that DN correlates with pK_a.

Correlating and comparing parameters from different acid-base scales not only provides a way to estimate missing or unavailable parameters in a system, but also emphasizes and strengthens the premise of quantitatively scaling chemical properties so as to predict equilibrium or kinetic constants via LFERs. Such comparisons also expose the limitations of any particular system.

4.7 METAL ION BINDING CONSTANTS

Metal ions form complexes with biochemicals in the familiar 1:1 stoichiometry. They can, however, also form complexes in other ratios, occasionally involving hydronium ions. We can use this information to quantify the different equilibrium constants as stability constants, reflecting the fact that they measure the tendency of a biochemical to form a particular complex.

In general, metal ion complex stability constants are defined as follows:

$$qM + pL + rH = M^q L^p H^r \tag{4.52}$$

$$\beta_{pqr} = \frac{M_q L_p H_r}{M^q L^p H^r} \tag{4.53}$$

In these equations, M, L, and H stand for the metal ion, biochemical, and hydronium ion activity, respectively. Often, $q = p = 1$ and $r = 0$ so that the stability constant is equal to a 1:1 association equilibrium constant.

Besides temperature, an important parameter that should be considered when using stability constants is the salt concentration. The types and concentrations of salts in the aqueous solution can affect the activity coefficients. Various spectroscopic and wet chemical techniques are used to determine stability constants (11).

4.7.1 Nucleic acids

In single-stranded polynucleic acids, all atoms are accessible to metal ions. In contrast, the only double-stranded nucleic acids sites that engage in specific interactions are the N7 atom in guanidine and adenine and the O4 atoms in thymine and uracil. The N7 site is the most reactive.

Nitrogen and oxygen atoms in nucleic acids act as electron donors for metal ions. Metal ions that bind nitrogen atoms preferentially to oxygen atoms follow the following order:

Mn(II) > Fe(II) > Zn(II) > Co(II) > Cd(II) > Ni(II) > Cu(II)

Soft metal ions, such as Hg(II), Ag(I), Pt(II), and Pd(II), also have a high affinity for nitrogen. Hard metal ions, such as Na(I), Ca(II), Mg(II), Al(III), Mn(II), and Fe(III), have greater affinity for the oxygen atoms in the phosphate group of nucleic acids. While borderline metal ions, such as Fe(II), Co(II), Cu(II), and Zn(II), and some soft metal ions, such as Cd(II) and Pb(II), have a strong affinity for nitrogen atoms on the pyridine or pyrimidine nucleic base. They also bind to phosphate groups.

4.7.2 Amino acids

A large body of data is available on the stability constants of amino acids and a variety of metal ions. Table 4.13 summarizes 1:1 stability constants.

One important conclusion that can be gleaned from these data is that copper ions generally have the highest affinity for amino acids among all transition metal ions. Moreover, histidine has the highest stability constant. It is thus not surprising that immobilized metal ion affinity chromatography (IMAC) uses

Table 4.13 Tabulation of 1:1 Stability Constants ($\log K_1$) for Various Amino Acids with Divalent Metal Ions

Amino Acid	Mg(II)	Ca(II)	Mn(II)	Fe(II)	Co(II)	Ni(II)	Cu(II)	Zn(II)
α-Alanine	2.0	1.2	3.1		4.4	5.5	8.2	4.6
Arginine	1.3	2.2	2.6	3.2	4.0	5.2	7.9	4.1
Asparagine					4.6		7.9	
Aspartic acid	2.4	1.6	3.7		5.9	6.7	8.6	5.8
Cysteine			4.6	6.2	9.3	9.8		9.2
Glutamic acid	1.9	1.4	3.3	4.6	4.5	5.9	7.9	5.5
Glycine	1.3	1.4	2.9	4.1	4.6	5.7	8.1	5.0
Histidine			3.4		6.8	8.9	10.2	6.7
Leucine			2.8	3.4	4.6	5.6	7.9	4.9
Lysine			2	4.5	3.6	5.5	7.6	
Methionine			2.8		4.4	5.2	8.0	4.4
Phenylalanine			2.4	3.7	4.1	5.6	8.3	4.4
Proline			2.8		5.0	6.2	8.9	5.1
Serine		1.4	3.9		4.5	5.4	7.9	4.7
Tryptophan			2.9	2.9	4.6	5.7	8.3	5.2
Tyrosine	2	1.5	2.4		4.0	5.1	7.9	4.2
Valine			2.8		4.6	5.3	8.1	5.0

Source: Adapted from Dawson RMC, Elliot DC, Elliot WH, Jones KM. Data for biochemical research. Oxford: Claredon Press, 1986:486–488.

immobilized copper ions to recover proteins that are expressed with cleavable tails containing six histidine residues. Measurements of stability constants with other metal ions, especially soft metal ions, will help determine how protein and amino acid selectivity can be manipulated through a judicious choice of metal ions.

4.8 SUMMARY

In bioseparation processes, process selection and design depends upon the prediction or measurement of chemical and physical properties such as solubility, charge, chemical interactions, and diffusivity. It is also useful to understand general trends in these properties based on molecular information, such as structure, sequence, or the presence of chemical functional groups. Recognizing these trends will aid in the evaluation of process choices. This chapter introduced many concepts that can be used to guide bioseparation process selection, such as hydrophobicity scales, linear free-energy relationships, molecular thermodynamics, and Lewis acid-base scales.

As analytical methods continue to improve our knowledge of biological structure and function, bioseparation process design will become even more sophisticated at the molecular scale. Thermodynamics may provide a framework with which to organize this information so as to minimize the number of experiments needed to characterize a system. Molecular intuition and a thermodynamic approach therefore constitute powerful tools for bioseparation process design.

Table 4.14 Retention Times of Carboxylic Acids

Carboxylic Acid	Retention Time* (min)
Acetic acid	19.0
Ascorbic acid	13.1
Citric acid	11.0
Formic acid	17.5
Fumaric acid	19.8
Lactic acid	16.0
Malic acid	12.9
Oxalic acid	9.0
Quinic acid	13.3
Shikimic acid	15.5
Tartaric acid	11.7

*Using a Supelco C-610 H column, mobile phase 0.1% H_3PO_4, 0.5 mL/min at 30 °C.
Source: Adapted from Supleco Chromatography Products Catalogue. 1996:154.

4.9 PROBLEMS

4.1 In extracting a carboxylic acid into an organic phase, an extractant molecule, E, dissolved in the organic phase will form a complex with the acid HA in the organic phase only when the acid is in its un-ionized form:

$$HA_{org} + E_{org} \leftrightarrow HA - E_{org} \tag{4.54}$$

Assuming that that this reaction can be expressed with an equilibrium constant $K_{complex}$, derive an expression for the effective dissociation constant K_D. K_D is defined as follows:

$$K_D = \frac{[\text{acid in the organic phase}]}{[\text{acid in the aqueous phase}]} \tag{4.55}$$

This constant is a function of pH, $K_{complex}$, and any other variable. Assume that the carboxylic acid is so hydrophilic that it will not partition into the organic phase unless it forms a complex with the extractant molecule.

4.2 You need to keep a lysine-rich peptide in solution. What steps would you take to ensure that it does not precipitate? Number and explicitly state in single-sentence form the steps you would take.

4.3 Calculate the charge versus pH diagram for a dipeptide of glutamic acid and lysine. To perform this calculation, you must assume that the amino and side chains that are ionizable in this peptide have the same pK_a values, even though they are now in the same polymer chain. Is this assumption valid? Why or why not? (Hint: Remember that a peptide bond is formed.)

4.4 When separating amino acids using ion-exchange chromatography with a sulfonated polystyrene packing, the following elution order is observed.

Starting at pH 3.25:

asparagine < threonine < serine < glutamic acid < proline

When the pH is changed to 4.25:

glycine < alanine < cysteine < valine < methionine < isoleucine < leucine
tyrosine < phenylalanine

Ending with pH 5.28:

lysine < histidine < arginine

Explain the observed retention time, taking into account any hydrophobic and ionic interactions with the negatively charged sulfonated polystyrene packing used in the analytical method.

4.5 In ion-exclusion chromatography, organic acids can be separated at low pH based in part on their acidity because the packings have strong acid functional groups. Using the information in Table 4.14, interpret the retention time data by graphing the retention time versus a property that you feel most strongly characterizes the acidity of these carboxylic acids. Should you group the data to improve your prediction for the retention time of organic acids not listed in Table 4.14?

4.6 A computer algorithm for computing the isoelectric point of proteins (Genetics Computer Group, Inc. 1993) uses the following equation:

$$\begin{bmatrix} \text{Net} \\ \text{charge} \end{bmatrix} = \begin{bmatrix} \text{\# of positively} \\ \text{charged residues} \end{bmatrix} - \begin{bmatrix} \text{\# of negatively} \\ \text{charged residues} \end{bmatrix} + \begin{bmatrix} \text{\# of protonated} \\ \text{amino termini} \end{bmatrix}$$
$$- \begin{bmatrix} \text{\# of deprotonated} \\ \text{carboxy termini} \end{bmatrix} \tag{4.56}$$

For each amino acid residue type (lysine, histidine, and so on) found in a protein, the number of protonated residues is found with the following equation:

$$N(p) = N(t) \frac{[H^+]}{[H^+] + K(N)} \tag{4.57}$$

$N(p)$ is the number of protonated residues, $N(t)$ is the total number of residues of a specific type, $[H^+]$ is the hydrogen ion concentration, and $K(N)$ is the amino acid dissociation constant.

(a) Using the algorithm and the discussion in the chapter, calculate the net charge of a peptide that contains two lysine residues and one proline residue.

(b) Derive the equation given for the number of protonated residues.

(c) Review the equations given in the chapter for amino acid charge as a function of pH, and explain how the equation for the number of protonated residues compares with them.

4.7 Rapid methods for recovering very small quantities of DNA are important in biological preparations and forensic science. One method involves precipitating DNA using the large dye molecule N,N′-Bis[3,3′-(dimethylaminopropylamine)-3,4,9,10-perylenetracarboxylic diimide, which is shortened to the simple name of DAPER (29). This molecule has tertiary amine and imide chemical functional groups on the periphery; its hydrophobic core contains four benzene rings and one cyclohexane ring. It has a molecular weight of 560 daltons.

Figure 4.11 shows the recovery of 310 base-pair DNA by precipitation using isopropanol or DAPER. Isopropanol was added until the final volume percentage

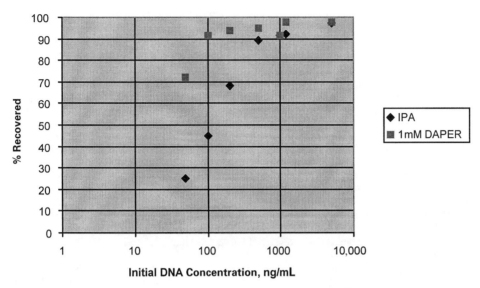

Figure 4.11 Recovery of 310 base-pair DNA by precipitation using isopropanol or DAPER (29).

was 50%; sodium acetate was added until the final concentration reached 0.3 M. For the DAPER precipitation, one-fifth volume of 1 mM DAPER is added to the DNA solution. Both precipitation methods were conducted at room temperature. Determine the solubility of DNA in isopropanol and DAPER for the conditions given.

4.8 One method for understanding the interactions between metal ions and amino acids is to use linear free-energy relationships to relate various properties. For example, one approach involves the use of the following equation:

$$\log K_1 = a\left(\frac{z^2}{r}\right) + b(\text{hydrophobicity}) + c(\text{basicity}) \qquad (4.58)$$

In this equation, z is the charge of the metal ions, r is the ionic radius, "hydrophobicity" is a measure of the amino acid, "basicity" is a measure of the donor properties of the amino acid, and a, b, and c are constants.

(a) Try to fit data for each amino acid (assume $b = c = 0$) where this chapter has provided such data. You will need to look up ionic radii for the metal ions in standard chemistry references such as *Lang's Handbook of Chemistry*.

(b) Determine if the data for all of the amino acids collected can be correlated using Equation 4.58. You will need to choose hydrophobicity and basicity scales from those provided in the chapter.

(c) Describe some of the difficulties or accomplishments associated with your correlation from part (b).

4.10 REFERENCES

1. Wells PR. Linear free energy relationships. London: Academic Press, 1968.
2. Smith JM, Van Ness HC. Introduction to chemical engineering thermodynamics, 4th ed. New York: McGraw-Hill, 1987.

3. Wolfenden J, Andersson L, Cullis PM, Southgate CCB. Affinity of amino acid side chains for solvent water. Biochemistry 1977;20:849–855.
4. Cornette J, Cease KB, Margalit H, Spouge JL, Berzofsky JA, DeLisi C. Hydrophobicity scale and computational techniques for detecting amphipathic structures in proteins. J Mol Biol 1987; 195:659–685.
5. Bailey PD. An introduction to peptide chemistry. New York: John Wiley and Sons, 1990.
6. Bezkorovainy A. Basic protein chemistry. Springfield: Charles C. Thomas, 1970.
7. Kiss T. Complexes of amino acids. In: Burger K, ed. Biocoordination chemistry: coordination equilibria in biologically active systems. New York: Ellis Horwood Limited, 1990.
8. Martell AE, Smith RM. Critical Stability Constants. Vol. 1. New York: Plenum Press, 1974.
9. Martell AE, Smith RM. Critical Stability Constants. Vol. 5. New York: Plenum Press, 1982.
10. Lehninger AL. Principles of biochemistry. New York: Worth Publishers, 1982.
11. Martell AE, Motekaitis RJ. Determination and use of stability constants, second edition. New York: VCH Publishers, 1992.
12. http://expasy.hcuge.ch/ch2d/pi_tool.html
13. Nozaki Y, Tanford C. The solubility of amino acids and two glycine peptides in aqueous ethanol and dioxane solutions. J Biol Chem 1971;246:2211.
14. http://cherubino.med.jhu.edu/~raj/Research/Linus/polars.html
15. http://www.eng.uci.edu/students/hslee/aainfo/hydro.htm
16. Pearson RG. Hard and soft acids and bases. Chem Brit 1967;3:103–107.
17. Pearson RG. Hard and soft acids and bases, HSAB. Part I. J Chem Ed 1968;45:581–587.
18. Pearson RG. Hard and soft acids and bases, HSAB. Part II. J Chem Ed 1968;45:643–648.
19. Drago, RS. Physical methods for chemists, 2nd ed. Fort Worth: Saunders, 1992.
20. Fuchs R, Abraham MH, Kamlet MJ. Solute–solvent interactions in chemical and biological systems. IV. Correlations of ΔG, ΔH, and $T\Delta S$ of transfer of aliphatic and aromatic solutes. J Phys Org Chem 1989;2:559.
21. Grinberg AA. An introduction to the chemistry of complex compounds. London: Pergamon Press, 1962.
22. Mulliken RS. Molecular compounds and their spectra. III. The interactions of electron donors and acceptors. J Phys Chem 1952;7:801–822.
23. Mulliken RS. Molecular compounds and their spectra. J Am Chem Soc 1952;74:811–824.
24. Weiss J. The formation and structure of some organic molecular compounds. J Am Chem Soc 1942:245–252.
25. Chatt J, Wilkins RG. Bis-ethylene-dichloroplatinum. Nature 1950;165:859–860.
26. Chatt J, Duncanson LA, Venanzi LM. Directing effects in inorganic substitution reactions. Part I. A hypothesis to explain the trans-effect. J Am Chem Soc 1955:4456–4460.
27. Pitzer KS, Catalano E. Electronic correlation in molecules. III. The paraffin hydrocarbons. J Am Chem Soc 1956;78:4844–4846.
28. Pitzer KS. London force contributions to bond energies. J Chem Phys 1955;23:1735.
29. Liu ZR, Rill RL. N,N'-Bis[3,3'-(dimethylaminopropylamine)-3,4,9,10-perylenetracarboxylic diimide a dicationic perylene dye for rapid precipitation and quantitation of trace amounts of DNA. Anal Biochem 1996;236:139–145.
30. Rose GD, Geselowitz AR, Lesser GJ, Lee RH, Zehfus MH. Hydrophobicity of amino acid residues in globular proteins. Science 1985;229:834–838.

5

Biocolloidal Interactions and Forces

Because they are far larger than simple molecules, biopolymers and cells respond to their surroundings in a unique fashion. The term *colloid* is generally used to describe these large biopolymers and cells, as well as any particle with a diameter of approximately 1 μm. Biological colloids (also known as biocolloids) behave in a similar fashion in aqueous solution as other types of colloidal particles. Aqueous colloids can be affected by a variety of forces, such as long-range forces associated with the hydrophobic effect, electrostatic forces, and van der Waals interactions. Paramagnetic and magnetic colloids, by virtue of their intrinsic characteristics, interact through long-range magnetic forces. Short-range forces, which originate from chemical and hydrogen bonds, become significant only when a colloid approaches a surface or another colloid to within molecular distances.

Although biocolloids share many aqueous properties with other colloidal particles, interactions involving these compounds can be somewhat more complex. For example, proteins undergo relatively slow conformational changes when they bind to a surface. Cell receptors can change the shape and chemistry of the cell surface, depending on the presence of other solutes. These and other complex biocolloidal phenomena lie beyond the scope of this chapter.

The environment-sensitive or "smart" responses of colloids do play a role in biocolloidal behavior. This chapter treats biopolymers and cells as colloids, and then highlights how the unique nature of biocolloids affects their behavior. As always, our focus is on understanding and designing separation processes, and we must recognize that models that describe colloidal effects are important to the design of cell, protein, and DNA separation and purification processes.

5.1 SHORT-RANGE INTERACTIONS

Table 5.1 provides an overview of the various bonds that can form between molecules and gives the interaction energy associated with each type of bond. Biopolymers and cells possess chemical functional groups on their accessible surfaces that are capable of hydrogen bonding and forming dipole-dipole interactions. These types of chemical interactions depend upon electron orbital effects that arise only when the chemical groups come in close proximity to one another—typically to within the length of a chemical bond. By solvating chemical functional groups, water molecules play an active role in these processes.

Small molecules can collide with many degrees of orientational freedom, which can help them form bonds. In contrast, the specific chemical groups on col-

Table 5.1 Range of Bond Energies Between Molecules

Interaction	Bond Energy (kcal/mol)
Ionic bond	140–250
Covalent bond	15–230
Metallic bond	25–85
Dipole-dipole interaction	<9
Dipole-induced dipole interaction	<5
Dispersion forces	<10

loidal particles are tethered to a surface that possesses only a limited ability to form chemical or hydrogen bonds with its counterpart on another colloid. Brownian motion roughly dictates the speed and frequency with which colloidal particles undergo collisions, except when external electrical or magnetic fields are present. Particle curvature and molecular-scale or larger surface roughness, however, prevents all possible acid-base or dipole interactions from taking place between colloids.

Because hydrogen bonds form between water molecules, the aqueous environment represents a somewhat unusual case. In this type of solution, large numbers of hydrogen-bonded water molecules form ordered clusters, which can extend the range of functional group interactions. The formation of an aqueous layer on some functional groups, for example, can increase the effective distance of usually short-range ion-ion or acid-base interactions. In the next section, we will see how molecular attractions caused by van der Waals forces and electrostatic interactions can become transformed into long-range actions in a condensed medium. We will also examine a long-range interaction exclusive to water—the hydrophobic effect.

5.2 LONG-RANGE INTERACTIONS

5.2.1 Van der Waals forces

All molecules can be attracted to each other by becoming dipoles for a fleeting moment. Because these dipoles are not permanent, the strength of the ensuing interaction does not depend upon perfect alignment of the molecules. The forces due to this interaction are known as van der Waals forces (or London dispersion forces). They explain why all molecules undergo phase transitions and why interfacial tension exists between two phases (1). In 1930, London derived the potential energy of interaction through dispersion forces between two identical atoms:

$$W_D = -\frac{\lambda}{r^6} \tag{5.1}$$

In this equation, r is the distance between the atoms and λ is a constant given by the following equation:

$$\lambda = \frac{3}{4}\alpha^2 h \nu_0 \tag{5.2}$$

In Equation 5.2, α is the polarizability of the atom and $h\nu_0$ is the characteristic energy for the quantized energy at frequency ν_0. From a practical standpoint, $h\nu_0$ is also known as the ionization potential.

While Equation 5.1 applies to the interaction between two molecules, this form cannot be readily used to describe the interaction between particles on the macroscopic scale. As we shall see, because of surface roughness on the scale of 0.1 mm, macroscopic bodies do not approach one another closely enough to make London dispersion forces significant. Also, Brownian motion cannot keep these large bodies in solution if they have densities close to, or greater than, the solvent density. To apply these molecular interactions to molecules in macroscopic bodies, we need to consider that each molecule on one body can interact with each molecule on the other body.

When an atom closely approaches a flat body, the force that results reflects the interactions with the atoms in the flat body. To correctly sum the contributions to the overall force, we can integrate the energy of dispersion, W_D. The force varies with distance. As a first approximation, we can assume that the overall force felt by each atom is additive. Thus the following integration will give the total force:

$$F_{12} = \int_v \frac{dW_D}{dr} dV \tag{5.3}$$

This equation adds the forces by taking into account the interaction of one atom with all atoms on the other body. Figure 5.1 depicts this integral as performed by two researchers through the use of geometry (2). Drawing a cone from atom A to the atoms on the other flat body gives, on the base of the cone, the molecules that interact with the same force; these molecules are equidistant from atom A. Integrating using a solid ring of thickness dx and depth $d\rho$ gives a differential volume

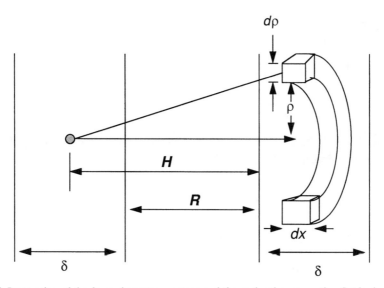

Figure 5.1 Integration of the forces between an atom and the molecules on another flat body.

that correctly accounts for all forces created by this infinitesimal volume of molecules that act on atom A.

Some final assumptions are important in interpreting the results from this model:

- The body has a uniform density of atoms.
- The dispersion forces are additive.
- The body is treated as a continuum of atoms, contrary to the quantum mechanical approach.

Using a ring of differential thickness dx with a constant density of atoms, the volume in the ring is given by the following equation:

$$V = dx\pi\left((\rho + d\rho)^2 - \pi\rho^2\right) \tag{5.4}$$

Letting q be equal to the number of atoms per unit volume, and assuming that $d\rho^2 \ll 2\rho d\rho, \pi\rho^2$, the following relationship for the number of atoms in the differential ring is obtained:

$$\text{Number of atoms} = 2\pi q\rho d\rho dx \tag{5.5}$$

The force component perpendicular to the ring is given by the following equation:

$$F_x = F_D\cos(\theta) = \frac{6\lambda}{r}\cos(\theta) \tag{5.6}$$

Thus the total force between a single atom and all atoms on a flat body separated by a distance R can be obtained through the following double integral:

$$F_{0-2} = \int_0^\infty \int_R^{R+\delta} \frac{12\pi\lambda}{r^7} q\rho\cos(\theta)dxd\rho \tag{5.7}$$

The integrals are evaluated by converting r to terms involving x and ρ using the definition of $\cos\theta$,

$$\cos(\theta) = \frac{x}{r} = \frac{x}{\sqrt{\rho^2 + x^2}} \tag{5.8}$$

which yields the following integral:

$$F_{0-2} = 12\pi q\lambda \int_0^\infty \rho \int_R^{R+\delta} \frac{x}{(\rho^2 + x^2)^4} dxd\rho \tag{5.9}$$

The final results for both the interaction force and energy between one atom and a flat body are as follows:

$$F_{0-2} = \frac{\pi q\lambda}{2r^4} \tag{5.10}$$

$$w_{0-2} = -\frac{\pi q\lambda}{6r^3} \tag{5.11}$$

We can easily extend this result to the case in which two flat bodies approach one another [shown in Figure 5.3(b)]:

Table 5.2 Dispersion Force Equations for Colloidal Particles of Various Geometries

Situation	Interaction Energy
Two spheres	$w = -\dfrac{Aa_1a_2}{6R(a_1+a_2)}$
Sphere–flat body	$w = -\dfrac{Aa}{6R}$
Two parallel chains of molecules	$w = -\dfrac{3\pi\lambda L}{8a^2R^5}$
Two cyclinders	$w = -\dfrac{AL(a_1a_2)^{0.5}}{12\sqrt{2}R^{1.5}(a_1+a_2)^{0.5}}$
Two crossed cylinders	$w = -\dfrac{A\sqrt{a_1a_2}}{6R}$
Two flat bodies	$w = -\dfrac{A}{12\pi R^2}$

$$F_{1-2} = \frac{\pi q^2 \lambda}{4r^3} \tag{5.12}$$

$$w_{1-2} = \frac{\pi q^2 \lambda}{12r^2} \tag{5.13}$$

More complex geometries, such as those involving two spheres, have also been solved (2); these results are listed in Table 5.2. The most important result from this exercise is that dispersion forces are shown to act across a longer distance when colloidal bodies are involved. In addition, this analysis gives rise to a new term, known as the Hamaker constant:

$$A = \pi^2 \lambda \rho_1 \rho_2 \tag{5.14}$$

The Hamaker constant indicates that the interaction potential and atomic densities of two materials dictate the attractive dispersion force between two colloidal particles.

Several modifications to this rather simple model of additive forces has been developed. For example, at larger separation distances (on the order of 10 nm), dispersion forces become retarded by a phase difference between the instantaneous dipole on opposing bodies. The diminished van der Waals forces demonstrate that dispersion forces decrease more rapidly for larger separation distances. We can account for this case by using a correction factor (2).

5.2.1.1 Medium effects

The continuous phase in which the colloids are suspended plays an important role in moderating dispersion forces. Because the medium itself interacts with the colloidal particles, a model for determining the conditions that induce two colloidal particles to "touch" must account for the medium-generated dispersion forces. One simple method uses a pseudochemical model to determine the energy of interaction.

Figure 5.2 Tracking the creation and destruction of surfaces to determine the energy change when two spherical bodies come together.

Figure 5.2 illustrates the interaction energy between two spherical particles. The following equation gives the net change in the free energy for the process in which the liquid-colloid surfaces are lost and a colloid-colloid surface is gained:

$$\Delta G_{2\text{-}1\text{-}2} = w_{1\text{-}1} + w_{2\text{-}2} - 2w_{1\text{-}2} \qquad (5.15)$$

In terms of the Hamaker constant, this process allows for the determination of a new constant, $A_{2\text{-}1\text{-}2}$:

$$A_{2\text{-}1\text{-}2} = A_{1\text{-}1} + A_{2\text{-}2} - 2A_{1\text{-}2} \qquad (5.16)$$

In molecular thermodynamics, a frequently used assumption to determine cross-constants such as $A_{1\text{-}2}$ is to take the geometric mean of the individual Hamaker constants $A_{1\text{-}1}$ and $A_{2\text{-}2}$. We can then calculate $A_{2\text{-}1\text{-}2}$ as follows:

$$A_{2\text{-}1\text{-}2} = A_{1\text{-}1} + A_{2\text{-}2} - 2\sqrt{A_{1\text{-}1}A_{2\text{-}2}} \qquad (5.17)$$

With rearrangement and simplification, the final result takes the following form:

$$A_{2\text{-}1\text{-}2} = \left(A_{1\text{-}1}^{1/2} - A_{2\text{-}2}^{1/2}\right)^2 \qquad (5.18)$$

Because the Hamaker constant is positive, a net attraction always exists between two colloidal bodies in a medium (or even in a vacuum). The interaction energies shown in Table 5.2 are negative, so a positive Hamaker constant allows for a negative interaction energy—the hallmark of a spontaneous process. Note also that the overall Hamaker constant ($A_{2\text{-}1\text{-}2}$) decreases as the Hamaker constant of the medium ($A_{1\text{-}1}$) approaches that of the colloidal particle.

5.2.2 Electrostatic interactions and DLVO theory

Another type of molecular interaction that is additive over colloidal distances derives from ionic charges. To understand electrostatic interactions between colloidal particles, we can first consider a pseudocapacitor model that describes the interplay between mobile ions dissolved in the medium and fixed charges on the surface of a colloid. As this structural model shows, colloids begin to exert electrostatic forces upon one another when they are close enough that the mobile ions in their diffuse double layer overlap (Figure 5.3). When the surfaces are not highly charged, with a surface potential (ψ_0) less than 25 mV, the thickness of the diffuse double layer (κ^{-1}) is given by the Debye-Hückel approximation:

$$\frac{1}{\kappa} = \left(\frac{1000\varepsilon kT}{8\pi e^2 N_A I}\right)^{1/2} \qquad (5.19)$$

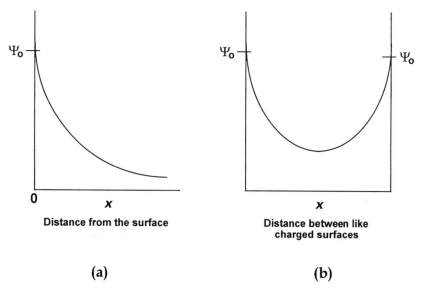

Figure 5.3 (a) Variation of electrostatic potential over distance from a surface. (b) Overlap of the electrical double layer of like charged surfaces.

In this equation, I is the ionic strength, N_A is Avogadro's number, ε is the dielectric constant of the medium, e is the charge on an electron, and k is Boltzmann's constant. For example, the double-layer thickness for a standard biological solution containing 0.15 M NaCl in 0.2 M pH 7 phosphate buffer at room temperature is quite small—less than 1 nm. Thus biocolloids usually remain quite well shielded from one another in terms of electrostatic interactions.

Moving beyond our simple capacitor model, the diffuse layer can be better described using the Poisson-Boltzmann equation. A detailed discussion of the Guoy-Chapman theory and Derjaguin's derivation of the interaction of flat surfaces and spherical particles of like charges at low surface potentials can be obtained from several sources (2–4). For our purposes, we can simply proceed to the combined theory of electrostatic interactions and dispersion forces, known as DLVO theory (named after Derjaguin, Landau, Verwey, and Overbeek, the framers of this model). For two flat bodies separated by a distance R, DLVO theory gives the following expression:

$$w_{\mathrm{DLVO}} = 64 n_0 k T \gamma_0^2 \exp(-\kappa R) - \frac{A}{12\pi r^2} \tag{5.20}$$

Here n_0 is the number of molecules and γ_0 is given by Equation 5.21:

$$\gamma_0 = \frac{\exp\left(\dfrac{zF\psi_0}{2n_0 kT}\right) - 1}{\exp\left(\dfrac{zF\psi_0}{2n_0 kT}\right) + 1} \tag{5.21}$$

DLVO theory for two spheres is given by the following equation:

$$w_{\text{DLVO}}^{\text{spheres}} = \frac{64\pi a n_0 k T \gamma_0^2}{\kappa^2} \exp(-\kappa R) - \frac{Aa}{12r} \tag{5.22}$$

In this expression, a is the diameter of the like spheres and R is the distance between the spheres' surfaces.

For two like, flat bodies with low surface potentials, a schematic representation of the DLVO equation shows how double-layer repulsion and van der Waals attraction gives rise to an interactive energy curve with two minima. At low salt concentrations, the secondary minimum shows that colloidal particles remain more than 5 nm from one another, thereby producing a stable suspension. Increasing the salt concentration decreases the surface potential and leads to rapid coagulation, as repulsive forces decline. The repulsive force, which is felt as much as 60 nm away from the surface in purified water, is undetectable at salt concentrations greater than 0.01 M KCl.

DLVO theory and experimental data clearly show that, at the physiologic conditions normally found in biological fluids, long-range electrostatic repulsion is a weak effect. This relationship suggests that cells and proteins can survive in an environment that enables association. Although this knowledge can be exploited to promote tissue formation and assembly of complex structures, aggregation such as clotting and precipitation can prove deleterious to an organism.

The next section discusses hydrophobic interactions, another type of attractive force that can act over even longer ranges and thus must be taken into account in biological separations.

5.2.3 Hydrophobic effects

So far, the physics of colloidal interactions in aqueous salt solutions suggest that biocolloids in buffers and physiological fluids can approach one another with a mild attractive force. An added, and possibly quite long-range, interaction is called the hydrophobic effect (5, 6).

Several researchers (7–9) have developed equations to describe the data obtained using two curved mica cylinders that have been coated with a long alkyl chain hydrocarbon. Some of the experimental information suggests that attractive forces can be felt as far away as 70 nm (Figure 5.4). This finding is somewhat surprising, as this force is generated by water molecules, which tend to create well-integrated, hydrogen-bonded clusters. The forces caused by the hydrophobic effect are also quite large, exceeding 100 mN/m (Force/Diameter) at short (less than 3 nm) separation distances.

By comparison, van der Waals attractive forces reach only 10 mN/m under similar conditions (2); indeed, repulsive forces are less than 1 mN/m for double-layer repulsion of spheres with moderate surface potentials ($\psi_0 < 100$ mV). The hydrophobic effect is quite formidable, making it is easy to understand why bacteria with hydrophobic membrane surface layers (such as mycobacteria) readily aggregate to form cords.

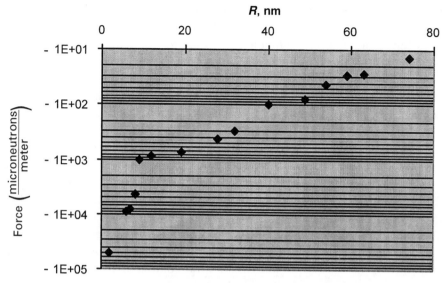

Figure 5.4 Force–distance curve for long-range hydrophobic attraction. (Source: Adapted from Claesson HK, Christenson PM, Berg J. Forces between fluorocarbon surfactant monolayers: salt effects on the hydrophobic interaction. J Phys Chem 1989;93:1472.)

5.2.4 Magnetic interactions

Most biochemical procedures that involve external fields deal only with electrical or gravitational field effects. Magnetic and paramagnetic materials, however, have utility in bioseparation procedures. Paramagnetic particles, which range from 50 nm to several micrometers in diameter, can be used to isolate specific cell subtypes in medical applications; they can also be exploited to recover bacteria in environmental applications. Analysis of magnetic field effects and the magnetic force's influence on the distribution of components, therefore, is an important consideration when designing separation procedures.

Paramagnetic particles are created by embedding iron oxide grains within a polymer matrix, such as polystyrene. The iron oxide grains are relatively uniformly distributed throughout the polystyrene matrix. Because the grains have no long-range order, they become magnetic only in the presence of a magnetic field. As they have a completely reversible magnetization, paramagnetic particles can prove quite useful because they can be readily dispersed and collected.

When an external magnetic field is applied, a single paramagnetic particle develops a dipole whose strength is given by the following equation:

$$\mu = \frac{4}{3}\pi a^3 \mu_0 X H \tag{5.23}$$

Here a is the particle radius, X is the magnetic susceptibility, μ_0 is the magenetic permeability of a vacuum, and H is the external field strength. Two particles have an interaction energy that depends upon their orientation to one another (10):

$$w_{\mathrm{mag}} = \frac{\mu^2(1 - 3\cos^2(\theta))}{4\pi\mu_0 R^3} \tag{5.24}$$

In this case, R is the distance between particles, measured from center to center. Equation 5.24 must be used with caution, however. The interaction energy used here ignores contributions from multipole effects, and the equation cannot explain the many-body problem (a typically intractable problem in classical physics).

It is important to contrast magnetic interactions with interactions caused by the electrical double layer. Magnetic forces arise from the entire volume of the bodies; in contrast, electrical double-layer interactions are solely caused by charges on the surface. This difference leads to a much stronger dependence of a sixth power on particle radius in the case of magnetic forces. Double-layer interactions have only a first-power dependence. In fact, van der Waals interactions have only a first-order dependence on the particle radius.

Another important distinction is that double-layer interactions between particles drop off exponentially with distance, with the exponential decay being controlled by the double-layer thickness. Magnetic interactions have a shorter range than van der Waals interactions, because their span of influence is determined by the inverse of the separation distance raised to the third power. Van der Waals interactions, however, fall off with the inverse first power of the separation distance.

The velocity dependence on particle size for particles moving under the influence of an electric field also differs from that of particle movement under a magnetic field. Chapters 12 and 13 examine electrophoretic and magnetophoretic mobilities in detail. Here, we will simply show the practical consequences of having the particle radius strongly depend on the strength of magnetic forces. Compared with electrophoretic mobility, and assuming Stokes' law for particle drag, magnetophoretic mobility (11) has a stronger dependence on particle radius. The function $f(C_p)$ accounts for the effect of particle-particle interactions that themselves depend upon particle concentration:

$$\mu_{\text{magnetophoresis}} \propto \frac{r_p^2 \Delta X}{\mu(1 + f(C_p))} \tag{5.25}$$

$$\mu_{\text{electrophoresis}} \propto \frac{\zeta}{\mu} \tag{5.26}$$

This strong dependence on particle radius has important implications for magnet design and separation time. To keep separation times with reasonable limits, separation of small paramagnetic particles will require the use of a much more powerful rare-earth magnet.

5.3 SUMMARY

Colloidal forces attributable to physical and chemical phenomena play an important role in cell suspension and protein solution behavior. Short-range forces (less than 0.2 nm) dictate biological interactions. Long-range forces, which include van der Waals interactions, electrical double-layer effects, and the hydrophobic effect, allow cells and proteins to "feel" each other's presence as far as 60 to 70 nm away.

In this chapter, we derived a model of van der Waals interactions between flat surfaces in some detail. This derivation illustrates the methodology of converting forces between molecules into a net force between micrometer-scale bodies. The cumulative effect of many individual molecules from one body interacting with those of another body increases the range of action associated with van der Waals forces for colloidal particles.

In the case of electrostatic interactions, the counter-ion layer dispersed from the ions on the colloidal surface creates an interaction energy. The scope of this energy is dependent on the surface charge density as well as the properties of the medium in which the counter ions are dissolved. Electrical double-layer forces depend on the surface property of the colloidal particle and not on the interior molecular structure of the particle.

Another long-range force that can be introduced in bioseparations arises from magnetic interactions. Both magnetic and electrical fields can be exploited in bioseparations by varying the mobility of biological molecules and cells.

5.4 PROBLEMS

5.1 Using Figure 5.5 as a guide, derive the equation for the van der Waals attraction between a sphere and a cylinder. What is the equation that results when the cylinder radius approaches infinity?

5.2 Derive an equation for the van der Waals interaction between a sphere and a chain of molecules when the two bodies are relatively similar in dimension, as shown in Figure 5.6. Discuss the utility of studying this geometry in biology.

5.3 Consider two spherical particles that are paramagnetic and have an electric charge. Taking magnetic, double-layer, and van der Waals interactions into account simultaneously, sketch the dependence between distance and the interaction energy for the following cases:

(a) Variation of the particle radius
(b) Variation of the ionic strength of the solution

Note and comment on any secondary minima that can occur in these cases.

5.4 In which of the following phenomena do colloidal phenomena play a role?

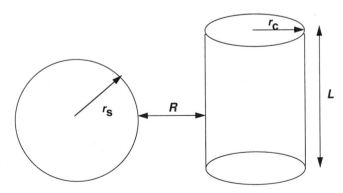

Figure 5.5 Details of the interaction between a sphere and a cylinder.

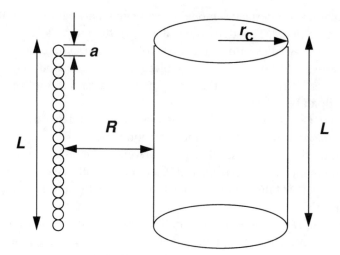

Figure 5.6 Details of the interaction between a chain of molecules and a cylinder.

(a) Coagulation
(b) Cell adhesion
(c) Enzyme-substrate binding
(d) Phagocytosis
(e) Virulence
(f) Biocompatibility

Discuss why colloidal effects are important or are not important in these biological events.

5.5 Calculate the net force as a function of separation distance between two colloids with a diameter of $1\,\mu m$ that are found in $0.2\,M$ phosphate-buffered saline if the surface potential of each of the colloids is $100\,mV$. Assume that DLVO theory for spheres applies. Where are the minima? Do these colloidal particles adhere to one another? What is the energy needed to separate these colloids? Assume that the drag force on these particles is given by Stokes' law: $F = 6\pi\eta aU$, where η is the viscosity of the solution (which is approximately the same as the viscosity of water). If the fluid can be swirled at a velocity of $1\,mm/s$ at room temperature, would this force be adequate to separate the particles?

5.5 REFERENCES

1. Van Oss CJ. Interfacial forces in aqueous media. New York: Marcel Dekker, 1994.
2. Israelachvilli J. Intermolecular and surface forces, 2nd ed. London: Academic Press, 1991.
3. Hiemenz PC. Principles of colloid chemistry, 3rd ed. New York: Marcel Dekker, 1997.
4. Adamson AW. Physical chemistry of surfaces, 5th ed. New York: John Wiley and Sons, 1990.
5. Yoon RH, Ravishankar SA. Long-range hydrophobic forces between mica surfaces in alkaline dodecylammonium chloride solutions. J Colloid Int Sci 1996;179:403.
6. Kurihara K, Kato S, Kunitake T. Very strong long-range attractive forces between stable hydrophobic monolayers of a polymerized ammonium surfactant. Chem Lett 1990;9:1555.
7. Eriksson JC, Ljunggren S, Claesson M. A phenomenological theory of long-range hydrophobic attraction forces based on a square-gradient variational approach. J Chem Soc, Faraday Trans 1989;85:163.

8. Claesson PM, Malmsten M, Lindman B. Forces between hydrophobic surfaces coated with ethyl(hydroxyethyl)cellulose in the presence of an ionic surfactact. Langmuir 1991:1441.

9. Claesson PM, Arnebrant T, Bergenstahl B. Direct measurement of the interaction between layers of insulin adsorbed on hydrophobic surfaces. J Coll Int Sci 1989;130:457.

10. Promislow JHE, Gast AP, Fermigier, M. Aggregation kinetics of paramagnetic colloidal particles. J Chem Phys 1995;102:5492.

11. Schramm LL, Clark BW. On the measurement of magnetophoretic mobilities.. Colloids and Surfaces. 1983;7:135–146.

6

Bioaffinity

To carry out life functions, biological systems undergo physical and chemical interactions that rely on variations in molecular selectivity and binding strength. The particular set of physical and chemical interactions in which structure can play a major role in shaping is referred to as bioaffinity. The specificity of noncovalent interactions (such as hydrogen-bonding, ionic-bonding, and hydrophobic interactions and interactions involving van der Waals forces) relates to the strength of the particular interaction. For example, these interactions may be specific to molecules in general or to clusters of functional groups or molecules (as depicted in Figure 6.1). As the strength of these interactions is relatively weak compared with the strength of covalent interactions, a large number of noncovalent reactions are needed to achieve the selectivity and strength found in biological systems. These nonspecific interactions must also occur at close proximity (i.e., less than 0.1 nm between the reactants), making the complementary structure between the ligand and receptor a critical parameter (1).

The individual contributions of these interactions to the overall binding energy has been an issue of great interest. Hydrogen bonds are estimated to add 2.1 to 7.5 kJ/mol of binding energy if polar charged groups are involved or 12.6 to 25.1 kJ/mol of such energy if uncharged groups are involved. In addition, hydrophobic interactions have been estimated to generate binding energy of 10 to 21 kJ/mol per square nanometer of contact. The considerable contributions from this source become apparent when we account for the size of the contact area, which can reach 10 nm^2.

In this chapter, we will review the methodology behind the major bioaffinity interactions that have been employed in bioseparation processes. Specifically, this chapter contains an in-depth discussion of the molecular recognition process and the most important interactions involved, including a review of the structure and function of specific receptor-ligand complexes.

6.1 MOLECULAR RECOGNITION PROCESSES

From a molecular perspective, the binding process between a receptor and its ligand can be viewed as moving from electrostatic interaction to solvent displacement, then to steric selection, and finally to charge and conformational rearrangement (2).

As the ligand approaches the receptor from 20 to 1 nm, Coulombic attractive/ repulsive forces dominate. In this region, both the receptor and ligand reorient themselves so as to maximize the attractive interaction. If this interaction is stronger than the repulsive forces, the process continues with the goal of stabilizing the

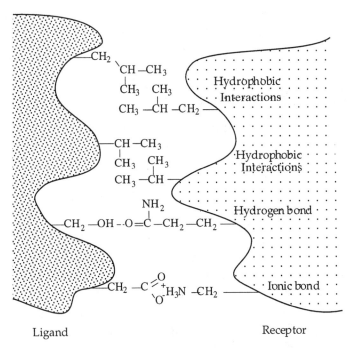

Figure 6.1 Receptor-ligand interactions.

complex. Although these electrostatic interactions might seem less important in molecular recognition, molecular modeling studies comparing the complex stability of structurally different ligands have provided new insight into their roles. Ligands that provide a better steric fit to the receptor have been shown to produce a complex as stable as that formed with less complementary ligands—as long as both ligands provide the same level of electrostatic interaction with the receptor.

The second step in molecular recognition, which occurs from 10 to 1 nm, involves the displacement of solvent molecules. Most biomolecules are water-soluble, which means that they contain a predominantly hydrophobic interior and a hydrophilic exterior. The hydrophilic exterior is composed of hydrated polar groups (i.e., polar groups surrounded by hydrogen-bonded water molecules). The displacement of water from the biomolecule surface can render a sterically hindered ligand more accessible for interaction through short-range forces (e.g., dipole-dipole, charge transfer). According to classical thermodynamics theory, water close to the surface has a much higher fugacity than in bulk solution. Its displacement therefore adds to the solvent entropic effect that is associated with surface reduction, which in turn contributes to the binding energy.

The third step in the molecular recognition process involves steric fit and the matching of groups, which allows for attractive interactions. In this step, the ligand fits into the receptor in a "lock and key" arrangement, reflecting various conformational changes in both the ligand and the receptor molecules.

The final step in the molecular recognition process comprises charge redistribution and conformational rearrangement. These processes eventually bind the ligand to the receptor. Redistribution of the valence electrons in the ligand takes place when atoms in the receptor become perturbed after solvent displacement—a process that modifies the electron cloud. The changes induced by the electrostatic and steric effects result in conformational changes, yielding a stable complex.

6.2 RECEPTOR-LIGAND INTERACTIONS

As mentioned earlier in this chapter, receptor-ligand interactions involve noncovalent interactions such as hydrogen-bonding, ionic-bonding, and hydrophobic interactions, as well as van der Waals forces (see Figure 6.1). Table 6.1 provides some general information about these interactions.

6.2.1 Ionic bonds

Ion interactions can occur between the net charge of a ligand and the counter-ions in the receptor. Consider, for example, the case of amino groups that are protonated at physiologic pH. These positively charged groups can interact with the negative charge of carboxy groups on amino acids such as aspartate (the ionized form of aspartic acid) or glutamate (the ionized form of glutamic acid).

Metal ions, especially members of the transition series, can also bind to certain charged groups. This interaction is described qualitatively by the principle of "hard and soft acids and bases," which is based on the definition of Lewis acids and bases. According to this principle, hard acids prefer to coordinate with hard bases, and soft acids prefer to coordinate with soft bases. Hard acids are defined as small acids with a high positive charge and no unshared pairs of electrons in their valence shells. These properties produce high electronegativity and low polarizability. Soft acids are large in size, carry a low positive charge, and contain unshared pairs of electrons (p or d) in their valence shells. They are therefore characterized by high polarizability and low electronegativity. Thus soft acids form stable complexes with highly polarizable bases. Hard acids will usually form stable complexes with hard bases, with polarizability playing a minor role. Acids and bases can thus be classified according to these guidelines as hard, soft, or borderline.

6.2.2 Hydrogen bonds

The formation of hydrogen bonds requires not only a certain proximity (as discussed previously), but also other chemical, electronic, and geometric conditions. The chem-

Table 6.1 Main Types of Receptor-Ligand Interactions

Interaction	Equilibrium Separation (nm)	Dissociation Energy (kJ/mol)
Ionic bond	0.23	670.0
Ion-dipole	0.24	84.0
Dipole-dipole (hydrogen bond)	0.28	20.0
Hydrophobic	0.30	4.0
van der Waals	0.33	0.25

ical conditions include the presence of a small, highly electronegative donor and an acceptor atom that possesses an unshared pair of nonbonding electrons. The electronic conditions require that both interacting molecules be at ground state with closed valence shells. The geometric constraints limit the distance between the donor and acceptor to between 0.18 and 0.3 nm. Such conditions occur between hydroxyl groups on the ligand and amino groups on the receptor.

6.2.3 Hydrophobic interactions

Hydrophobic interactions occur between nonpolar groups in an aqueous solution when the hydrogen-bonded structure of water is altered. These complex forces reduce the number of water molecules found in close proximity to the polar groups. Such interactions can increase by taking advantage of dispersion forces between the planar surfaces of a pair of aromatic groups, in what are known as stacking interactions.

We can explain these effects by noting that the driving force for the hydrophobic interaction is entropic, while the enthalpy change opposes this tendency. Thus $\Delta H > 0$ when two or more initially solvated, hydrophobic groups clump together inside a biomolecule. An increase in temperature will favor the hydrophobic interaction if ΔH remains positive.

6.2.4 Van der Waals forces

Van der Waals forces are attractive forces between neutral molecules that originate from electrical interactions. Although several such forces are generated from electrical interactions (i.e., dipole-dipole interactions), the most important in the formation of a receptor-ligand complex are those operating between nonpolar molecules.

The van der Waals forces that predominate in the formation of a receptor-ligand complex are known as London dispersion forces. They result from the polarization of one molecule by quantum fluctuations in the charge distribution in a second molecule, and vice versa. These fluctuations create temporary dipoles on an atom, which distorts the electron cloud of a second atom, thereby creating a second dipole. The dipoles then attract one another, even when they form in two different molecules. Compared with dipole-dipole forces, London forces are much weaker. The atoms forming the dipoles must be almost touching before these forces become significant (3).

6.3 THEORETICAL ASPECTS OF RECEPTOR-LIGAND AFFINITY

This section will present two different approaches for analyzing the receptor-ligand complex formation. The first, termed the thermodynamic approach, relies on the free-energy contributions of the different steps that result in the formation of the receptor-ligand complex. The second, an equilibrium approach, uses equilibrium binding data to determine receptor-ligand binding affinities.

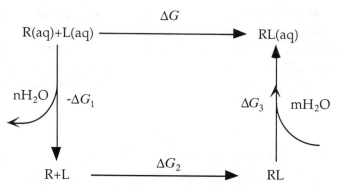

Figure 6.2 Calculational path for the free energy of binding, ΔG.

6.3.1 Thermodynamic approach

In the thermodynamic approach, we construct a theoretical path. We then calculate parameters that do not depend upon this path, but rather are state functions of the receptor-ligand complex formation process. Each step is analyzed in terms of its energy contributions to the process, and each contribution is then added to determine the overall change in the energy state of the system (4).

Receptor-ligand complex formation involves three main steps: 1) water displacement; 2) the actual receptor-ligand interaction; and 3) hydration after the complex is formed. Figure 6.2 illustrates a calculational path for this process. The overall free-energy change, ΔG, consists of the three steps' energy contributions:

$$\Delta G = -\Delta G_1 + \Delta G_1 + \Delta G_3 \tag{6.1}$$

In this equation, ΔG_1 and ΔG_3 correspond to the hydration of the separated R and L species and the hydration of the RL complex, respectively. The free-energy term for the receptor-ligand interaction can be divided into other specific contributions:

$$\Delta G_2 = \Delta U_{RL} + \Delta U_{conf} - T\Delta S \tag{6.2}$$

In Equation 6.2, ΔU_{RL} refers to the change in potential energy for the interaction of the receptor and ligand in their complex form, ΔU_{conf} is the potential energy increase for the receptor from its "low-energy" unbound state to the "high-energy" complexed state, and ΔS is the entropy change for the receptor-ligand interaction.

This model is related to changes in intermolecular forces. It does not include alterations in the arrangement of chemical bonds, which forms the basis of computational chemistry. This technique can be used to determine each of the contributions using the principles of quantum mechanics and statistical physics (these principles go beyond the scope of our discussion).

6.3.2 Equilibrium approach

The affinity interactions between a receptor (R) and its complementary ligand (L) can be described using the following equilibrium expression:

Table 6.2 Dissociation Constants of Typical Bioaffinity Interactions

Receptor-Ligand Pair	K_D (M)
Antibody-antigen	10^{-7}–10^{-11}
DNA-protein	10^{-8}–10^{-9}
Cell receptor-ligand	10^{-9}–10^{-12}
Enzyme-substrate	10^{-3}–10^{-5}
Avidin-biotin	$\sim 10^{-15}$
Streptavidin-biotin	$\sim 10^{-15}$
Lectin-monosaccharide	10^{-3}–10^{-5}
Lectin-oligosaccharide	10^{-5}–10^{-7}

$$R + L \underset{k_D}{\overset{k_A}{\rightleftharpoons}} RL \tag{6.3}$$

The dissociation constant K_D is the ratio between the dissociation (k_D) and association (k_A) rate constants:

$$K_D = \frac{k_D}{k_A} = \frac{[R][L]}{[RL]} \tag{6.4}$$

The values in brackets are concentrations, and RL represents the receptor-ligand complex. Low values of K_D represent a very stable complex, while high values denote a weakly associated complex. Ligands can be classified as highly specific or group-specific based on the strength of their interactions with the biomolecule of interest. For highly specific ligands, the dissociation constants will range from 10^{-3} M (for enzyme-substrates interactions) to 10^{-15} M (for the avidin-biotin interaction). Group-specific ligands have smaller dissociation constants, implying a stronger bond, as more than one particular molecule or biopolymer subgroup can be bound. Table 6.2 shows the dissociation constants for a variety of bioaffinity reactions.

To determine these dissociation constants experimentally, we can employ equilibrium dialysis. In this procedure, two compartments are separated by a dialysis membrane that is permeable only to the ligand. Known amounts of the receptor and the ligand are placed in different compartments. The system is left to equilibrate (with the time depending on the kinetics of the specific receptor-ligand system), allowing part of the ligand to become bound to the receptor. Because the membrane is permeable only to the ligand, the concentration of the receptor remains constant in its compartment. Thus only the ligand concentration needs to be determined, which can be accomplished by tagging the ligand with known labels (e.g., isotope, chromophore) and quantifying the amount with a technique appropriate for the specific label.

A control experiment is also run, in which no receptor is added, and thus nothing remains in the receptor compartment. An amount of labeled ligand added to the second compartment will be equal to the amount added in the actual experiment. If both compartments have the same volume and contain the same solvent conditions (i.e., pressure, buffer concentration), then the equilibrium ligand concentration in both compartments will be the same.

In the actual experiment, the amount of ligand bound to the receptor will be the difference between the ligand concentration in the two compartments. The higher the affinity of the receptor, the higher the amount of ligand bound.

Because the data from the experiments just described are gathered at equilibrium conditions, we can use the previously derived equilibrium expression. As the receptor concentration is a known constant, we can rewrite the equilibrium expression as follows:

$$K_D = \frac{[R][L]}{[RL]} = \frac{(n-r)c}{r} \tag{6.5}$$

In this equation, r is the ratio of the bound ligand to the receptor concentration, c is the free ligand concentration, and n is the valency of the receptor (i.e., the number of ligand binding sites per receptor molecule). Equation 6.5 can be rearranged to obtain

$$\frac{r}{c} = \frac{n}{K_D} - \frac{r}{K_D} \tag{6.6}$$

We can use Equation 6.6 to determine the dissociation constant and the receptor valency by plotting r/c versus c. This plot, known as a Scatchard plot, will be linear with a slope of $-1/K_D$ and n as the c-axis intercept. The fact that the plot is nonlinear indicates that the receptor population is not homogeneous, but rather includes receptors with different affinities.

Example 6.1

You are trying to determine which of two antibodies, called Ab 1 and Ab 2, has a higher affinity for a specific antigen (ligand). After performing repeated equilibrium dialysis with a constant concentration of antibody and varying the concentration of antigen, you plot your data as shown in Figure 6.3. Determine the following:
(a) The dissociation constant for each antibody
(b) Which antibody has a higher affinity for the antigen
(c) How many antigens each of the antibodies can bind
(d) Whether the population is homogeneous or heterogeneous for each of the antibodies

Solution
(a) Using Equation 6.6, it can be seen that:

$$\text{Slope} = -\frac{1}{K_D} \tag{6.7}$$

For Ab 1:

$$\text{Slope} = \frac{3-0}{0-2} = -\frac{3}{2} \tag{6.8}$$

$$K_D = \frac{2}{3} \text{ M} \tag{6.9}$$

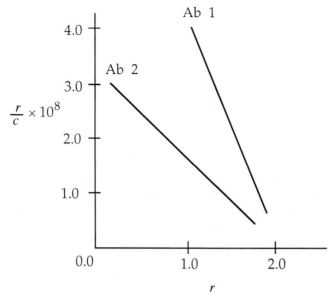

Figure 6.3 Scatchard plot.

For Ab 2:

$$\text{Slope} = \frac{4-0}{1-2} = -2 \tag{6.10}$$

$$K_D = \frac{1}{2}\,M \tag{6.11}$$

(b) The larger the value of K_D, the lower the affinity of the antibody for the antigen. Using this logic, Ab 1 has the higher affinity.

(c) When we extrapolate each line to the r-axis intercept, we see that the valency n is equal to 2. This result means that both antibodies can each bind two antigens.

(d) The binding data for both antibodies behaved linearly when plotted in a Scatchard plot. Thus the preparations of antibodies used in the experiment were homogeneous.

6.4 SPECIFIC INTERACTIONS

In this section, we will examine the interactions that result during the formation of different receptor-ligand complexes. We have selected complexes for discussion based on their applicability to the biotechnology industry. The complex formation sequence is analyzed step by step, including considerations of the specific bonds formed to the conformational changes involved (5).

6.4.1 Antibody-antigen interactions

In vertebrates, the immune system protects the organism from invading pathogens and cancer. This defense system generates proteins known as immunoglobulins or antibodies. In general, these proteins consist of two identical heavy chains and two identical light chains linked by disulfide bonds that recognize a specific region (epitope) on the foreign substance (antigen). Each of these chains contains a variable sequence in the amino-terminal, with the rest consisting of conserved sequences.

The first antibody to be characterized was immunoglobulin G (IgG). This antibody was briefly treated with the enzyme papain, a protease that cleaves the disulfide bonds holding the heavy chains together. This process yields three fragments: two identical 45 kD fragments, both of which are known as Fab, and a 50 kD fragment, known as Fc. The Fab fragments receive their name because they retain their antigen binding capacity; in contrast, Fc is named for its ability to crystallize during cold storage (6).

The heavy-chain constant region determines the class of antibody (that is, whether it is IgM, IgG, IgD, IgA, or IgE) and thus its function. The variable regions of both chains contain several hypervariable domains known as complementarity-determining regions (CDRs). These CDRs form the antigen binding site, which in turn determines the specificity of the antibody.

Another important aspect of the antibody structure relates to the hinge region in which the two heavy chains are united. This area exhibits conformational flexibility, which allows a relative lateral combining site motion. This motion, in turn, increases CDR availability for binding.

Each heavy and light chain includes several homologous domains, each of which incorporates approximately 110 amino acid residues. Each domain consists of a loop formed by the amino acids with an intrachain disulfide bond (Figure 6.4). The light chain contains one variable (V_L) and one constant domain (C_L). The heavy chain contains one variable (V_H) and three or four constant domains (C_{H_1}, C_{H_2}, C_{H_3}, C_{H_4}), depending on the antibody class.

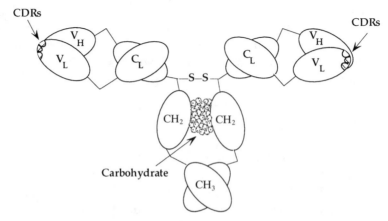

Figure 6.4 Schematic diagram of the immunoglobulin domains.

High-resolution X-ray crystallography has provided new insights into the structure of both the V_H and V_L domains. Specifically, it has illuminated the differences found when binding large, globular protein antigen or binding a smaller antigen. The antibody binds a rather flat, undulating surface of the globular protein antigen. The binding site on the antigen contains protrusions and depressions that are complemented by protrusions and depressions on the antibody. Antibodies that bind smaller antigens contain deep pockets within the CDR region, in which most of the antigen becomes buried.

Other studies of antigen binding have revealed that more residues on the heavy-chain CDRs bind antigen than do residues in the light-chain CDRs. Thus the V_H domain seems to be more intensely involved in antigen binding than the V_L domain. In addition, the formation of the antibody-antigen complex often produces conformational changes in one or both of the biomolecules. These changes tend to lower the entropy and thus facilitate a stronger interaction. The functions of the constant region in the antibody are mostly to influence conformational changes and other biological functions in the immune system (7).

6.4.2 DNA-protein interactions

Deoxyribonucleic acid (DNA), arguably the most important molecule in living cells, contains all of the information that specifies the cell function. DNA and ribonucleic acid (RNA) are macromolecules that consist of linear polymers built up from simple subunits called nucleotides. Nucleotides have three components:

*   A cyclic five-carbon (pentose) sugar (deoxyribose for DNA, and ribose for RNA)
*   A nitrogenous base of either purine or pyrimidine classification, covalently attached to the 1'-carbon atom of the sugar by an *N*-glycosylic bond
*   A phosphate attached to the 5' carbon of the sugar by a phosphodiester bond

The purines making up the nitrogenous base are either adenine (A) or guanine (G). The pyrimidines are either cytosine (C) or thymine (T) for DNA, and uracil (U) for RNA.

The nucleotides of DNA are called deoxyribonucleotides, because they contain the sugar deoxyribose. The nucleotides of RNA are called ribonucleotides, because they contain the sugar ribose instead. Each nucleotide contains a specific and a non-specific region. The phosphate and sugar groups represent the nonspecific portion of the nucleotide, while the purine and pyrimidine bases make up the specific portion.

Nucleotides join to one another linearly by chemical bonding between atoms in the nonspecific regions, which forms polynucleotides. This type of linkage, called a phosphodiester bond, occurs between a phosphate group and a hydroxyl group on the sugar component. The most important feature of DNA is that it usually consists of two complementary strands coiled about one another to form a double helix. Each of these strands comprises a polynucleotide.

The two chains are joined by hydrogen bonds between the purine-to-pyrimidine base pairs. Adenine, a purine, is always paired with thymine, a pyrimidine; and guanine, a purine, is always paired with cytosine, a pyrimidine. As a result, chemical analysis of the molar content of the bases in DNA assumes that the amount of adenine always equals the amount of thymine, and that the amount of guanine always equals the amount of cytosine. This base pairing provides stabilization of the DNA through hydrogen bonding between complementary bases.

Three forms of this double helix exist: DNA-A, DNA-B, and DNA-Z. X-ray diffraction data reveal that the A and B forms are right-handed, while the Z form is left-handed. Also, the A and B forms have 10 or 11 nucleotides on each chain at every turn of the helix; in contrast, the Z form has 12 nucleotides at these locations. Each chain makes a complete turn every 32 Å (A form), 34 Å (B form), or 36 Å (Z form).

These three forms of DNA are created by the macromolecule's efforts to balance the attractive and repulsive forces that result from ionic interactions between phosphodiester and other ionic groups; other sources of these forces include hydrogen bonds and base stacking. Because DNA-B is associated with a higher water activity and a lower ionic strength, it is thought to be the predominant form in living systems, as these characteristics are the prevailing conditions encountered in such organisms (8).

The specificity provided by base pairing permits the transmission of genetic information from one generation to another. When cell growth and differentiation occur, the DNA double helix unwinds. Two new DNA strands are then formed that are complementary to the original DNA strands. In this way, one newly synthesized strand becomes incorporated into the DNA double helix. The sequence of bases (A, G, T, and C) in a strand of DNA specifies the order in which amino acids are assembled to form proteins. The genetic code is the collection of base sequences that correspond to each amino acid codon. As DNA contains only four bases and protein includes 20 amino acids, each codon must contain at least three bases.

Proteins known as transcription factors regulate gene expression, generally by binding to a control region of the gene. These regions usually form the major groove of the DNA double helix. The DNA-protein complex is thought to form through nonspecific electrostatic interactions with the phosphate backbone, which are followed by weak, noncovalent, specific interactions between amino acids in the proteins and the base pairs in the major groove of DNA-B. To initiate a specific interaction, the protein may slide until it reaches the specific domain; alternatively, proteins may be transferred from nonspecific sites to specific sites in a process called intersegment transfer.

Domains (also called motifs) on the DNA that are used for binding proteins include helix-turn-helix, leucine zipper, β-ribbons, TATA box binding, and zinc finger protein domains.

The helix-turn-helix domain is found in some bacterial regulatory proteins. It consists of two α-helices that are connected by an amino acid chain, which constitutes the turn. Each α-helix comprises a linear sequence of amino acids that folds into a right-handed helical path. The α-helix with the carboxyl terminal fits into the major groove of DNA.

The leucine zipper is found in many eukaryotic genes. Located next to the DNA binding region in the protein, it is formed when an α-helix, created from a leucine heptad repeat, joins another α-helix from another protein to form a dimer in a coiled-coil or "zipper" structure. The two helices are held together by hydrophobic interactions between the leucines. In the Y-shaped structure created by their joining, each arm (an α-helix) binds a specific motif in the major groove of DNA.

The β-ribbon domain consists of a two-stranded, antiparallel β-sheet that binds to the major groove of DNA. This β-sheet forms when an amino acid chain folds upon itself, with each section of the chain running in a direction opposite to that taken by the neighboring chains. Although the β-ribbon domain does undergo binding, the α-helix forms a more stable complex when bound to DNA.

The TATA box is a sequence in the promoter region in the DNA of many eukaryotic cells that specifies where transcription begins. This motif is bound by all three classes of eukaryotic RNA polymerases using a large β-sheet. The specific subunit of the polymerase that binds the TATA box, known as TFIID, resembles a saddle that fits over the DNA double helix. This interaction is unique in one sense— for this interaction to occur, the TATA box motif must undergo a severe deformation before it can accommodate the TFIID. In this deformation, two kinks in the double helix become separated by a partially unwound DNA strand.

The zinc fingers motif is found in many DNA-binding proteins. It comprises an α-helix that binds a zinc atom. This structure can be repeated multiple times in the protein. For example, an α-helix and a β-sheet may be held together by a zinc atom. In most cases, this structure is clustered in such a way that the α-helix can interact with the major groove of DNA, creating a very strong and specific interaction. Another type of zinc fingers structure consists of a dimer in which two α-helices interact with the major groove of DNA, much like the helix-turn-helix motif.

Although these complexes represent only some of the DNA-protein interactions discovered so far, we nevertheless make some generalizations about this type of interaction. The specific interaction occurs between a short segment of DNA (located in the major groove) and an α-helix, or between a short segment of DNA and a β-sheet in the protein. By interacting with the phosphate backbone in DNA, the protein appears to become reoriented so as to provide a better fit.

6.4.3 Cell receptor-ligand interactions

Cells communicate either through direct contact or via the secretion of chemical substances that are recognized by a receptor in the target cell. In the latter case, the cell receptors can appear on the surface of the cell or inside the cell. If the receptor is located on the surface of the cell, then the receptor-ligand complex forms there. If the receptor is inside the cell, however, the ligand must diffuse through the cell membrane or be transported by vesicles. The cell receptors act as signal transducers, converting the binding event into signals inside the cell that modulate its behavior. In this section, we will discuss the structure and function of the principal classes of extracellular receptors, which include ion-channel receptors, G-protein-linked receptors, and enzyme-linked receptors. Figure 6.5 illustrates each of these classes of cell-surface receptors.

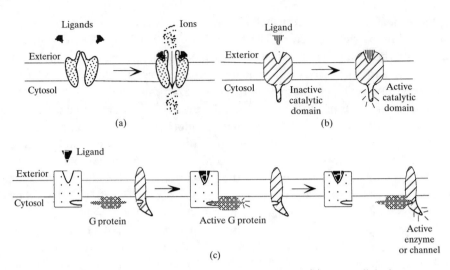

Figure 6.5 Classes of cell-surface receptors: (a) ion-channel receptor; (b) enzyme-linked receptor; (c) g-protein-linked receptor.

An ion-channel receptor is a transmembrane protein complex that forms a water-filled hole that becomes permeable to specific ions across the cell membrane when its is bound by specific ligands. For example, the ligands g-aminobutyric acid (GABA) and glycine activate Cl^- channels, while acetylcholine, serotonin, and glutamate activate Na^+ and K^+ channels.

G-protein-linked receptors activate trimeric GTP-binding protein (G protein), which then activates a membrane-bound enzyme or ion channel. Activation of an enzyme alters the concentration of at least one intracellular mediator, while activation of an ion channel changes the ion permeability of the plasma membrane. The change in mediator concentration then modifies other proteins. Structurally, these receptors consist of seven hydrophobic transmembrane domains.

An enzyme-linked receptor can either become enzymatically active or activate other enzymes after being bound by its ligand. Most such enzymes are protein kinases, which phosphorylate specific proteins. They are involved in the activation of promoters through a cascade of intracellular reactions.

6.4.4 Enzyme-substrate interactions

Enzymes are catalysts that have high affinity for specific substrates. They bind to these specific substrates in such a way so as to reduce the activation energy necessary for a particular reaction in which covalent bonds are rearranged. The substrate generally becomes covalently bound to the binding site on the enzyme. It then reacts with a second molecule on the enzyme surface, breaking the covalent bond just formed. The enzyme is then regenerated to repeat the cycle again.

Approximately two-thirds of known enzymes need nucleotide coenzymes to perform their catalytic activity. These coenzymes bind the enzyme so as to provide the additional reactive groups necessary for the catalytic reaction. Figure 6.6 depicts

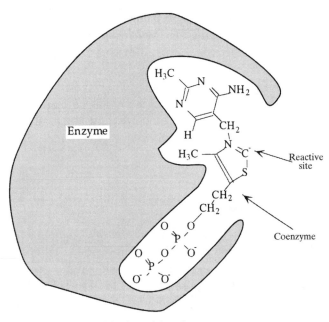

Figure 6.6 Binding of the coenzyme thiamine triphosphate to an enzyme.

an enzyme-coenzyme interaction. Coenzymes include the nucleotides AMP, ATP, NAD, and NADP, as well as iron-containing heme groups in hemoglobin, cytochromes in enzymes that are involved in aldehyde-group transfer, and vitamins that promote carboxyl-group transfer (e.g., biotin).

6.4.5 Biotin-avidin/streptavidin interactions

In recent years, biotechnologists have become greatly intrigued by the strong interaction between the vitamin biotin with the egg-white avidin (for which $K_D = 10^{-15}$ M). This interest results from the need to manipulate biomolecules in biotechnology settings. The biotin-avidin interaction could satisfy this need, as biotin-labeled biomolecules can be isolated, almost irreversibly, using immobilized avidin. Because avidin is glycosylated and highly alkaline, it can also bind other molecules, such as the sugar-binding lectins. To minimize these nonspecific interactions, researchers have attempted to remove the carbohydrate residues. As yet, these approaches have not borne fruit, as avidin is not susceptible to glycosidases.

Another strategy for minimizing nonspecific interactions relies on an analogous molecule known as streptavidin. This molecule, which is obtained from the bacterium *Streptomyces avidinii*, is a tetramer with two-fold symmetry. It consists of four identical 14 kD subunits, each with a high affinity for biotin. The subunits are arranged as two pairs located on opposing faces of the molecule.

Streptavidin has approximately 40% homology with avidin, with the conserved residues being short homologous domains that form part of the binding sites. The binding sites comprise four tryptophanyl residues that interact hydrophobically with biotin. The tryptophanyl residues in the binding site do not bind biotin directly, but rather stabilize other residues on the site (by hydrogen bonds), which then directly bind the biotin. The binding site consists of a β-barrel structure formed by a curved β-sheet, with the biotin site being found inside this β-sheet.

This bacterial protein displays remarkable similarity with avidin in terms of its biotin-binding properties. On the other hand, it is not glycosylated and is not as basic as avidin. Nevertheless, streptavidin not only provides a high affinity binding site, but also a high capacity for binding biotin or biotinylated molecules, thanks to its tetrameric structure (9).

6.4.6 Lectin-carbohydrate interactions

A lectin is a type of protein that contains at least two binding sites for specific carbohydrates. Its general structure consists of an oligomer composed of different subunits, each of which usually includes one carbohydrate binding site. Lectins are classified according to the monosaccharide that they bind—for example, D-mannose, D-galactose, N-acetyl-D-glucosamine, N-acetyl-D-galactosamine, L-fructose, or N-acetylneuraminic acid. These monosaccharides, which are all part of animal glycoconjugates, appear on the surface of cells. Most known lectins also contain metal ions—specifically Mg^{2+} and/or Ca^{2+}—which are needed to bind sugars. In the case of Ca^{2+}, binding of the metal ion appears to alter the environment of the transition metal site, making it possible for the lectin-sugar complex to form and then stabilize. Although many enzymes also bind carbohydrates (e.g., kinases, mutases, glycosidases, and transferases), lectins are monovalent, with each binding only one type of carbohydrate.

The lectins that bind monosaccharides are not only specific for a sugar, but also specific to a particular isomer. They bind either an α- or β-isomer of a specific monosaccharide. The nature of the glycoside on the sugars is also an important consideration in binding, with aromatic glycosides (such as phenyl or p-nitrophenyl groups) binding more strongly than aliphatic ones (such as ethyl or methyl groups). Based on this characteristic, hydrophobic interactions appear to be an important part in the formation of the complex.

Certain lectins demonstrate a higher affinity for oligosaccharides than for monosaccharides. In such a case, the oligosaccharide will contain the monosaccharide for which the lectin is specific. Although this monosaccharide usually appears in the nonreducing end of the oligomer, some lectins can also recognize them in internal positions.

6.5 SUMMARY

The term "bioaffinity" is used to describe the physical and chemical interactions between various biomolecules—namely, ligands and receptors. A ligand will bind to its receptor counterpart through noncovalent means such as ionic bonds, hydrogen bonds, hydrophobic interactions, and van der Waals forces. These binding relationships can be mathematically described using either the thermodynamic approach or the equilibrium approach. In specific interactions, antibodies will react with antigens, proteins with DNA, ligands with cell receptors, enzymes with substrates, biotin with avidin or streptavidin, and lectins with carbohydrates. These specific interactions can be exploited in bioseparation design, as in affinity chromatography.

6.6 PROBLEMS

6.1 You are trying to develop an immunoassay for a hormone that occurs in the blood at a concentration around 10^{-7} M. You have narrowed your search for the antibody to three antisera, whose affinity for the hormone you have determined by equilibrium dialysis. The results of this analysis are shown in the Scatchard plots in Figure 6.7.

(a) The average dissociation constant for a heterogeneous population is determined by calculating the dissociation constant when one-half of the binding sites are occupied (i.e., $r = 1$). Determine the dissociation constant for each antiserum.

(b) What is the valence of each antibody?

(c) Which of the antisera is homogeneous, and which is heterogeneous?

(d) Which antisera would you use for your immunoassay? Explain your rationale.

6.2 Explain why different receptor-ligand complexes have different affinities, while still retaining a high selectivity.

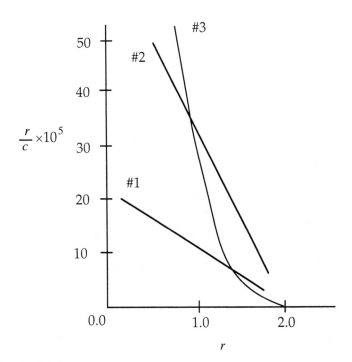

Figure 6.7 Scatchard plots.

Table 6.3 Dissociation Constants

Drug	K_D (M)
A	0.016×10^{-6}
B	8.114×10^{-6}
C	0.316×10^{-6}

Table 6.4 Supplied Data

Receptor	K_D (M)	K_A ($M^{-1} s^{-1}$)
A	1.2×10^{-9}	2.0×10^{7}
B	2.8×10^{-6}	2.0×10^{7}

6.3 Explain why bioaffinity interactions are important in the biotechnology industry. Give examples of their application.

6.4 Why do most of the receptor-ligand complexes lack covalent bonds?

6.5 Explain why coenzymes are important in enzymatic reactions.

6.6 Describe the process by which receptor-ligand complexes form.

6.7 Why is water displacement important during the receptor-ligand complex formation?

6.8 You are testing different drugs to measure their affinity to a certain receptor. You have determined dissociation constants for the drugs being tested, as shown in Table 6.3.

    (a) Which of these drugs has the greatest affinity for the receptor?

    (b) Which has the weakest affinity for the receptor?

6.9 You are examining the binding of a certain hormone to two different receptors. The data in Table 6.4 were supplied to you. What would be the value of the rate constant k_D for the release of the hormone from each receptor?

6.7 REFERENCES

1. Voiculetz N, Motoc I, Simon Z, eds. Specific interactions and biological recognition processes. Boca Raton: CRC Press, 1993.
2. Alberts B, Bray D, Lewis J, Raff M, Roberts K, Watson JD. Molecular biology of the cell, 3rd ed. New York: Garland, 1994.
3. Meyers RA, ed. Molecular biology and biotechnology. New York: VCH Publishers, 1995.
4. Tinoco I, Sauer K, Wang JC. Physical chemistry: principles and applications in biological sciences. Englewood Cliffs: Prentice Hall, 1995.
5. Cantor CR, Schimmel PR. Biophysical chemistry. San Francisco: WH Freeman, 1980.
6. Kubi J. Immunology, 2nd ed. New York: WH Freeman, 1994.
7. Padlan, EA. Antibody-antigen complexes. Austin: RG Landes, 1994.
8. Voet D, Voet JG. Biochemistry, 2nd ed. New York: John Wiley and Sons, 1995.
9. Wilchek M, Bayer EA, eds. Methods in enzymology series: avidin-biotin technology. Vol 184. New York: Academic Press, 1990.

PART III

BIOSEPARATION METHODS

7

Crystallization and Precipitation

Crystallization—the formation of solid particles of defined shape and size from a homogeneous liquid phase—is the oldest and most widely used type of purification. It occurs when the liquid phase becomes supersaturated because of cooling, solvent evaporation, or the addition of a nonsolvating diluent. Most bulk pharmaceutical and organic chemicals are marketed as crystalline products because crystals of exceptional purity can be manufactured. In addition, the use of a crystalline-form product facilitates the finishing steps in the production process, such as filtering and drying. Apart from conferrring these advantages, crystallization also improves the product's appearance—an important aspect of consumer acceptance.

Precipitation and crystallization can be analyzed as a single process, as no clear distinction can be drawn between the two stages. Precipitates form when a solid phase is created rapidly, and they generally do not have a regular morphology. Crystals, on the other hand, form more slowly and have regular shapes. In both cases, growth rates and solid- and liquid-phase balances can be calculated, which allows us to treat both processes with the same approach.

Although bench-scale laboratory crystallization is a simple technique, the scale-up from this level to industrial crystallization processes can be complicated. Large-scale crystallization processes suffer from an ill-defined geometry, with simultaneous mass and heat transfer taking place in a multiphase, multicomponent, thermodynamically unstable system. The resulting crystals can be profoundly affected by trace impurities. Today, industrial crystallization remains more of an art than a science.

7.1 SATURATION AND SUPERSATURATION

Saturation conditions exist when a thermodynamically stable solution contains the maximum concentration of solute. Saturation results from phase equilibrium, in which the chemical potentials of the solid crystal phase and the surrounding solution (commonly referred to as the mother liquor) are equal. In Chapter 4, we examined equilibrium data and developed theoretical models for predicting solubility. In reality, however, a solution can contain a higher concentration of solute than predicted during the formation of a solid phase. Such supersaturated solutions are thermodynamically unstable.

The supersaturated state can be divided into three loosely defined zones (Figure 7.1). The first zone is called the stable zone, where no crystallization occurs. Next comes the metastable zone, where growth of existing crystals and formation of new

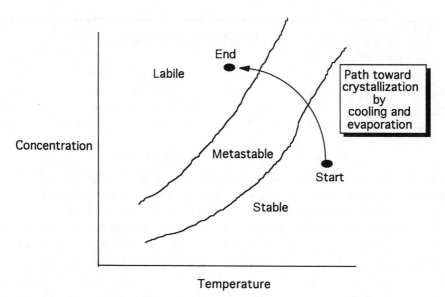

Figure 7.1 Illustration of the three characteristic zones exhibiting different crystallization behavior.

nuclei occur simultaneously. The third zone is the labile zone, where new nuclei form spontaneously from a clear solution.

It is useful to define the degree of supersaturation using the following equations:

$$S_D = \frac{C_A}{C_A^*} \tag{7.1}$$

$$\Delta C_D = C_A - C_A^* \tag{7.2}$$

$$\tilde{C}_A = \frac{C_A}{C_A - C_A^*} \tag{7.3}$$

In these equations, S_D is the degree of saturation, C_A is the concentration of solute A in solution, and C_A^* is the concentration of A at the saturation point.

The rate at which precipitation and crystallization occur is directly related to the degree of supersaturation, as this factor typifies the driving force for solid-phase formation. The maximum value of the degree of supersaturation, however, is often dictated by the type of nucleation observed.

7.2 NUCLEATION PHENOMENA

The solid phase can be created by homogeneous, heterogeneous, secondary, or attritive nucleation. In industrial practice, the most unlikely of these processes is homogeneous nucleation, which requires a perfectly clean vessel with no rough surfaces. In heterogeneous nucleation, the process is affected by the vessel's wall morphology or the presence of insolubles, such as dust particles in the mother liquor. Seeding a crystallizer can be the best method in commercial operations, as the growth on the seed crystals, known as secondary nucleation, leads to large, well-formed crystals. In a highly agitated vessel, crystals can break; the fragment formed then serve as surfaces for growth by attritive nucleation.

Homogeneous nucleation occurs when molecules cluster and undergo a phase transition, thereby creating a solid surface. This process is easily analyzed on a thermodynamic and transport basis, and this assessment can then be used to build models for the more complex heterogeneous, secondary, and attrition nucleation phenomena. The free-energy change upon crystal formation during homogeneous nucleation is the sum of the free energy from surface formation and the entropically driven process of molecular aggregation to form a solid phase:

$$\Delta G_{\text{homogeneous}} = \Delta G_{\text{surface formation}} + \Delta G_{\text{clustering}} \tag{7.4}$$

The free energy of surface formation for a spherical particle is given by the product of the interfacial area and the surface tension of the solid-liquid interface:

$$\Delta G_{\text{surface formation}} = 4\pi r^2 \gamma_{\text{sl}} \tag{7.5}$$

In this equation, r is the particle radius and γ_{sl} is the surface tension of the solid-liquid interface.

The free energy associated with the creation of a cluster of solid material can be defined as the energy of transferring n moles from the liquid to the solid phase at a common standard state. Assuming that the ratio of the activity coefficients is unity, the clustering free energy is given by

$$\Delta G_{\text{clustering}} = -nkT \ln\left(\frac{C}{C^*}\right) = -\left(\frac{4}{3}\pi r^3\right)\frac{\rho RT}{M_{\text{w}}} \tag{7.6}$$

where R is the universal gas constant and T is the absolute temperature. The product of the Boltzmann constant (k) and the number of moles undergoing the phase change (n) can be replaced by the solid density ρ, the molecular weight M_{w}, and the volume of the spherical solid formed. Putting these terms together, we realize that if C is greater than C^*, the two terms have opposite signs. Thus a maximum value of the free energy must exist for a given particle size. To ensure that a solid phase is created, the radius of the solid particle must exceed the critical radius, r_{c}, dictated by the free-energy maximum. This free-energy maximum is given by the following equation:

$$\Delta G_{\text{max}} = \frac{4}{3}\pi r_{\text{c}}^2 \gamma_{\text{sl}} = \frac{16\pi \gamma_{\text{sl}}^3 M_{\text{w}}^2}{3\rho^2 R^2 T^2 (\ln(S_{\text{D}}))^2} \tag{7.7}$$

Because the free energy must be supplied as work, we can see that nucleation is analogous to reaction kinetics. Thus we can use an Arrhenius-type expression to obtain the rate of nuclei formation:

$$B^0 = \frac{dN}{dt} = A\exp\left(-\frac{\Delta G_{\text{max}}}{RT}\right) = A\exp\left(-\frac{16\pi \gamma_{\text{sl}}^2 M_{\text{w}}^2}{3\rho^2 R^2 T^2 (\ln(S_{\text{D}}))^2}\right) \tag{7.8}$$

In Equation 7.8, B^0 is the rate of nuclei formation, A is the Arrhenius preexponential constant, and N is the cumulative number of crystals per unit volume.

Equation 7.8 is not useful for describing industrial crystallization because it accounts for only homogeneous nucleation. Rather, it can be employed to estimate the rate of nucleation. Beyond the carefully controlled lab-scale crystallization process, heterogeneous and secondary nucleation correlate with the most important and easily measured quantity, ΔC. Based on Equation 7.8, we can see that a power law function is likely an appropriate starting point to describe industrial crystallization. In fact, we can use the following equation to obtain B^0:

$$B^0 = k_N (\Delta C_{\max})^i \tag{7.9}$$

Empirical descriptions are also used to deal with issues such as the hydrodynamic environment in the crystallizer, the presence of impurities, and the direct effect of temperature—all of which can be difficult to quantify. In one approach to these problems, we add a term to deal with the effect of secondary nucleation. For example, we might account for the magma density, or slurry concentration:

$$B^0 = k_B M_T^j (\Delta C_{\max})^i \tag{7.10}$$

To use Equation 7.10 to deal with temperature and changes in impeller speed, we make k_B a function of T and impeller tip speed. Typically, the exponent i ranges from 0.5 to 2.5 for secondary nucleation; j is usually lower and often less than 1.

To validate the mathematical models applied to crystallization models, we can track the actual composition of the solution or quantify the size and number of crystals created. It is usually more convenient to use the linear growth rate G, which is equal to the rate of change of a characteristic dimension of a crystal, L. We use the chain rule

$$\frac{dN}{dt} = \frac{dN}{dL}\frac{dL}{dt} = \frac{dN}{dL} G = n^0 G \tag{7.11}$$

to obtain the following equation:

$$n^0 \equiv \frac{dN}{dL}\bigg|_{L=0} \tag{7.12}$$

In Equation 7.12, the term n^0, which is called the population density of nuclei, is used to define the change in the number of crystals per unit crystal length. Conveniently, the linear growth rate is generally proportional to the supersaturation. Through these substitutions, the nucleation rate can be described by the following equation:

$$B^0 = k_R M_T^j G^i \tag{7.13}$$

7.3 GROWTH OF CRYSTALS

Before we can model the processes occurring in the crystallizer, we must understand how crystals grow after nuclei form. In Section 7.2, we stated quite matter-of-factly that this growth rate is generally proportional to the degree of supersaturation. We can begin to understand why this relationship holds by studying the sequence of steps in crystal growth (Figure 7.2).

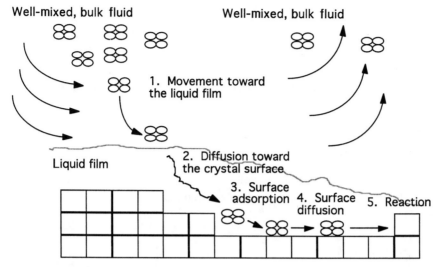

Figure 7.2 Idealized representation of some major events during crystal growth. (Four circles = molecule in liquid state; square = molecule in solid state.)

As it turns out, it is not possible to faithfully account for each of these phenomena and generate a model that unambiguously determines each of the adjustable parameters. Instead, we must rely on the power law function once again:

$$G = k_g \Delta C^g \tag{7.14}$$

The exponential factor g usually lies between 0 and 2.5, and commonly has a value of 1. This value suggests that a linear mass transfer process (such as film transfer) or a first-order surface reaction is the rate-limiting step in the overall process given in Figure 7.2. We must also emphasize that Equation 7.14 assumes that the linear growth rate is independent of crystal size. When this assumption is not valid, we can rely on an empirical relationship known as the Abegg-Stevens-Larson equation:

$$G = G^0 (1 + aL)^b \tag{7.15}$$

Here G^0 is the nuclei growth rate and a and b are constants (1).

7.4 BATCH CRYSTALLIZATION

Now that we have useful equations for expressing nucleation and growth phenomena, we can develop an analysis methodology for describing crystallizer performance. Two material balances can be employed for crystallizer analysis: the solution balance and the crystal-phase balance. These measures are two sides of the same coin. Depending on our goal and the crystallizer instrumentation, however, one may prove more useful than the other.

7.4.1 Solution balance

Because the equations for nucleation and growth account for the degree of super-saturation, it is useful to measure the rate of change in supersaturation as a function of time in the crystallizer. For a well-mixed, constant-liquid-volume crystallizer, we can derive an equation describing the decrease in the supersaturation with time from the following species balance:

$$\begin{bmatrix} \text{Volumetric rate of} \\ \text{solute disappearance} \end{bmatrix} = \begin{bmatrix} \text{volumetric rate of} \\ \text{nuclei formation} \end{bmatrix} + \begin{bmatrix} \text{volumetric rate of} \\ \text{crystal growth} \end{bmatrix} \quad (7.16)$$

Using the rate of change of supersaturation to track the rate of solute disappearance, and then dividing by the constant crystallizer liquid volume, yields Equation 7.17:

$$-\frac{d\Delta C}{dt} = k_R M_T^j \Delta C^i + k_g A_T \Delta C^g \quad (7.17)$$

This nonlinear equation is usually solved numerically, as both the magma concentration and crystal area are functions of time. After solving this equation, the results can be plotted as a desupersaturation curve (ΔC versus t). When the production process uses an isothermal, seeded crystallizer, the rate of desupersaturation typically shows a maximum at low values of time.

The most useful application of solution balances involves temperature control during cooling-induced crystallization. Belter (2) provides an example of how the solution balance can be simplified to generate a polynomial function describing the temperature needed to control the crystallizer so as to increase crystal size (3):

$$T = T_0 - \frac{3Gt}{L_{\text{seed}}} \frac{W_{\text{seed}}}{\frac{dC^*}{dt}} \left(1 + \frac{3Gt}{L_{\text{seed}}} + \frac{1}{3} \left(\frac{Gt}{L_{\text{seed}}} \right)^2 \right) \quad (7.18)$$

We can easily see that the solution tracking method is quite complex—many of the terms are hard to quantify directly, and the equation to be solved is nonlinear. For this reason, it is more common to track the solid phase rather than the solution phase. The remainder of this chapter focuses on tracking solid-phase formation and growth, based on population balances. In these processes, it is assumed that super-saturation is maintained throughout the crystallization by such means as evaporation or cooling, both of which are typically used in industrial processes to form and grow crystals as quickly as possible.

7.4.2 Solid-phase balance

A methodology that can successfully track the solid phase requires the use of measurements that can be readily performed. We can easily track the size distribution by sieving the resultant crystals and plotting the cumulative number or weight of crystals versus the crystal size. A more convenient method is to track the population density n, which is the slope of the curve of the cumulative number of crystals (N) versus the crystal linear size (L).

The population density monitors the number of crystals within a small size range, just as a demographer might track how the number of people within a specific age group changes with time. Thus crystals enter and leave the specific size range based on their growth rate. A population balance for a batch crystallizer can be written as follows:

$$\begin{bmatrix} \text{Number of} \\ \text{crystals initially} \\ \text{within range} \end{bmatrix} + \begin{bmatrix} \text{number of} \\ \text{crystals growing} \\ \text{into range} \end{bmatrix} = \begin{bmatrix} \text{number of} \\ \text{crystals at end} \\ \text{within range} \end{bmatrix} + \begin{bmatrix} \text{number of} \\ \text{crystals growing} \\ \text{out of range} \end{bmatrix}$$

(7.19)

$$Vn_{\text{initial}}\Delta L + VG_1 n_1 \Delta t = Vn_{\text{final}}\Delta L + VG_2 n_2 \Delta t \tag{7.20}$$

Dividing by ΔL and Δt, and then letting ΔL and Δt go to 0, results in the following differential equation:

$$\frac{dn}{dt} + \frac{d(Gn)}{dL} = 0 \tag{7.21}$$

Here the volume of crystals and solution are assumed to remain constant.

One strategy for solving this equation is to assume that the growth rate, G, is constant. Although this assumption is not entirely accurate, because the degree of supersaturation changes with time, it nevertheless provides a reasonable starting point for the analysis. After solving this equation, we can introduce another concept, known as the growth rate dispersion, to derive a more meaningful result from this mathematical model. Using the growth rate dispersion equation, we solve for the population balance once again to see what happens to it when the growth rate bcomes distributed over the course of operating a batch reactor.

The goal of the population balance is to determine how time and growth rate affect the total weight of crystals and the crystal size distribution. More extensive analyses of this concept are provided in the references listed at the end of this chapter.

Assuming that the growth rate is constant, the population balance becomes

$$\frac{dn}{dt} + G\frac{dn}{dL} = 0 \tag{7.22}$$

At this point, we can apply the method of Laplace transforms (see Appendix A) to solve Equation 7.22. As part of this calculation, we use one of two sets of initial and boundary conditions, shown in Equations 7.23–7.28:

Set 1

at $t = 0, n = 0$ (7.23)

at $L = 0, n = n^0 \delta(t)$ (7.24)

as $L \to \infty, n$ is finite (7.25)

Set 2

at $t = 0, n = 0$ (7.26)

at $L = 0$, $n = \dfrac{B^0}{G}$ (7.27)

as $L \to \infty$, n is finite (7.28)

The conditions in Set 1 are meant to simulate the seeding of a batch crystallizer with tiny crystals of uniform size. The impulse function $\delta(t)$ (described in detail in Appendix A) accounts for the fact that these seeds appear only at the beginning of the crystallizer operation. The conditions in Set 2 are relevant for operations that do not involve seeding but rather continual nucleation of crystals; the formation rate for such crystals is therefore given by the rate of nuclei formation.

The answers for each set of boundary conditions given in Laplace form after finding the inverse transform (see Appendix A) are as follows:

Set 1

$$\bar{n} = n^0 \exp\left(-\frac{Ls}{G}\right)$$ (7.29)

$$n = n^0 \delta\left(t - \frac{L}{G}\right)$$ (7.30)

Set 2

$$\bar{n} = \frac{B^0}{Gs} \exp\left(-\frac{Ls}{G}\right)$$ (7.31)

$$n = B^0 u\left(t - \frac{L}{G}\right)$$ (7.32)

7.4.3 Crystal size distribution

Two quantities are especially relevant when examining experimental data: N, the cumulative number of crystals per unit volume L, and M, the cumulative mass of crystals per unit volume of size L. Of these two quantities, M is the most readily obtained. As it is simply the integral of the population density with respect to L, N is easily calculated for Set 1 and Set 2 boundary conditions:

Set 1

$$\bar{N} = n^0 \int_0^L \exp\left(-\frac{Ls}{G}\right) dL = \frac{n^0 G}{s}\left(1 - \exp\left(-\frac{Ls}{G}\right)\right)$$ (7.33)

$$N = n^0 t\left(1 - u\left(t - \frac{L}{G}\right)\right)$$ (7.34)

$$N_{L \to \infty} = n^0 G t$$ (7.35)

Set 2

$$\bar{N} = \frac{B^0}{Gs} \int_0^L \exp\left(-\frac{Ls}{G}\right) dL = \frac{B^0}{s^2}\left(1 - \exp\left(-\frac{Ls}{G}\right)\right)$$ (7.36)

$$N = B^0\left(t - \left(t - \frac{L}{G}\right)u\left(t - \frac{L}{G}\right)\right)$$ (7.37)

$$N_{L \to \infty} = B^0 t$$ (7.38)

Through this calculation, we find that the total number of crystals is proportional to time and either the growth rate or the nucleation rate, depending on whether seeding takes place. Note that the equations for N contain the unit step function $u(t - L/G)$, where $u = 0$ from $t = 0$ to $t = L/G$, and $u = 1$ when t is greater than L/G (see Appendix A). Although this conclusion applies only to this highly idealized model of a batch crystallizer, it seems intuitively correct as a first approximation. Note also that it is more useful to perform mathematical operations such as taking limits and integrating using the Laplace form of the population density equation.

To obtain the easily measured quantity M (the cumulative mass of crystals per unit volume), we calculate the mass of a crystal using the density of the crystal, ρ_c, and a shape factor k_v:

$$M = \text{crystal mass} = \rho_c k_v L^3 \tag{7.39}$$

To determine the cumulative mass, we then employ the population balance in the integral shown in Equations 7.40–7.44, where W is defined as the total mass for all crystal sizes:

$$M = \rho_c k_v \int_0^L n L^3 dL \tag{7.40}$$

Set 1

$$M = 6\rho_c k_v n^0 G \left(\frac{G^3 t^3}{6} - \left(\frac{G^4 \left(t - \frac{L}{G}\right)^3}{\Gamma(3)} + \frac{LG^3 \left(t - \frac{L}{G}\right)^2}{\Gamma(2)} + \frac{L^2 G^2}{2} \left(t - \frac{L}{G}\right)^2 \right.\right.$$

$$\left.\left. + \frac{L^3 G}{6} \right) u\left(t - \frac{L}{G}\right) \right) \tag{7.41}$$

as $L \to \infty$, $M = W = k_v \rho_c n^0 G^4 t^3$ \tag{7.42}

Set 2

$$M = 6\rho_c k_v \frac{B^0}{G} \left(\frac{G^4 t^4}{24} - \left(\frac{G^4 \left(t - \frac{L}{G}\right)^4}{\Gamma(4)} + \frac{LG^3 \left(t - \frac{L}{G}\right)^3}{\Gamma(3)} + \frac{L^2 G^2 \left(t - \frac{L}{G}\right)^2}{2\Gamma(2)} \right.\right.$$

$$\left.\left. + \frac{L^3 G}{6}\left(t - \frac{L}{G}\right) \right) u\left(t - \frac{L}{G}\right) \right) \tag{7.43}$$

as $L \to \infty$, $M = W = \frac{1}{4} k_v \rho_c B^0 G^3 t^4$ \tag{7.44}

Although complex, these equations suggest that the cumulative mass can be modeled as a function of crystal length, using a polynomial expression. Also, the total mass depends on time and growth rate as a power law expression. To the extent that these results are borne out by experiment, they are highly useful. The following sub-sections explore how these results can be applied to control crystallizer conditions.

Example 7.1
Write a solute mass balance for a batch crystallizer with no seeding. Use this mass balance to determine the time-temperature relationship needed to maintain constant supersaturation through control of the cooling curve.

Solution
The rate of change in solute concentration in the crystallizer volume is equal to the rate of change of solid concentration:

$$\frac{dC}{dt} = -\frac{dW}{dt} \tag{7.45}$$

Using Equation 7.42, we can substitute for W:

$$\frac{dC}{dt} = -k_v \rho_c B^0 G^3 t^3 \tag{7.46}$$

As our goal is to keep the degree of supersaturation constant, the change in temperature must be proportional to the change in liquid concentration:

$$\frac{dC}{dt} = k_T \frac{dT}{dt} = -k_v \rho_c B^0 G^3 t^3 \tag{7.47}$$

Integrating with the temperature at which crystals start to form (defined as T_0), we find

$$T_0 - T = \frac{k_v \rho_c B^0 G^3 t^4}{4 k_T} \tag{7.48}$$

We can then create a ratio of the temperature difference at any time to the total temperature difference:

$$\frac{T_0 - T}{T_0 - T_f} = \left(\frac{t}{t_f}\right)^4 \tag{7.49}$$

This ratio is frequently called the "unaccomplished temperature." Maintaining this temperature programming may stabilize the degree of supersaturation obtained, which is believed to lead to the production of more uniform crystals.

7.4.4 Organic solvent and salt precipitation

Just as supersaturation can be created by cooling, the solubility of a biosolute can be lowered by adding an ionogenic component to the aqueous phase to "salt out" the solute (as discussed in Chapter 4). It is also possible to precipitate proteins or biochemicals by adding a water-soluble, polar organic solvent. Both of these effects can be modeled with the following expression:

$$\ln\left(\frac{C}{C_0}\right) = a - bD \tag{7.50}$$

In this equation, D is the concentration of the ionogenic agent or polar organic solvent introduced, and a and b are experimentally determined constants. Maintaining a constant degree of supersaturation requires that the rate of D addition remain proportional to the rate of change in the degree of supersaturation. To realize this goal, we must use the following equation:

$$\frac{dC}{dt} = -bC_0 \exp(a - bD)\frac{dD}{dt} \tag{7.51}$$

Using the same approach as employed in Example 7.1 yields the following:

$$\exp(a - bD)\frac{dD}{dt} = \frac{k_v \rho_c B^0 G^3 t^3}{bC_0} \tag{7.52}$$

Solving this expression for the concentration of D gives the following result:

$$D = a - \frac{1}{b}\ln\left(-\frac{k_v b \rho_c B^0 G^3 t^4}{4} + a\right) \tag{7.53}$$

7.4.5 Growth rate dispersion

It is unrealistic to assume that a batch crystallizer can maintain a constant growth rate because the degree of supersaturation changes over the course of a run. As a simple approach to deal with this effect, we can model the variation in the growth rate as a dispersion term:

$$\frac{dn}{dt} + G\frac{dn}{dL} - D_G\frac{\partial^2 n}{\partial L^2} = 0 \tag{7.54}$$

Now that it includes the growth rate dispersion term for a spiked seeding of the batch crystallizer, the population density exhibits a realistic distribution as a function of crystal length. Equation 7.54 can be solved using the following boundary and initial conditions:

at $t = 0$, $n = 0$ (7.55)

at $L = 0$, $n = n^0 \delta(t)$ (7.56)

as $L \to \infty$, n is finite   (7.57)

The resulting expression for n takes the form of a Gaussian or normal distribution with respect to time:

$$n = \frac{n^0}{\sqrt{\dfrac{4\pi L D_G}{G^3}}}\exp\left(\frac{(L - Gt)^2}{\dfrac{4 L D_G}{G^2}}\right) \tag{7.58}$$

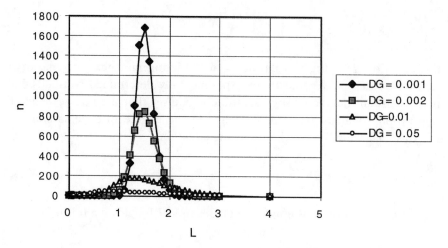

Figure 7.3 Dependence of n on L at various values of D_G ($G = 0.01$; $t = 15$).

This expression is obtained via Laplace transforms and van der Laan's theorem (see Appendix B). From the Laplace transform of the population density, we can readily determine the cumulative number distribution. Our purpose here is to assess how the cumulative number and mass distributions differ from the results previously obtained without the use of the growth rate dispersion term. Based on our definition of N, we need to calculate the following integral:

$$\frac{N}{n^0} = \int_0^L \sqrt{\frac{G^3}{4\pi L D_G}}\ \exp\left(\frac{(L - Gt)^2}{\dfrac{4LD_G}{G^2}}\right) dL \tag{7.59}$$

Figure 7.3 provides a graph of n as a function of L for various values of D_G. The total area under each curve in this figure represents N values as L goes to infinity at specific times and for specific growth rates and dispersion values. Although the graph of n versus L is bell-shaped, it is not a symmetric, Gaussian distribution at higher values of D_G. Note that the Gaussian distribution is obtained by plotting n versus t. In this graph, we see a noticeable tailing off at higher values of L, a decline that is exacerbated by increasing values of D_G. At low values of D_G, the curve is nearly symmetric and provides a result similar to the case with no growth rate dispersion. In addition, the maximum in the n versus L curve shifts toward smaller values of L. This trend has important implications in the calculation of dominant crystal size, which is discussed later in this section.

The cumulative mass of crystals can also be calculated using the following integral:

$$\frac{M}{\rho_c k_v n^0} = \int_0^L \sqrt{\frac{G^3}{4\pi L D_G}}\ L^3 \exp\left(\frac{(L - Gt)^2}{\dfrac{4LD_G}{G^2}}\right) dL \tag{7.60}$$

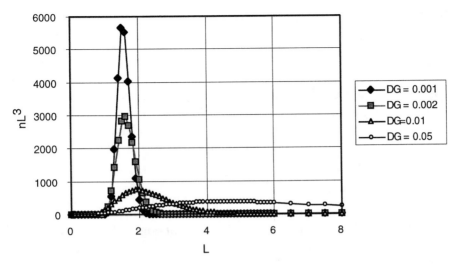

Figure 7.4 $n*L^3$ ($G = 0.01; t = 15$).

As L approaches infinity in Equation 7.60, the total weight of crystals can be found. As Figure 7.4 shows, low values of D_G increase the cumulative mass of crystals. At higher values of D_G, the curve develops pronounced tailing, showing asymmetrical behavior. This tailing indicates a broad distribution of crystal sizes.

The dominant crystal size without growth rate dispersion occurs at the mode, which is the maximum of the curve of M versus L. For a continuous crystallizer (see Section 7.5), the dominant crystal size occurs at $L_d = 3G\tau$. For a batch crystallizer with very low values of D_G, the dominant crystal size is $L_d = Gt$. With growth rate dispersion, the increasing values of D_G no longer allow for a clear maxima. Hence, instead of a dominant crystal size, the process yields a large crystal size range, with all sizes being essentially represented equally. This distribution lowers the overall crystal quality, as many physical properties—such as dissolution rates, solids flow rates, and dust formation from the presence of fines—can be undesirably affected in a broad size distribution.

Introducing the growth rate dispersion term into the calculations for batch crystallizers reveals how experimentally observed phenomena affect the production process—in particular, it demonstrates that crystals of different sizes grow at different rates. In the next section, we will see why a process vessel that continuously produces crystals is desirable not only for maintaining a high volumetric productivity, but for controlling growth rate dispersion because crystals will have a controlled, finite residence time in the vessel. Although growth rate dispersion can still occur in these vessels, continuous crystallizers can be operated in a manner that closely approximates the process that results in a small incremental change in crystal size. This small, differential crystal growth method allows for the measurement of the linear growth rate. Size-dependent growth and growth rate dispersion factors can also be considered as part of the analysis, thereby enabling us to model industrial crystallizers.

7.5 CONTINUOUS CRYSTALLIZATION

We have already derived the population balance for a batch crystallizer, so we can readily derive a continuous crystallizer solid-phase balance. A continuous crystallizer provides a stream of crystals at steady state. Because its input stream helps to maintain a constant supersaturation, this type of crystallizer can provide a constant growth rate much more easily than a batch operation can.

We can easily extend our population balances for modeling a batch crystallizer to the continuous system. The population balance for a steady-state continuous crystallizer is as follows:

$$
\begin{bmatrix} \text{Number of} \\ \text{crystals growing} \\ \text{into range} \end{bmatrix} + \begin{bmatrix} \text{number of} \\ \text{crystals entering} \\ \text{range by flow} \end{bmatrix} = \begin{bmatrix} \text{number of} \\ \text{crystals growing} \\ \text{out of range} \end{bmatrix} + \begin{bmatrix} \text{number of} \\ \text{crystals leaving} \\ \text{range by flow} \end{bmatrix}
$$

$$(7.61)$$

Assuming that the crystallizer liquid volume remains essentially constant, as do the growth and flow rates, we can take limits to obtain the following differential balance:

$$
VG\frac{dn}{dL} + Qn = 0 \tag{7.62}
$$

It is assumed that no crystals are entering, as happens in continual seeding, and that no growth rate dispersion occurs. This expression can be restated in terms of the residence time, $\tau = V/Q$:

$$
\frac{dn}{dL} + \frac{n}{G\tau} = 0 \tag{7.63}
$$

The population density can thus be solved quite easily:

$$
n = n^0 \exp\left(-\frac{L}{G\tau}\right) \tag{7.64}
$$

In this equation,

$$
n = n^0 \quad \text{when} \quad L = 0 \tag{7.65}
$$

Applying the analysis previously performed for the batch crystallizer, the cumulative mass of crystals can be expressed as in Equation 7.66:

$$
M = 6k_v\rho_c n^0 G\tau\left(G^3\tau^3 - \left(G^3\tau^3 + G^2\tau^2 L + \frac{1}{2}G\tau L^2 + \frac{1}{6}L^3\right)\exp\left(-\frac{L}{G\tau}\right)\right) \tag{7.66}
$$

$$
\text{as } L \to \infty, M = W = 6k_v\rho_c n^0 G^4\tau^4 \tag{7.67}
$$

Although continuous crystallizers are desirable for determining growth rates and other kinetic parameters, they are not a popular choice in real-world bioseparations. For more detailed analyses of these types of crystallizers, the reader is directed to the reference list at the end of this chapter.

7.6 YIELD

Before concluding a process analysis of crystallization, it is important to calculate the process's yield. Crystals, and especially protein precipitates, can carry water. In most cases, water is considered to be an essential part of the product. With most precipitates, however, it is considered undesirable and even an impurity. It is also useful to directly measure the amount of solid generated as part of the yield calculation, as well as the change in the liquid-phase concentration.

Tavare (1) has tabulated equations for theoretical yields in crystallization depending on the method used to achieve supersaturation. These equations apply to both batch and steady-state continuous crystallization. (Note that the initial concentration is the feed concentration and the final concentration is the concentration exiting the crystallizer.)

7.6.1 Removal of solvent and diluent

7.6.1.1 Complete removal of solvent and diluent

Before crystallization begins, a liquid phase can contain both solvent and diluent; the latter is a liquid phase added to modify the properties of the mother liquor and solubility of the product. During the evaporative crystallization process, both solvent and diluent can be fully removed. In this case, the theoretical crystal yield, or the mass of crystal product, can be calculated as follows:

$$Y = S_0 R m_i \tag{7.68}$$

S_0 is the mass of the solvent, R is the ratio of the hydrous crystal salt's molecular weight to the anhydrous salt's molecular weight (equal to 1 if the crystal does not have a hydrous form), and m_i is the ratio of the mass of the anhydrous crystal to the mass of solvent. (The preferred mass unit is kilograms.) In Equation 7.68, the solvent mass refers to "free" solvent, meaning that this quantity of solvent is not entrained within the crystal. Ensuring that nearly all of the solvent and diluent are removed from the crystal also keeps crystal purity high.

7.6.1.2 No loss of solvent and diluent

In cooling crystallization, the solvent and diluent are not removed. Hence the crystal yield is given by the following equation:

$$Y = S_0 R(m_i - m_f) \tag{7.69}$$

Here m_f is the mass of crystal exiting the crystallizer or remaining at the end of a batch crystallization.

7.6.1.3 Diluent addition with no loss of solvent

Adding a diluent that does not solubilize the product in the liquid phase can create supersaturation, which then leads to crystal formation. The yield can be calculated in this case as follows:

$$Y = S_0 R(m_i - m_f(1 + V_d)) \tag{7.70}$$

V_d is the ratio of the mass of diluent added to the mass of free solvent.

7.6.1.4 Partial loss of solvent

In evaporative crystallization, only a portion of the solvent is typically removed. The crystals are recovered from the mother liquor by filtration or centrifugation. In this case, the yield can be obtained as follows:

$$Y = S_0 R(m_i - m_f(1 + V_d - V_s)) \tag{7.71}$$

V_s is the ratio of the mass of solvent removed to the mass of free solvent.

7.7 SUMMARY

This chapter has emphasized the development of analytical tools that incorporate an understanding of the phenomena of crystallization and precipitation; these tools are intended to guide the operation of batch and continuous crystallizers. We have not given a detailed analysis of specific conditions for crystallizing certain types of biomolecules. Rather, our discussion has focused on the general methods through which a state of supersaturation can be created. These methods include solvent evaporation, cooling, and diluent addition, which are used in nearly all industrially relevant crystallization and precipitation strategies. We also examined the implications of the crystallizer type selected and the phenomena of growth rate dispersion, which must guide design and experimental data analysis.

Crystalline product sales are based on mass rather than the number of crystals. Thus one of the most important factors in industrial crystallization is crystal size distribution with respect to crystal mass.

Other strategies for analyzing growth rate differences among crystals can be found in the extensive process engineering literature covered by Tavare (1) and others listed in the references at the end of this chapter. Here we have focused on the use of Laplace transforms and moments analyses using van der Laan's theorems. These mathematical tools are employed in Chapter 10 as well, and the reader is encouraged to compare and contrast the results provided in both of these chapters.

7.8 PROBLEMS

7.1 Graph the equation for N versus t in a batch crystallizer with growth rate dispersion, and compare the result with the batch equation for no growth rate dispersion. Use the values of $G = 0.01$ cm and $D_G = 0.003$. Explain the significance of the differences between these two results in terms of process outcomes.

7.2 Derive the equation for the total mass W in a batch reactor with growth rate dispersion. What does it suggest to you about operating a batch reactor? Provide processing strategies to improve yield and size distributions.

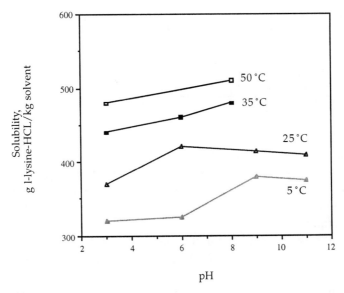

Figure 7.5 Solubility of l-lysine as a function of pH.

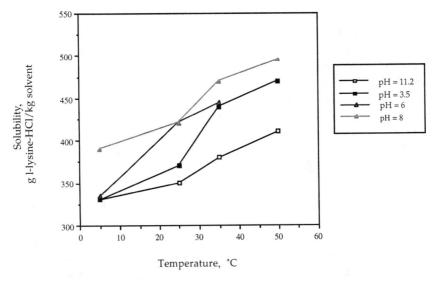

Figure 7.6 Solubility of l-lysine as a function of temperature.

7.3 Your technicians perform the crystallization work and provide the following data:

Crystallization of l-lysine from a mother liquor containing 650 g/L of l-lysine-HCl was conducted in a 10 L batch crystallizer by cooling the 50 °C solution to 10 °C at a rate of 0.3 °C to 1 °C per minute, and then holding the solution in the crystallizer for 30 minutes. After filtering the solids, the crystals appeared brown and translucent when wet, and became opaque when dried. The crystallization step gave an l-lysine-HCl product with a purity of 98% or higher. The mother liquor contained approximately 420 g/L of l-lysine-HCl after crystallization. Figures 7.5 and 7.6 give

Table 7.1 Crystal Formation Data

	Time to Start of Crystal Formation (min)	Temperature at Start of Crystal Formation (°C)
With seeding	40	40
Without seeding	90	27

solubility data for l-lysine as a function of pH and temperature. A brief study on the effectiveness of seeding with l-lysine-HCl during crystallization indicated that seeding increased the crystallization rate but had no effect on yield, purity, or crystal size distribution. Crystal formation also occurred earlier and, consequently, at a higher temperature. (See Table 7.1.)

(a) What is the crystal yield in this case?

(b) What process conditions (pH, T, and so on) will provide the best product production rate?

(c) Assuming that you start with a 650 g/L lysine solution, what would the crystal yield be based on the conditions you determined in part (b)?

7.4 For a continuous crystallizer, assume that the growth rate can be increased by 5% by removing a crystal habit modifier from the mother liquor. Alternatively, the nucleation population density n^0 can be increased by 25% by lowering the crystallizer temperature. Either of these process changes introduces an additional cost in the final product of approximately $1/kg.

(a) Which process modification is the most cost-effective in terms of an improved production rate (kg/day)?

(b) If the dominant crystal size is decreased 1% by increasing the nucleation population density, which process modification will be more cost-effective?

(c) What, if any, are the pros and cons of either process modification with respect to product characteristics? Give supporting formulae and reasoning.

7.5 Derive the population density for a continuous crystallizer exhibiting growth rate dispersion. How are N and M affected compared with the case involving no growth rate dispersion?

7.6 Using the Abegg-Stevens-Larson equation where $b = 2$, determine the population density for a batch crystallizer. Note that specifying the dependence of G on L means that you do not need to use the growth rate dispersion model.

7.7 What is the protein precipitate yield when 50 g/L of protein in 2 L of an aqueous solution is combined with 5 L of ethanol, resulting in a final concentration (ethanol-water mixture) of 0.5 g/L protein?

7.8 Plot the effect of the solid-liquid interfacial tension γ_{sl} on the maximum free-energy change in nucleation for the l-lysine crystallization described in Problem 7.3. Assume that the crystallization was performed at a pH of 6 and temperature of 10 °C, with the mother liquor containing a starting concentration of 630 g/L. A reasonable estimate of γ_{sl} for most solid-liquid interfaces is about 10 dynes/cm, considering that the surface tension of one of the more energetic surfaces, the air-water interface, is about 72 dynes/cm at room temperature. What is the crystal radius that must develop to begin nucleation?

7.9 In separate graphs, plot N and M as functions of the crystal size L for a batch crystallizer having a perfect, infinitesimal pulse of seed crystals and no growth dispersion. Explain how you could use these plots to estimate N and M when a finite pulse of seed crystals is introduced. A finite pulse is one that occurs over a time interval and can be mathematically described as $n = n^0 u(t - t_0)$, where t_0 is the pulse duration.

7.10 A plug flow crystallizer (uniform velocity across the pipe diameter) can be a useful process vessel during cooling crystallization. Assuming that the mother liquor and crystals in this crystallizer have a uniform velocity of U, derive a population balance for this crystallizer based on the following initial and boundary conditions. Also, assume that there is a constant growth rate, G.

Initial condition:
at $x = 0, n = 0$ (7.72)

Boundary conditions:
at $L = 0$, $n = n^0\delta(t)$ (7.73)
as $L \to \infty$, n is finite (7.74)

(a) Solve the population balance for n.
(b) How does this setup compare with the batch crystallizer?

7.11 In a continuous crystallizer with constant growth rate, derive an expression for the size range of the crystals by mass for those crystals that are within one standard deviation from the dominant crystal size.

7.12 Reviewing the scientific literature on crystallization in space, describe the effect of microgravity on a batch crystallizer. If a biotechnology facility is created in space, will it have any operational benefits in terms of crystal size distribution, yield, or growth rates? Which of the following methods for supersaturation would be more attractive in space?

(a) Evaporation
(b) Cooling
(c) Diluent addition

7.9 REFERENCES

1. Tavare NS. Process simulation analysis and design. New York: Plenum, 1995.
2. Belter PA, Cussler EL, Hu WS. Bioseparations: downstream processing for biotechnology. New York: John Wiley and Sons, 1988.
3. Wankat PC. Rate-controlled separations. London: Elsevier, 1990.

8

Membrane Filtration

The separation of molecules through the use of a barrier is commonly known as membrane filtration. The composition of this barrier, which can be either an insoluble liquid phase or a solid phase, is a critical factor in membrane separations. Commonly used membranes consist of cellulose, Teflon, or nylon polymers that have been formed into sheets or tubes.

Solid-liquid separations are difficult to perform and scale up, especially when the product must be recovered from a fermentation broth. Although the filtration of well-defined crystalline solids is a straightforward process, the physical characteristics of a broth can complicate the filtration for several reasons. Broth characteristics such as small particle size, wide particle-size distribution, high solids compressibility, increasing viscosity as the suspension becomes more concentrated, the introduction of non-Newtonian fluid behavior, and minimal density differences between the solid and liquid phases can all create difficulties in separating the components of a fermentation broth. To alleviate the problems associated with the compressibility of the accumulated solids on the filter surface, one may add precoat materials and filters aids to the filtration broth; these materials increase the cake's porosity and reduce its resistance to flow.

8.1 MEMBRANE MATERIALS

Membrane materials are classified according to their pore size, porosity, and mechanical stability. They range from cellulose and parchment for caustic recovery, to polyvinyl chloride for acid recovery processes.

Dialysis membranes made from cellulose or modified polyacrylonitrile materials are available in hollow-fiber form. Membranes made from cellulose esters or other polymers (such as nylon, polyvinyl chloride, or acrylonitrile) have a porosity of approximately 80%, and their pore sizes range from 0.1 to 1.0 μm.

A second type of membrane is made from nonporous films of polymers such as polypropylene. These polymers have a porosity of 35% and a thickness of 0.3 cm.

A third membrane type, made by exposing nonporous films of mica or polycarbonate to α radiation, has a 3% porosity. It is ten times thinner than other membranes, however, which results in a permeability similar to that obtained with competing membranes.

Controlling membrane porosity and pore size is important because these characteristics define the selectivity and transfer rate across the membrane. To manipulate these properties, one can add swelling agents, stretch the membrane, or initiate

146

chemical reactions (as the acetylation of cellophane, for example). In addition, the material must be durable and both physically and chemically stable if it is to be used in commercial bioseparations. For example, membranes made of cellulose esters and other polymers with similar structures remain highly stable over wide ranges of pH, temperature, and organic solvent concentrations.

8.2 DRIVING FORCES IN MEMBRANE SEPARATIONS

Three fundamental driving forces are commonly used for membrane separations: pressure, concentration, and electrical potential. Table 8.1 compares these membrane separation processes.

Pressure-driven membrane processes can be classified into microfiltration (MF), ultrafiltration (UF), and reverse osmosis (RO) separations. MF membranes reject particulates of different sizes depending on the membrane pore size, which can range between 0.02 and 10 mm. UF membranes reject macromolecules with molecular-weight values ranging between 1000 and 1,000,000. The manufacturer typically specifies the molecular-weight cut-off (MWCO) of an UF membrane. RO membranes prevent the passage of low-molecular-weight solutes such as salts, and their performance is evaluated in terms of the percentage of salt rejected.

Concentration-driven membrane processes include dialysis and processes that use liquid membranes. In these processes, the transfer of the solute across the membrane occurs by diffusion.

In electrically driven membrane processes, the transfer of the solute across the membrane occurs through electromigration of an electrolyte solution.

8.3 GENERAL THEORY OF MICROFILTRATION

In this section, we will provide a mathematical description of the theory behind microfiltration. First, we must consider Darcy's law, which relates the flow rate through a porous bed to the pressure drop:

$$v = \frac{k\Delta P}{\mu l} \tag{8.1}$$

Table 8.1 Classification of Membrane Separation Processes

Filtration Process	Driving Force	Pore Size
Microfiltration	Pressure (0–1 bar)	0.02–10 μm
Ultrafiltration	Pressure (0–10 bar)	0.001–0.2 μm
		MW 10^3–10^6
Reverse osmosis	Pressure (0–100 bar)	Nonporous
		MW < 10^3
Dialysis	Concentration difference	1–3 μm
Electrodialysis	Electrical potential	MW < 200
Electro-osmosis	Electrical potential	
Liquid membranes	Concentration difference	

In this equation, v is the liquid velocity, ΔP is the pressure drop across the bed, k is the Darcy's law permeability of the bed, μ is the liquid viscosity, and l is the bed thickness. For a batch filtration process, we can represent the liquid velocity by the following expression:

$$v = \frac{1}{A}\frac{dV}{dt} \tag{8.2}$$

Here V is the total volume of the filtrate, t is time, and A is the cross-sectional area of the bed. From Darcy's law, the ratio $\frac{k}{l}$ represents resistance:

$$\frac{k}{l} = r_m + r_c \tag{8.3}$$

In Equation 8.3, r_m is the resistance of the filter media and r_c is the resistance of the cake, or the accumulated biomass. A filter cake comprises the accumulation of solute on the filter medium or membrane.

Combining Equations 8.1, 8.2, and 8.3 gives a differential equation for filtration at a constant pressure drop. This general equation for microfiltration applies to both compressible and incompressible cakes:

$$\frac{dV}{dt} = \frac{\Delta PA}{\mu(r_m + r_c)} \tag{8.4}$$

8.3.1 Incompressible cakes

For an incompressible cake, the cake thickness is directly proportional to the filtrate volume and inversely proportional to the filter area. Thus the cake's resistance can be described by the following equation:

$$r_c = \alpha\,\frac{W}{A} = \alpha\,\frac{\rho_0 V}{A} \tag{8.5}$$

In this equation, W is the total weight of cake or filter, a is the average specific resistance of the cake, and $\Delta\rho_0$ is the mass of the solid cake per filtrate volume. Combining Equations 8.4 and 8.5 gives the following expression:

$$\frac{dV}{dt} = \frac{\Delta PA}{\mu\left(r_m + \alpha\,\dfrac{\rho_0 V}{A}\right)} \tag{8.6}$$

Integrating Equation 8.6 from $V_0(t = 0)$ to $V(t)$ gives

$$V^2 + 2VV_0 = Kt \tag{8.7}$$

where

$$K = \left(\frac{2A^2}{\alpha\rho_0\eta}\right)\Delta P \tag{8.8}$$

and

$$V_0 = \left(\frac{r_m}{\alpha\rho_0}\right)A \tag{8.9}$$

leading to the Ruth equation:

$$\frac{t}{V} = \frac{1}{K}(V + 2V_0) \tag{8.10}$$

In Equation 8.10, V is the total volume of filtrate.

K and V_0 are determined by plotting t/V versus V, and then measuring the resulting curve's slope $1/K$ and intercept $2V_0/K$. K is a function of ΔP and the properties of the cake, whereas V_0 is a function of the filter medium's resistance (which is usually negligible). Thus the Ruth equation can be expressed in a simpler form:

$$t = \left(\frac{\alpha\rho_0\mu}{2\Delta P}\right)\left(\frac{V}{A}\right)^2 \tag{8.11}$$

8.3.2 Compressible cakes

Because most biological-material cakes are compressible, we must account for this effect. The compressibility of a cake increases the pressure drop across the cake, thereby strengthening the cake's resistance. This relationship is best represented by an exponential equation:

$$\alpha = f(\Delta P) \tag{8.12}$$

$$\alpha = \alpha'(\Delta P)^a \tag{8.13}$$

where α is the compressibility of the cake, which can vary from 0 for an incompressible cake to 1 for a compressible cake. Values of α usually range from 0.1 to 0.8. α' is a constant that relates to the size and shape of the biological particles forming the cake. Values of α and α' can be determined by performing a least-square analysis of $\log\alpha$ versus $\log(\Delta P)$. This analysis gives a straight line of slope α with an intercept of $\log\alpha'$. The compressibility of the cake does change the expression of the cake resistance because the pressure drop increases across the cake, which affects K but not V_0.

When the biological materials are highly susceptible to compressibility effects, it is recommended to treat the feed with filter aids. Filter aids reduce the pressure drop across the cake, thereby describing resistivity. Common filter aids include 85% to 90% SiO_2 and 10% to 15% MeO_2, where Me represents a metal such as Al, Fe, Ti, Ca, or Mg.

8.4 MICROFILTRATION

Microfiltration and ultrafiltration are both pressure-driven membrane processes that use the membrane's permeability to differentiate between solutes of different sizes. The two processes can be distinguished based on the size of the solutes each technique can separate. However, a considerable overlap occurs in terms of the size ranges that can be accommodated. The following discussion about microfiltration pertains to ultrafiltration as well, as these processes are analyzed in a similar fashion.

Conventional filtration usually involves large particles or microorganisms, which accumulate as a filter cake. Because microfiltration involves such small microorganisms, however, no cake can form. Instead, the filtration process is carried out by applying suitable hydrodynamic forces—originated through pumping or stirring— to force the suspension across the membrane at high tangential velocities in a cross-flow pattern. Because of these velocities, a relatively slow flow moves perpendicular to the membrane. The large ratio of cross flow to perpendicular flow works to minimize cake formation.

Microfiltration can be used to concentrate an insoluble material. The performance achieved with this system depends on the applied hydrodynamic conditions, characteristics of the membrane, and fluid parameters such as particle size and distribution. In most cases, the microfiltration feed takes the form of a dilute suspension; as the liquid is removed, this suspension becomes more concentrated. To maintain the tangential flow across the membrane surface, we must apply mechanical agitation or pumping. Mechanical agitation, as in rotating systems, is commonly used for inorganic or shear- and heat-insensitive materials. Pumping or periodic purging is a necessity for sustaining cross flow. Because the discharged suspension usually contains more liquid than a conventional filter cake, an additional centrifugation or conventional filtration step is required to further concentrate the insolubles.

The microfiltration process can, for example, serve as a method for cell harvesting after fermentation. Such harvesting is necessary if the product is produced intracellularly, as is the case for some enzymes. With this technique, it is not necessary to produce a highly dewatered cell mass, and a volume reduction to approximately 40% wet cells is sufficient to apply further procedures such as disruption and purification.

The characteristics of the microfiltration membrane can play a very important role in determining the overall performance achievable with the filtration process. This membrane is usually thin and microporous, containing small and highly monodispersed pores. It retains the particles being filtered, but allows liquid and smaller solutes to pass through quickly. As a result of its small pores, the microfiltration membrane has low permeability and high resistance to flow.

Microfiltration membranes can handle suspensions in which very small density differences exist between the solution and the solute, as in microbial suspensions. As only a negligible filter cake is formed in this process, any resistance must be due solely to the membrane. As we will see, this case does not hold for ultrafiltration, where resistance arises from both the cake formed and the membrane.

Because the flow per area through a microfiltration membrane is slow, the filter's geometry is an important factor in determining how well the process works. To assure an adequate flow rate, membranes are designed with a large filter area per filter volume. Three designs are common: plate-and-frame, spiral-wound modules (Figure 8.1), and fiber modules. These filters are resistant to fouling and are easy to clean.

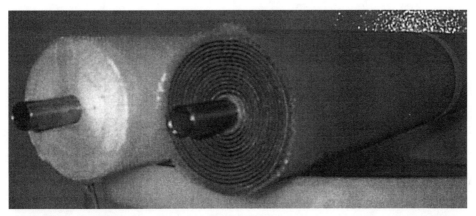

Figure 8.1 Spiral-wound membrane cartridges. The cartridge on the left is new; the cartridge on the right has been used in cell filtration. (Courtesy of Genencor International, Palo Alto, California.)

8.4.1 Staging in microfiltration

The combination of several microfiltration membranes in a filtration system is often advantageous, as such an arrangement can sometimes produce more concentrated or purer products than a single membrane. The treatment of such a system is merely an extension of the single membrane case, as outlined in Example 8.1.

Example 8.1
Discuss the merits of having three versus two membranes in series, as illustrated in Figures 8.2 and 8.3, respectively. Assume that membrane size is sufficiently large so that no product concentration takes place.

Solution
The mass balances for each membrane unit give the following values:

$$100C_0 + 50C_2 = 150C_1 \tag{8.14}$$

$$30C_1 + 50C_3 = 80C_2 \tag{8.15}$$

$$30C_2 + 30 \times 0 = 60C_3 \tag{8.16}$$

To solve for C_1, C_2, and C_3, we need to solve three equations for three unknowns, from which we obtain $C_1 = 0.8C_0$. Fractional product recovery is determined as follows:

$$\frac{120C_1}{100C_0} = 0.96 \tag{8.17}$$

Also, the dilution factor is

$$\frac{C_1}{C_0} = 0.80 \tag{8.18}$$

For the two-stage process, mass balances for each filter give the following:

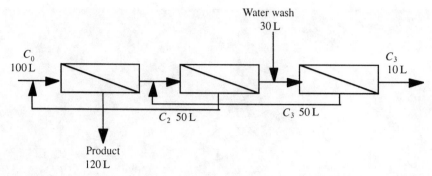

Figure 8.2 Three-stage microfiltration.

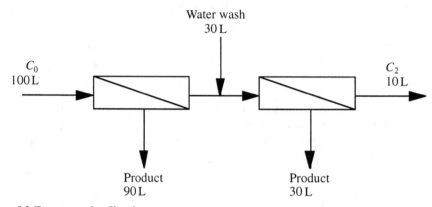

Figure 8.3 Two-stage microfiltration.

$$100C_0 = 100C_1 \tag{8.19}$$

$$10C_1 = 30 \times 0 = 40C_2 \tag{8.20}$$

Thus $C_0 = C_1 = 4C_2$. Fractional product recovery is determined as follows:

$$\frac{90C_1 + 30C_2}{100C_0} = 0.975 \tag{8.21}$$

Also, the dilution factor is

$$\frac{C_1 + C_2}{C_0} = 1.25 \tag{8.22}$$

Comparing the performance of the two processes shows that the two-stage process, in addition to its simplicity, gives a better fractional product recovery and dilution factor than the three- or one-stage process.

8.5 ULTRAFILTRATION

Ultrafiltration is a pressure-driven process that can separate high-molecular-weight solutes from a solvent. The membranes used in this process have a much smaller

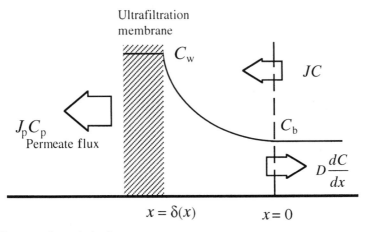

Figure 8.4 Concentration polarization.

pore size of solvent than microfiltration membranes, with a relatively small pore area per membrane surface area. Consequently, ultrafiltration membranes permit a lower flux of solvent than microfiltration membranes.

Ultrafiltration (UF) processes take advantage of a high cross flow and depend on the geometry of the membrane. Since UF separates solutes based on differing molecular size, the separation can become hindered if large molecules accumulate at the membrane. Such a buildup reduces the flux and influences the solute retention characteristics. The accumulation of the solute at the membrane surface produces a higher solute concentration at the membrane surface than in the bulk of the macromolecular solution, a phenomenon known as concentration polarization. The formation of the filter cake in this way can retard the filtration process and increase the solute concentration in the product solvent. To reduce the concentration polarization effect, a large flow can be applied across the membrane surface, perpendicular to the flow through the membrane. Unfortunately, this technique yields a lower product-to-feed ratio.

We can define concentration polarization in mathematical terms by applying a mass balance to the solute:

$$\begin{bmatrix} \text{Solute carried toward} \\ \text{membrane by solvent} \end{bmatrix} = \begin{bmatrix} \text{solute moving away from} \\ \text{wall by diffusion} \end{bmatrix} + \begin{bmatrix} \text{solute permeating} \\ \text{the membrane} \end{bmatrix}$$

(8.23)

$$JC = -D\frac{dC}{dx} + JC_{p}$$

(8.24)

In Equation 8.24, J is the flux of solute carried by the solvent toward the membrane, D is the solute diffusion coefficient, x is the coordinate that is used to measure boundary-layer thickness, and C_p is the concentration of solute in the permeate. In the system depicted in Figure 8.4, C_W is the concentration at the membrane wall, $\delta(x)$ is the boundary-layer thickness, and C_b is the concentration in the bulk solution.

Assuming that a negligible amount of the solute passes through the membrane, the mass balance becomes

$$JC = -D\frac{dC}{dx} \tag{8.25}$$

Integrating from C_w at $x = 0$ to C_b at $x = d(x)$ gives us

$$\int_{C_w}^{C_b} \frac{dC}{C} = \int_0^{\delta(x)} \frac{J}{D}dx \tag{8.26}$$

$$J(x) = \frac{D}{\delta(x)} \ln\left(\frac{C_w}{C_b}\right) \tag{8.27}$$

Equation 8.27 describes the solute flux in concentration polarization. Concentration polarization increases as the flux increases and the ratio $D/\delta(x)$ decreases. The decrease in $D/\delta(x)$ reflects the low solute diffusivity, which can result from the presence of a high-molecular-weight solute, high solution viscosity, or a large boundary-layer thickness (created by low turbulence). The most simple case arises when concentration polarization is very low:

$$\frac{D}{\delta(x)J} \gg 1 \tag{8.28}$$

In this case, $C_w = C_b$.

8.5.1 Ultrafiltration process application

Generally, separation of proteins requires that the protein be concentrated, then washed to remove salts. This application of ultrafiltration is discussed in Example 8.2.

Example 8.2
A 10 L batch of protein A in a 5 mg/L suspension is to be concentrated by a factor of 5 using ultrafiltration, with a retention coefficient (R) of 0.98. Mathematically model this process.

Solution
Performing a mass balance for the batch ultrafiltration process gives the following equation:

$$[\text{rate of loss of solute}] = [\text{solute flux through membrane}] \tag{8.29}$$

$$-\frac{d}{dt}(CV) = JAC_p = JAC(1 - R) \tag{8.30}$$

In Equation 8.30, V is the retention volume, C is the retentate concentration, C_p is the permeate concentration, J is the flux through membrane, A is the membrane area, and R is the retention coefficient.

Note that the salt concentration does not change. Therefore, a mass balance describes the protein concentration:

$$-C\frac{d}{dt}(V) - V\frac{d}{dt}(C) = JA(1 - R)C \tag{8.31}$$

For a batch ultrafiltration process,

$$\frac{dV}{dt} = -JA \tag{8.32}$$

Integrating from V_0 ($t = 0$) to V ($t = t$) gives

$$\int_{V_0}^{V} dV = -\int_0^t JA\,dt \tag{8.33}$$

$$V = V_0 - JAt \tag{8.34}$$

Replacing V with $V_0 - JAt$, replacing $-dV/dt$ with JA, and integrating the mass balance from C_0 ($t = 0$) to C_{final} ($t = t$) gives the following:

$$\frac{C_{final}}{C} = \left(\frac{V_0}{V_{final}}\right)^R \tag{8.35}$$

Therefore C_{final} is 24.2 mg/L. The amount of protein lost is $V_0C_0 - V_{final}C_{final} = 1.6$ mg.

The second step in this ultrafiltration process involves the removal of the salt from the concentrated protein solution. This washing is performed at a constant volume by adding make-up water at the same rate as the material is being removed through the membrane—a process known as diafiltration, or dilution mode filtration. In diafiltration, the mass balance describes the salt, not the protein. Also, because none of the salt is retained, $R = 0$. Thus the mass balance can be expressed as follows:

$$-V\frac{dC}{dt} = JAC \tag{8.36}$$

When integrated from C_0 ($t = 0$) to C_{final} ($t = t$):

$$C_{final} = C_i \exp\left(-\frac{JAt}{V}\right) \tag{8.37}$$

or

$$C_{final} = C_i \exp\left(-\frac{\text{volume of flush}}{\text{volume in vessel}}\right) \tag{8.38}$$

If the protein A solution originally contained 12 mg/L of salts, how long would it take to reduce this concentration to 0.6 mg/L, if the water flow rate is 1 L/h and the solution volume is maintained at 1.2 L? Using the previously derived expression of the final concentration, C_{final},

$$0.6 = 12\exp\left(-\frac{t}{1.2}\right) \tag{8.39}$$

we find that $t = 3.6$ h.

8.5.2 Ultrafiltration membrane application and modification

Therapeutic proteins are usually produced by extraction from human and animal tissue or extraction from continuously cultured mammalian cells (1,2). These techniques introduce a risk of contamination, as the final product may potentially contain a virus or viruslike particles. Even tests performed on the final product to check for the presence of a certain virus cannot completely eliminate this risk, as many of these assays are not sensitive enough to detect various viruses that may cause medical concern.

To overcome this problem, manufacturers of therapeutic proteins have adopted different techniques to reduce the risk of virus transmission. In the case of downstream purification, physical removal and inactivation operations have been employed for viral clearance. The clearance factors associated with these operations have proved effective in protecting against viral transmission. On the other hand, many inactivation operations are based on heat or chemical treatment, which may denature the product protein. Consequently, stabilization procedures are required to minimize the loss of the protein.

Although ultrafiltration membranes are commonly used in virus removal, most of them cannot discriminate between different-sized viruses because of their large pore size distribution. This pore distribution may satisfy the requirements for most protein concentration and diafiltration applications, but it does not suffice for virus removal. In addition, ultrafiltration membranes are prone to developing mechanical defects that can enlarge the pore openings. These defects or imperfections occur randomly and cannot be very well controlled.

Taking these factors into account, the preferred option for virus removal applications is a membrane structure with a narrow pore size distribution and no large pore defects. An example of such a membrane system is the Viresolve/70. This membrane is composed of polyvinylidene fluoride, which is rendered hydrophilic and low-protein-binding by the addition of hydroxy-propyl acrylate. The Viresolve/70 system includes an asymmetric, finely porous active layer of 5 to 10 microns, supported by a preformed microporous membrane. The resultant refined physical separation barrier will remove virus particles on the basis of their size. The protein-sieving characteristics of this membrane are nearly equivalent to those of conventional 100,000 Da MWCO ultrafiltration membranes. Because it is based on the actual sieving properties of the membrane rather than random defects, the minimum virus clearance achievable with this filtration system is reproducible and predictable. The membrane's performance in both single- and two-stage systems shows that overall removal of virus particles is possible with simultaneous high recovery of product protein.

8.6 REVERSE OSMOSIS

The concept of osmotic pressure can be illustrated by considering a semipermeable membrane that separates a macromolecular solution from a solvent. Because of the size of the species and the chemistry of the membrane, the solvent can flow through the membrane and into the macromolecular solution; the solute, however, cannot pass through the membrane. This dilution takes place until an equal concentration is achieved on both sides of the membrane. The pressure needed to halt this dilution is called the osmotic pressure, ΔP. We can measure the osmotic pressure experimentally by observing the increase in the solution volume or the rise in the capillary level.

If we apply a pressure sufficient to overcome the osmotic pressure, then the flow will reverse, resulting in the concentration of the macromolecular solution. This procedure is known as reverse osmosis. It can be used to filter pure solvents from solutions containing salts and dissolved organic molecules. A typical applied pressure across the membrane is in the range of 0 to 100 bar. Most reverse osmosis membranes are composed of cellulose, acetate, or polyamide, and are called asymmetric membranes. In addition, composite membranes—made from a dense, thin polymer coating on a polysulfone support film—may be used.

Osmotic pressure is a property of the solution, not the membrane. Essentially, it is a function of the solution concentration at a standard temperature. To derive an expression for osmotic pressure, we can consider the equilibrium between the liquids on each side of the membrane. As we are assuming equilibrium conditions, the chemical potential of the pure solvent must be equal to the chemical potential of the solvent in the macromolecular solution:

$$\mu_{\text{pure solvent}} = \mu_{\text{solvent in macromolecular solution}} \tag{8.40}$$

The chemical potential of a species i includes all forms of energy acting on the solute, and can be represented by the following equation:

$$\mu_i = \mu_i^0 + \overline{S}_i T + \overline{V}_i P + RT \ln(C_i) + ZF\psi + \dots \tag{8.41}$$

In this equation, $\overline{S}_i$ is the partial molar entropy, $\overline{V}_i$ is the partial molar volume, Z_i is the charge of species i, and Ψ is the electrical potential. In reverse osmosis, the chemical potential of the solvent on both sides of the membrane can written as follows:

$$\mu_s^0 + P^0 \overline{V}_s = \mu_s^0 + P\overline{V}_s + RT \ln(x_s) \tag{8.42}$$

In Equation 8.42, μ_s^0 is the chemical potential of the standard state, $\overline{V}_s$ is the partial molar volume of the solvent, x_s is the mole fraction of the solvent, and $P - P^0 = \Delta\Pi$, the osmotic pressure. Thus

$$P - P^0 = -\frac{RT}{\overline{V}_s} \ln(1 - x_m) \tag{8.43}$$

$$\Delta\Pi = -\frac{RT}{\overline{V}_s} \ln(1 - x_m) \tag{8.44}$$

where $\Delta\Pi$ is the osmotic pressure and x_m is the mole fraction of the macromolecular solution. For a dilute macromolecular solution, we can expand the logarithm using a Maclaurin series for x_m:

$$\Delta\Pi = -\frac{RT}{\overline{V}_s}(-x \dots) \tag{8.45}$$

Therefore,

$$\Delta\Pi = RTc_m \tag{8.46}$$

Equation 8.46, which is known as Van't Hoff's law, shows the dependency of the osmotic pressure on the concentration of the macromolecular solution, c_m. Van't Hoff's law provides only an approximation, because it assumes that the solution is ideal and the macromolecules are present in low concentration.

8.7 FLUX EQUATIONS

As previously mentioned, the flow of the solvent across the membrane in reverse osmosis arises because a hydrostatic pressure is applied. Consequently, the flux representing the amount of solvent passing through the membrane is directly proportional to the excess applied hydrostatic pressure. The flux across the membrane can be approximated as follows:

$$J_{solvent} = L_p(\Delta P - \Delta \Pi) = \frac{k}{\eta L}(\Delta P - \Delta \Pi) \tag{8.47}$$

In the equation, ΔP is the applied hydrostatic pressure, $\Delta \Pi$ is the osmotic pressure RTc_m, L_P is the membrane solvent permeability coefficient $k/\eta L$, k is the hydraulic permeability, η is the viscosity of the solvent, and L is the thickness of the membrane.

From Equation 8.47, we see that the solvent flux is inversely proportional to the membrane thickness. The flow of the solute from one side of the membrane to the other, however, remains independent of the membrane thickness. The solute flux is basically a function of a concentration driving force—that is, the difference in the solute concentration on the two sides of the membrane. It can be approximately written as

$$J_{solute} = L_s \Delta c \tag{8.48}$$

where Δc is the concentration driving force (i.e., the difference in the solute concentration across the membrane) and L_s is the solute permeability coefficient.

8.8 ELECTRODIALYSIS

Electrodialysis, an electrically driven membrane process, is commonly used to desalt protein solutions to which salts were added to precipitate a portion of the proteins. In this filtration process, the electrolyte solution concentration changes as it electromigrates through the membrane. The feed solution is pumped through stacks of narrow compartments (either horizontal or vertical), and the terminal compartments are separated by alternating cation-exchange and anion-exchange membranes. These membranes, therefore, are selectively permeable to positive and negative ions, respectively.

Consider, for example, a solution containing a protein and a small electrolyte, such as sodium chloride. The large protein cannot pass through the membrane, but the electrolyte is drawn to the electrode terminals. The sodium ions of the sodium chloride are drawn to the anode and the chloride ions are drawn to the cathode, leaving behind a pure protein solution. Upon the application of an electric field, the electromigration process begins. The salt is depleted from the center compartment and becomes concentrated in the neighboring compartments. During the migration, the cation permeate membrane acts as a barrier to keep out the anions of the electrolyte, as the anion membrane acts as a barrier against the cations.

To reduce the costs associated with electrodialysis, the system can be designed to minimize DC energy consumption. In the case of the protein solution containing sodium chloride, sodium ions will migrate to the anode and cause a water molecule to dissociate, producing NaOH. At the same time, the chloride ions will migrate to the cathode, where they react to form hydrogen chloride and an electron. In real-world applications of electrodialysis, pure water is fed into alternating compartments. The protein solution is desalted as before, with the salt existing as sodium chloride. As a result, the sodium chloride cannot react to form NaOH and HCl, which minimizes the amount of current needed to desalt the protein solution.

8.9 EMULSION LIQUID MEMBRANES

The emulsion liquid membrane (ELM) process is a concentration-driven membrane process. In general, it includes two miscible aqueous phases, separated by an organic phase called the membrane phase. The membrane phase represents the essential feature of the ELM process. Unlike a conventional solid membrane, this liquid membrane phase consists of an organic solvent that contains a sufficient amount of an appropriate detergent, known as a surfactant. The surfactant, when present at concentrations of 5% or more, functions to stabilize the membrane phase.

The membrane system is generally performed in a two-step process (Figure 8.5). First, the emulsification of the inner aqueous phase in the organic membrane phase takes place at very high shear rates, up to 10,000 rpm. This step yields an emulsion that remains very stable thanks to the presence of microscopic drops of the aqueous phase. Second, the emulsion globules become dispersed in the outer aqueous phase at low mixing rates, from 200 to 300 rpm. This step yields emulsion globules that have diameters ranging between 0.5 and 2 mm; their actual sizes depend on the viscosity of the emulsion, the pressure of the surfactants, and the intensity of mixing.

The principle underlying the ELM process relies on the selective transport of one or more solutes from the outer aqueous phase, through the emulsion globules, and into the inner encapsulated aqueous phase. This mass transfer process is main-

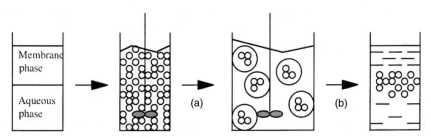

Figure 8.5 Forming a membrane system. (a) Emulsification of the inner aqueous phase in the organic membrane phase. (b) Dispersion of the emulsion globules in the outer aqueous phase.

tained by a concentration driving force, with the transfer of the solute usually taking only a few minutes. The rapid speed of this transfer is attributable to the large interfacial area between the outer aqueous phase and the emulsion globules; the large surface area of the inner aqueous phase that is encapsulated in the emulsion globules also works to boost the rate of transfer. After the mass transfer ends, the emulsion globules separate and become disrupted. The enriched inner aqueous phase undergoes further processing. In this case, the organic membrane phase can be reused.

Two mechanisms affect the transport of the solute across the liquid membrane.

The first mechanism maximizes the concentration driving force by modifying the chemistry of the inner aqueous phase. This modification ensures that the outer aqueous phase reacts upon its arrival at the membrane-encapsulated phase interface. Ideally, the product of this chemical reaction should be nondiffusing in the membrane phase. The recovery of phenol from a continuous aqueous phase to a sodium hydroxide-encapsulated solution represents a good example of how this mechanism works.

In the second mechanism, a carrier species becomes embedded in the membrane phase. This carrier species forms a complex with the solute in the outer aqueous phase at the membrane–outer aqueous phase interface. Next, the complex diffuses through the membrane to the interface with the encapsulated aqueous phase. At this point, the complex dissociates, releasing the solute. An example of this mechanism in action is the extraction of copper.

The stability of the emulsion is a critical factor in determining the success rate for ELM processes. The emulsion should resist shear forces, temperature, and concentration changes. On the other hand, an unduly stable emulsion may be difficult to disrupt. Thus the emulsion should be stable at process conditions, but should permit easy separation and recovery of the solute from the encapsulated phase.

8.10 SUMMARY

The theory of filtration described in this chapter is useful for analyzing experimental data. In the design of large-scale purification processes, it gives a qualitative picture that must then be verified by laboratory experiments.

In actual practice, many biological suspensions can be difficult to separate because of their slow filtration rates and cake formation. A typical cake is compressible in nature, which hinders its processing. Therefore, to improve filtration processes, consideration should be given to combining filtration with other separation techniques such as centrifugation.

8.11 PROBLEMS

8.1 In a reverse osmosis unit, we can develop a flux of $10\,L/m^2h$ for a protein and salt solution. The goal is to concentrate the solution without losing the buffer salts, which stabilize the protein. Given a membrane solvent permeability coefficient of

0.8 L/m²h atm and a concentration in the macromolecular solution of 5 mole/L at room temperature, determine the applied hydrostatic pressure, ΔP.

8.2 For batch microfiltration, discuss whether using n filters in series would offer a different performance than using a single filter. Assume that the total membrane area is the same for the two systems, the membrane resistance is negligible compared to the cake resistance, and the cake is incompressible.

8.3 Filtration tests are conducted in the laboratory at room temperature with a constant pressure drop using a slurry of $CaCO_3$ in H_2O. The results of these tests are given in Tables 8.2 and 8.3. The filter area is 440 cm² and the mass of the solid cake per filtrate volume is 23.5 g/L.

(a) Determine α and r_m as a function of ΔP.

(b) Derive an empirical relationship between α and ΔP.

8.4 Dialysis involves the transfer of low-molecular-weight solutes across a membrane by diffusion, where the driving force consists of a concentration gradient. This process is most useful for exchanging buffer solutions in which proteins have been dissolved. Determine an expression for the flux using a solute balance for the transfer into the membrane volume. Assume that the volume of the liquid in the membrane is constant. How does your equation compare with the equation for solute flux in ultrafiltration?

8.5 You need to separate cells from two proteins, one of molecular weight 20,000 Da and the other of 80,000 Da. Provide a process design assuming that you have three membranes with molecular weight cut-offs of 5000, 50,000, and 1,500,000, respectively. To narrow down the possibilities, discuss the merits of using coupled membrane batch modules (Figure 8.6) versus using an individual train (Figure 8.7).

Table 8.2   Filtration Test Results

	Test				
	1	2	3	4	5
Pressure drop $(-\Delta P)$ (lb$_f$/in²)	6.7	16.2	28.2	36.3	49.1

Table 8.3   Filtration Test Results

Filtrate Volume, V (L)	Test				
	1	2	3	4	5
0.5	17.3	6.8	6.3	5.0	4.4
1.0	41.3	19.0	14.0	11.5	9.5
1.5	72.0	34.6	24.2	19.8	16.3
2.0	108.3	53.4	37.0	30.1	24.6
2.5	152.1	76.0	51.7	42.5	34.7
3.0	201.7	102.0	69.0	56.8	46.1
3.5		131.2	88.8	73.0	59.0
4.0		163.0	110.0	91.2	73.6
4.5			134.0	111.0	89.4
5.0			160.0	133.0	107.3
5.5				156.8	
6.0			182.5		

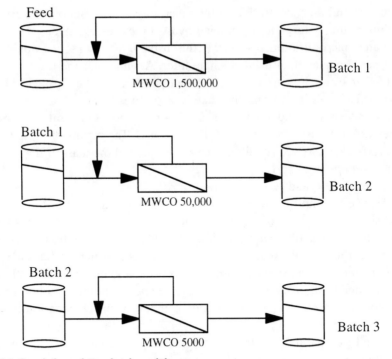

Figure 8.6 Coupled membrane batch modules.

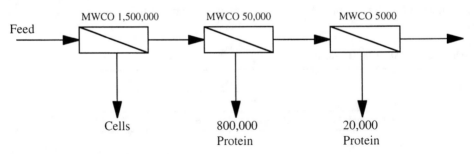

Figure 8.7 Individual membrane train.

Assume that the size exclusion is perfect ($R = 0$ or $R = 1$, depending on the molecular-weight cut-off).

8.6 A 400 L solution of a protein (MW = 3100 Da) at 0.05% needs to be concentrated to 1.5% by weight. Using an ultrafiltration membrane of surface area 500 cm^2, we can achieve a flux of 12×10^{-5} cm/s. The protein has a diffusion coefficient of 1.8×10^{-6} cm^2/s at 15 °C, and the pressure drop is 25 atm.

(a) Determine whether concentration polarization is an important external factor affecting the membrane's performance. Assume that the boundary layer thickness is 0.015 cm.

(b) Calculate the time needed to complete the filtration and achieve the desired final concentration.

8.7 What is the concentration on the surface of an ultrafiltration membrane if the flux is 2.4×10^{-4} cm/s and the bulk concentration is 6.3×10^{-5} g/cm^3? The solute diffusivity is 5.4×10^{-6} cm^2/s, and the boundary-layer thickness is 0.02 cm.

8.12 REFERENCES

1. Lubiniecki AS, Spier RE, Griffiths JB. Animal cell biotechnology. London: Academic Press, 1988.
2. Lubiniecki AS, May LH. Cell bank characterization for recombinant DNA mammalian cell lines. Dev Biol Stand 1985;60:141–6.

9

Centrifugation

Centrifuges are used primarily in solid-liquid or liquid-liquid separation processes, where the separation is based on density differences between the solid or liquid and the surrounding fluid. The higher-density phase will settle through the lower-density phase under the influence of the applied force created through centrifugal acceleration.

Centrifugation has long been employed as a separation technology. In 1878, Swedish engineer Gustaf de Laval developed a continuously operating centrifugal separator designed to separate cream from milk. In 1896, this design was used in the fermentation industry to recover yeast. Currently, centrifugation is used to separate such bioproducts as DNA, macromolecules, mammalian cells, and intracellular components. These applications have placed new demands on industrial-scale centrifugation equipment.

9.1 GOVERNING PRINCIPLES

A difference between the density of a molecule or particle and the density of the surrounding fluid can be exploited in a separation process. Solutions in which large density differences exist will often settle under the influence of gravitational acceleration alone. In industry, however, this technique is not feasible because of the inadequate density differences for the solution components, excessive processing times, and space required to house the process equipment (1). The modeling of a gravitational sedimentation process is simple, and it can easily be applied to centrifugation by replacing the gravitational force with a centrifugal force. Thus we will begin our mathematical analysis of centrifugation by deriving equations that describe a particle freely falling in a liquid.

When a particle settles in a condensed medium such as water, the opposing forces of drag and buoyancy exactly counter the gravitational force. Consequently, the particle moves at a constant settling velocity. As no net force is present, the acceleration of the particle is zero. A force balance results in the following expression:

$$F_{gravity} - F_{buoyancy} - F_{drag} = 0 \tag{9.1}$$

The objective of this force analysis is to determine the settling velocity of the particle. To resolve this issue, however, we must further define the buoyant and drag forces. The buoyant force derives solely from the density difference between the particle and the surrounding fluid; the drag force depends on the shape of the object.

164

Based on experimental fluid mechanics and theory, the drag force can be expressed in terms of a drag coefficient, C_D:

$$F_D = C_D A \frac{\rho_p U^2}{2} \tag{9.2}$$

In Equation 9.2, A is the cross-sectional area of the particle, ρ_p is the particle density, and U is the particle velocity. Drag coefficients for spherical particles in various flow regimes can be calculated as follows:

Stokes regime (Re < 0.4): $$C_D = \frac{24}{Re} \tag{9.3}$$

Allen regime (0.4 < Re < 500): $$C_D = \frac{10}{\sqrt{Re}} \tag{9.4}$$

Newton regime ($500 < Re < 2 \times 10^5$): $C_D = 0.44$ (9.5)

Re is the Reynolds number:

$$Re = \frac{\rho_p U d_p}{\mu} \tag{9.6}$$

In Equation 9.6, d_p is the particle diameter and μ is the viscosity of the suspension.

In general, the centrifugation of biological solutes occurs within the Stokes regime. Therefore, the drag force is given by Stokes' law (2):

$$F_{drag} = 3\pi R d \mu U \tag{9.7}$$

Both the gravitational and buoyancy forces may be written in terms of the particle volume and the particle and fluid density:

$$F_{gravity} + F_{buoyancy} = \frac{\pi}{6} d_p^3 g(\rho_p - \rho_f) \tag{9.8}$$

Here ρ_p is the particle density and ρ_f is the fluid density. Substituting the force definitions into Equation 9.8 and solving for the particle velocity provides an equation that gives the terminal velocity v_g for a small spherical particle:

$$v_g = \frac{d_p^2 g(\rho_p - \rho_f)}{18\mu} \tag{9.9}$$

In this equation, the gravitation constant g is equal to $9.8\,m/s^2$. Thus Equation 9.9 demonstrates that gravity is the driving force for a freely settling particle.

In contrast, an applied centrifugal force is used to accelerate the particle in a centrifuge. The magnitude of the centrifugal force is based on the rate of angular rotation in the centrifuge and the radial distance from the center of the centrifuge to the spherical particle. The rotational velocity expressions in the different regimes are as follows:

Stokes regime (Re < 0.4): $$v_\omega = \frac{\omega^2 R d_p^2(\rho_p - \rho_f)}{18\mu} \tag{9.10}$$

Allen regime (0.4 < Re < 500):
$$v_\omega = \left(\frac{4}{225} \frac{\omega^4 R^2 (\rho_p - \rho_f)^2}{\rho_p \mu} \right)^{1/3} d_p \quad (9.11)$$

Newton regime (500 < Re < 2 × 10⁵):
$$v_\omega = \left(3\omega^2 R (\rho_p - \rho_f) \frac{d_p}{\rho_f} \right)^{1/2} \quad (9.12)$$

In these equations, ω is the angular velocity (usually expressed in radians per second), R is the radial distance, and v_ω is the rotational velocity.

Although the centrifugation of proteins and whole cells occurs within the Stokes regime, which has a low Reynolds number, solutions of large crystals can have high Reynolds numbers. In the Stokes regime, centrifugation performance is affected by particle size, shape, and the density difference between the phases. Once the process moves into the Allen or Newton regime, however, performance is affected by additional phenomena, such as high particle velocities, wall effects, and hindered settling because of particle-particle interactions.

9.2 ADVANTAGES AND DISADVANTAGES OF CENTRIFUGATION

Centrifugation processes have short retention times—an important consideration in the processing of some labile and biologically active products. On the other hand, the operation itself generates heat that can damage some products, if they are exceptionally sensitive to thermal conditions. To facilitate the recovery of solid products, the process stream entering a centrifuge is normally pretreated to remove added contaminants, such as filter aids. Even if this additional step is not taken, the centrifugation process can nevertheless separate suspensions that are extremely difficult—if not impossible—to treat with filtration.

In industrial installations, centrifuges offer the advantage of being relatively compact, and the equipment can be operated in a continuous manner. When a process is continuous, lines can potentially become fouled by rapidly settling particles; the manufacturer must take care to avoid this problem. Preparation and start-up times for this process are relatively short, and automation of a centrifugation system can easily be achieved. When used to remove biologically hazardous materials, such as genetically altered microbes, the system can potentially achieve containment of these substances. Aerosol production, however, must be eliminated. In general, operational costs for a centrifuge are relatively low, but the machines themselves are expensive to purchase and maintain.

9.3 SELECTION OF CENTRIFUGES

Centrifuges have been designed and tailored to serve specific applications. Their performance in any given application will take into account the following considerations:

- Solid composition
- Particle size and distribution
- Particle shape, density, and rigidity

- Liquid composition, density, and viscosity
- Product stability

As might be expected, the efficiency of centrifugal separations rises when high density differences exist between the particle and the medium, particles are large, and liquid viscosities are low.

Density is an important parameter for centrifugation because the density of microorganisms is fairly close to that of water, making the separation of microbial suspensions by this method difficult.

Particle size affects how well the process works because the efficiency of throughput is related to the square of the particle diameter.

Although viscosity is rarely problematic, especially given the low viscosity of fermentation broths, viscous suspensions can present processing difficulties. When a centrifuge operates at high rotational speeds with low throughput, it may generate a significant amount of heat. As the temperature of the process stream rises, the viscosity will decrease, yielding a more efficient separation. In some disc-bowl centrifuges, the temperature of the effluent liquid may increase by 3 to 6 °C, while the temperature of the solid discharge can increase by 15 to 20 °C. This heating can harm thermosensitive products. Its effects can be mitigated through the use of double-jacketed covers, which circulate a coolant for heat removal and temperature control. In practice, the operating temperature is dictated by the thermal stability of the product, rather than centrifugation efficiency.

In some cases, disruption of cell debris may cause the viscosity of a suspension to rise, as fine particles are released into the mixture. With nucleic acids, for example, this type of viscosity increase can be mitigated through hydrolysis with deoxyribonucleases.

The designer of a centrifugal separation must consider the capacity required, the degree of separation desired, and the efficiency of the centrifuge. Typical biotechnological separations involve liquid phases with a viscosity of 1 to 2 cP, solid-liquid density differentials of $0.1 \, g/cm^3$ or less, and particle sizes ranging from $100 \, \mu m$ to less than $0.5 \, \mu m$. Limitations related to the release of cell debris or alterations in product thermal stability, as noted earlier, can complicate the process, making it difficult to achieve an adequate separation at sufficient process capacity. Indeed, high supernatant liquid clarity is often demanded in centrifugation processes—especially in protein purification, which includes the use of chromatographic techniques. Many of the newer centrifuge designs can provide increased centrifugal forces in conjunction with tighter temperature control.

The need for sterile, aseptic operation and either containment or exclusion also influences the design of a centrifugal separation. Biotechnology applications often mandate sterile conditions, especially when the products are intended for human use. In such cases, the centrifuge bowl must remain strictly isolated from bearings and other running parts. Hermetic seals or steam sterilization can be used to ensure that sterile conditions are maintained. Hermetic seals reduce protein denaturation and can increase separation efficiency by lowering turbulence in the feed zone. Steam sterilization dictates that the equipment include reinforced centrifuge covers that can withstand steam treatment at high temperatures and pressures. Ordinarily, this type of sterilization takes places when the process is off-line. When a process involves genetically manipulated, pathogenic, or otherwise toxic species, sterilizable and aerosol-preventing centrifuge designs are necessary. Most of the recent developments in this area focus on providing better clean-in-place (CIP) and steam sterilization capabilities, thereby permitting higher levels of containment and aseptic operations.

When considering the choice of a centrifuge, the designer can perform a series of preliminary laboratory-scale tests on the suspension to aid in the selection process. Such tests include gravity sedimentation, bottle centrifugation, and Buchner funnel filtration. In some cases, the designer might alter the suspension's properties prior to centrifugation. For example, particle size can be changed by flocculation, liquid viscosities can be decreased by increasing temperature, and solids concentration can be maximized through sedimentation. In the case of process streams that incorporate microbial cells, cell flocculation can enhance the centrifugation. After conducting these preliminary experiments, the designer can obtain relevant information from centrifuge manufacturers and users about equipment capability and requirements, respectively. The appropriate unit can then be selected based on economic concerns and reliability issues.

Industrial centrifuges are typically less efficient than laboratory centrifuges, as the larger units tend to generate lower centrifugal forces than the smaller systems. Frequently, two or more units of the same or different types may be required to carry out a separation efficiently.

9.4 TYPES OF CENTRIFUGES

A variety of rotor designs have been developed in an effort to avoid the problems of convection and zone detection. Convection problems can become amplified at high centrifugal forces. Likewise, zone detection is usually difficult with these systems; for example, it is problematic in a colloidal solution being centrifuged at a high angular velocity. Table 9.1 lists some common rotor types and their applications.

Several centrifugation techniques exist. Pelleting centrifugation (also known as differential pelleting), for example, takes advantage of variations in particle size to effect the separation. As might be expected, the larger particles settle more rapidly; the smaller particles follow suit as the centrifugal force increases. This procedure is characterized by low yields because the smaller particles tend to remain in solution.

Table 9.1 Types of Rotors and Their Application Range

	Pelleting	Rate-Zonal Sedimentation	Rate-Zonal Flotation	Isopycnic
Fixed-angle rotor	Excellent	Limited	Good	Variable
Near-vertical rotor	—	Poor	Good	Variable
Vertical rotor	—	Good	Good	Excellent
Swinging bucket rotor	Inefficient	Good	Excellent	Good
Zonal rotor	—	Excellent	Excellent	Good

Source: Adapted from Rickwood D, Ford T, Steensgaard J. Centrifugation: essential data. New York: Wiley, 1994:13.

In rate-zonal sedimentation, a density gradient is established in the solution. The sample is added as a narrow zone on top of this density gradient. Under centrifugal force, the particles will move at different rates depending on their mass, producing several "zones" of particles with similar mass.

Isopycnic centrifugation is employed to separate particles based on density. The process begins with a uniform mixture of particles and a density gradient solution, with centrifugal force then redistributing gradient and particles into their isopycnic positions (1).

Centrifuges are usually classified based on their speed and geometry, with common types being the tubular bowl, disc-type, and batch-basket systems.

9.4.1 Tubular bowl centrifuges

The tubular bowl centrifuge is based on a simple geometry [Figure 9.1(a)]. The process stream enters at the bottom of the centrifuge, and high centrifugal forces act to separate out the solids. The bulk of the solids will adhere to the walls of the bowl, while the liquid phase exits at the top of the centrifuge. As this type of system lacks a provision for solids rejection, the centrifuge must be stopped so that the solids can be removed from the walls. Although its simple configuration facilitates the process of disassembling and cleaning the centrifuge, the necessary interruptions to the process must be considered a limitation.

Tubular bowl centrifuges have dewatering capacity, but limited solids capacity. Foaming can be a problem unless the system includes special skimming or centripetal pumps. The main advantage of a tubular bowl centrifuge is its applicability to protein separation, as cooling is easily performed in these units (1, 3).

To mathematically describe the operation of a tubular bowl centrifuge, we will examine the position of a particle in the feed stream. As can be seen in Figure 9.1(a), a particle enters the centrifuge with a velocity in the z-direction. It then

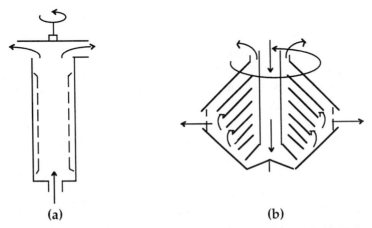

(a) **(b)**

Figure 9.1 Schematic of (a) a tubular bowl centrifuge and (b) a disc-type centrifuge. (Adapted from Belter PA, Cussler EL, Hu W-S. Bioseparations: downstream processing for biotechnology. New York: John Wiley and Sons, 1988:53.)

acquires an additional component of velocity in the r-direction as the centrifugal forces come into play. This derivation, performed by Belter (3), gives the following equation:

$$\frac{dz}{dt} = \frac{Q}{\pi(R_0^2 - R_1^2)}$$ (9.13)

Q is the feed flow rate, R_0 is the bowl radius, and R_1 is the distance between the liquid centerline and the surface of the accumulating solid. Equation 9.13 assumes uniform solids accumulation on the walls of the centrifuge and negligible gravitational effects.

Movement in the r-direction is related to the distance of the particle from the axis of rotation:

$$\frac{dr}{dt} = \frac{d_p^2}{18\mu}(\rho_p - \rho_f)r\omega^2$$ (9.14)

Particles entering the centrifuge at $r = R_1$ do not reach $r = R_0$ until the top of the centrifuge. Thus we can derive Equation 9.15 for the feed flow rate by rearranging our earlier equations:

$$Q = v_g\left(\frac{2\pi l R^2 \omega^2}{g}\right)$$ (9.15)

where v_g is the terminal sedimentation velocity and R is an average radius. As the conditions used to derive Equation 9.15 relate to the last particles removed from the system, this equation will predict the maximum solids removal at given conditions. Note, however, that the fluid in a standard tubular bowl centrifuge is highly turbulent in the feed zone. This high turbulence may disrupt some particles, thereby adversely affecting the separation.

9.4.2 Disc-type centrifuges

Disc-type centrifuges were originally developed for cream separation. Today, they are often used in bioprocessing to separate many biotechnology products including amino acids and single-cell proteins (1).

In this type of centrifuge, a solid bowl holds a stack of conical plates that act to decrease the settling distance while simultaneously increasing the settling area [see Figure 9.1(b)]. The stacking of the plates assists in the rapid sedimentation of the solids, and the stacking angle allows the solid particles to slide smoothly. The feed is introduced at the top of the bowl and then travels through discs. As they traverse through the unit, solids become deposited on the underside of the discs. Any remaining liquid is discharged through an annular slit at the top (1, 3). During rotation of the discs, an optical beam may be transmitted through the unit to detect different component zones.

Disc spacing depends on the particle size, solids concentration, and the properties of the solid sediment (commonly referred to as sludge). The usual ratio of disc spacing to particle size is 4 : 1, with at least 0.5 mm spacing needed to avoid clogging.

With gelatinous particles such as microbial cells, the spacing ratio can be as low as 2:1.

In the tubular bowl centrifuge, the position of a particle with time was a critical factor in describing the system's operation. For a disc-type centrifuge, the objective is the same—but the geometry of the centrifuge is more complex and hence the analysis is more difficult. To describe the position of a particle in a disc-type centrifuge, we will consider a particle at position (x, y), where x is the axis directed into the centrifuge along the edge of the outer disc and y is the axis normal to the outer disc. R_0 is the outer radius of the centrifuge, and R_1 is the inner radius. The disc stacking appears between the inner and outer radii (Figure 9.2).

The particle moves in the x-direction because of fluid flow and sedimentation. This motion can be represented by the following expression (3):

$$\frac{dx}{dt} = v_0 - v_\omega \sin\theta \tag{9.16}$$

In this equation, θ is angle of stacking, v_0 is the flow velocity of the particle, and v_ω is the settling velocity of the particle.

By definition, v_0 is described by the following expression (3):

$$v_0 = \left(\frac{Q}{n(2\pi rl)}\right) f(y) \tag{9.17}$$

Here Q is the total flow, n is the number of discs, r is the distance from the axis of rotation, and l is the distance between discs. $F(y)$ is a function of y representing the variation of v_0 in the y-direction. From Equation 9.17, we understand that at a constant total flow Q, v_0 is a function of the centrifuge radius. It therefore decreases as the fluid flows away from the axis of rotation. In addition, v_0 is a function of y, and is zero at the disc surface. Moreover, because the velocity due to fluid flow is larger than the sedimentation velocity $v_\omega \sin\theta$, we have

$$\frac{dx}{dt} = v_0 = \left(\frac{Q}{n(2\pi rl)}\right) f(y) \tag{9.18}$$

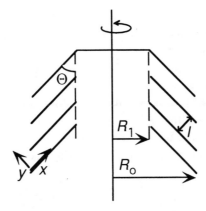

Figure 9.2 Schematic of a disc-type centrifuge, showing the stacking of discs, angle of inclination, inner and outer radii, and axes. (Adapted from Belter PA, Cussler EL, Hu W-S. Bioseparations: downstream processing for biotechnology. New York: John Wiley and Sons, 1988:57.)

The motion of the particle in the y-direction can be represented by the following expression:

$$\frac{dy}{dt} = v_\omega \cos\theta \qquad (9.19)$$

We express v_ω in terms of v_g and the angular velocity ω:

$$\frac{dy}{dt} = v_g \left(\frac{\omega^2 r}{g}\right) \cos\theta \qquad (9.20)$$

Using the chain rule, we develop an expression that describes the trajectory of the particle between the discs of the centrifuge:

$$\frac{dy}{dt}\frac{dt}{dx} = \frac{dy}{dx} = \left(\frac{2\pi n l v_g \omega^2}{Q g f(y)}\right) r^2 \cos\theta \qquad (9.21)$$

Note that in Equation 9.21, r can be replaced by $R_0 - x\sin\theta$. To ensure that all desired particles are removed, it is useful to focus on the particles that enter at the outer edge of the discs ($x = 0$, $y = 0$) and are removed at the inner edge of the discs ($x = (R_0 - R_1)/\sin\theta$, $y = 1$). Integrating the expression describing the trajectory, we can find an expression for total flow:

$$Q = v_g \left(\frac{2\pi n\omega^2}{3g}(R_0^3 - R_1^3)\cot\theta\right) \qquad (9.22)$$

To include the effect of disc spacing—an important consideration in disc centrifuges for which Equation 9.22 does not account—and to describe the effect of the shear and Coriolis forces on the velocity profile, we use a dimensionless hydrodynamic parameter λ:

$$\lambda = l\left(\frac{\omega\sin\theta}{\eta}\right)^{0.5} \qquad (9.23)$$

In this equation, η is the kinematic viscosity. For industrial disc centrifuges, λ is typically between 6 and 15, with average radial velocities varying from 0.05 to 0.3 m/s.

Vortices tend to arise in disc-type centrifuges because of the circumferential velocities that they generate, which creates turbulence and entrains particles in the supernatant. To suppress this circumferential velocity component, spacers may be included to hold the discs farther apart.

The two most widely used disc-type centrifuges are the nozzle and solid-ejecting centrifuges, both of which possess clean-in-place (CIP) capability. Depending on the type of separation undertaken, aqueous solutions of detergent, alkali, acid, or protease are passed through the system during shutdown. The cleaning solution can be fed through the nozzle in a nozzle centrifuge or processed through a sequence of solid-ejecting cycles in a solid-ejecting disc centrifuge.

9.4.3 Batch-basket centrifuges

Unlike the other centrifuges listed in Table 9.2, batch-basket centrifuges combine centrifugation with filtration. This type of centrifuge consists of a bowl with a

Table 9.2 Sedimenting Centrifuges

Centrifuge Type	Centrifugal Force (g)	Type of Operation	Particle Size (μm)	Applications
Tubular	20,000	Batch	0.1–100	Pilot-scale separation of animal, plant, and microbial cells
Single bowl	500–1500	Batch	0.5–20,000	Vaccine production Removal of *P. aureus* cells from proteins
Multiple bowl	5000–9000	Batch	0.5–20,000	Separation of human blood plasma components
Basic disc-type	5000–11,000	Batch	0.5–500	Production of extracellular glucose isomerase from *b. Coagulans*, soybean isolate
Nozzle disc-type	6000–9000	Continuous	0.5–500	Production of *S. cerevisia* for baking and brewing Cell recovery in tetracycline production
Ejector disc-type	5500–7500	Intermittent	0.5–500	Recombinant dna and Steroid production
Perforated bowl-basket	400–1300	Batch	10–1000	Adsorbents such as cellulose and agarose

Source: Adapted from Atkins B, Mavituna F. Biochemical engineering and biotechnology handbook, 2nd ed. New York: Stockton, 1991.

perforated side-sheet. A filter screen, usually consisting of a metal grid or woven cloth, is placed over this side-sheet to retain solids but allow liquids to pass through. In a batch-basket centrifuge, the suspension enters along the axis of the bowl. The solid is deposited on the walls of the basket, which is encompassed in a case that directs the flow of the isolated liquid to a tank for disposal or recycling (1, 3).

9.5 INDUSTRIAL-SCALE CENTRIFUGATION

The scale-up of centrifugation to an industrial scale depends on the availability of reliable laboratory data and the construction of an accurate mathematical model for the process taking place in the equipment. Two approaches are used to estimate scale-up (3): a quantitative method that uses a Σ factor, and a qualitative approach that uses an equivalent time concept, Gt.

In the Σ method, we must find an expression that defines the total feed flow Q for the different centrifuges. Generally, for sedimentation centrifuges, the volumetric flow rate is expressed by the following equation,

$$Q = \frac{\dfrac{d_p^2 \Delta \rho g}{18\mu}}{\dfrac{\omega^2 r V}{sg}} \tag{9.24}$$

or

$$Q = v_0 \Sigma \tag{9.25}$$

where d_p is the particle diameter, $\Delta\rho$ is the density difference between the particle and the liquid, η is the dynamic viscosity, ω is the angular velocity, r is the effective radius of rotation in the centrifuge, V is the liquid volume in the centrifuge, s is the effective distance of particle settling, g is the gravitational acceleration (expressed in CGS units), V_0 is the terminal particle velocity under gravitation, and Σ, the sigma factor, is equivalent to the surface area needed in gravitational settling to obtain the same separation as the centrifuge.

Equations 9.24 and 9.25 describe a spherical, rigid particle that is isolated from the bulk of the suspension. For other particles existing under more crowded conditions, we may adjust the expression for Q by defining v_h, the hindered settling velocity:

$$v_h = \frac{v_0}{(1 + A\varepsilon_p)} \tag{9.26}$$

In Equation 9.26, A is a constant and ε_p is the volume fraction of the particles in the suspension. The values for A and ε_p depend on the properties of the suspension (Table 9.3).

The effect of hindered settling has also been correlated with the following expression,

$$v_h = v_0 (1 - \varepsilon_p)^a \tag{9.27}$$

where a is a geometric constant. For a rigid spherical particle, a equals approximately 4.6, but it can range from 10 to 100 for nonrigid, nonspherical particles. The effect of hindered settling must always be considered in biotechnological applications, as microorganisms are easily compressible and nonspherical.

The quantitative approach for scaling up centrifuges employs the Σ factor. This factor helps determine which centrifuge is most applicable to a specific process. A higher Σ factor implies a higher centrifugal force, leading to a more efficient separation process. The Σ factor for the tubular bowl and disc-type centrifuge geometries was previously derived in the expression describing Q, and is given by:

$$Q = v_g \Sigma \tag{9.28}$$

Therefore we have the following expressions (3):

Tubular bowl: $\Sigma = \left(\dfrac{2\pi l R^2 \omega^2}{g} \right)$ \hfill (9.29)

Table 9.3 Values of A and ε_p for Varying Suspensions

Suspension	A	ε_p
Dilute	1–2	$\varepsilon_p < 0.15$
Spherical particles	$1 + (2.2)(\varepsilon_p)^{3.43}$	$0.2 < \varepsilon_p < 0.5$
Irregular particles	$1 + (3.05)(\varepsilon_p)^{2.84}$	$0.15 < \varepsilon_p < 0.5$

Disc bowl: $$\Sigma = \left(\frac{2\pi n\omega^2}{3g} (R_0^3 - R_1^3) \cot\theta \right) \tag{9.30}$$

Note that v_g is a function of the particles and essentially remains independent of the centrifuge. On the other hand, the Σ factor, which has the dimensions of length squared for both tubular and disc-type centrifuges, is not a function of the particle. Rather, it is a function of only the particular centrifuge geometry. Therefore, for the same suspension, the scale-up of liquid flow through the centrifuge is directly proportional to the Σ factor:

$$Q_2 = Q_1 \left(\frac{\Sigma_2}{\Sigma_1} \right) \tag{9.31}$$

In the qualitative method, we assess the difficulty of the separation by calculating Gt, the product of the centrifugal force G and the centrifugation time t. This value is given by the following equation (3):

$$Gt = \frac{\omega^2 R_0 t}{g} \tag{9.32}$$

In this equation, R_0 is the characteristic radius. Table 9.4 presents Gt values for a variety of biomass centrifugations.

A useful bench-scale laboratory technique for determining Gt involves the application of a Gyro tester. The Gyro tester is a bench-scale centrifuge, equipped with three interchangeable bowls: a tube holding unit, a disc bowl for liquid-liquid separation, and a disc bowl with a nozzle discharge for solid-liquid suspension separation (3). To measure Gt, the tubes of the tube holding unit are filled with 10 mL of suspension and the centrifugation time is varied between 5 and 30 seconds. The degree of settling is observed for each time. Gt is then easily calculated, as the centrifugal force for given speeds is known. This approach should be used with extreme caution, however, as it gives only a crude approximation.

In short, for selection purposes, we should choose a centrifuge that has the Σ value needed to satisfy the process requirements of v_g and Q. The Gyro tester must be used to find reliable values of v_g. And, of course, experience and judgment must be applied along with the two previously described approaches. Judgment comes from understanding the properties of the suspension and from available general performance data, such as that found in Table 9.5.

Table 9.4 Centrifugal Conditions

Solids	Centrifugation Conditions Gt (10^6 s)
Eukaryotic cells	0.3
Eukaryotic cell debris	2
Protein precipitates	9
Bacteria	18
Bacterial cell debris	54
Ribosomes	1100

Source: Adapted from Belter PA, Cussler EL, Hu W-S. Bioseparations: downstream processing for biotechnology. New York: John Wiley and Sons, 1988:63.

Table 9.5 Performance Data for Selected Biomass Centrifugations

Product	Microorganism	Size (µm)	Relative Throughput	Type of Separator
Baker's yeast	*Saccharomyces*	7–10	100	Nozzle
Brewer's yeast	*Saccharomyces*	5–8	70	Nozzle
Vaccines	*Clostridia*	1–3	5	Solid bowl
Enzymes	*Bacillus*	1–3	7	Nozzle
Single-cell Protein	*Candida*	4–7	50	Nozzle

Source: Adapted from Belter PA, Cussler EL, Hu W-S. Bioseparations: downstream processing for biotechnology. New York: John Wiley and Sons, 1988:65.

9.6 SUMMARY

Centrifugation, a very powerful method for removing insoluble matter from process streams, is commonly the first step in bioseparations. It is the only static field method, other than electrophoresis, that is commonly used for multicomponent separations.

The efficiency or throughput achievable with centrifugation improves with higher density differences between the particle and the medium, larger particle sizes, and lower liquid viscosities. The influence of the particle size is particularly important because efficiency (or throughput) is related to the square of the particle diameter. Centrifugal separation is an attractive method for solid-liquid as well as liquid-liquid separations as continuous processing is feasible, retention times can be short, and no filter aids are required.

9.7 PROBLEMS

9.1 In the recovery of yeast cells from a fermentation broth, the cells are assumed to be spherical with a density of $1.03 \, \text{g/cm}^3$ and a diameter of 50 mm. The broth has a density of $1.01 \, \text{g/cm}^3$ and a viscosity of 1.1 cp.

(a) Calculate the free settling terminal velocity of the yeast cells.

(b) Calculate the centrifugal velocity when the centrifuge is operated at 500 rpm, with a 3 cm radial distance from the center of the centrifuge to the particle.

(c) How does the terminal velocity in parts (a) and (b) change if we double the cell diameter? Decrease the viscosity by one-half?

9.2 A test-tube centrifuge is used to recover particles from a slurry. In this piece of equipment, the test tubes rotate perpendicularly to the axis of rotation. The tubes are 8 cm long, and during centrifugation the distance between the bottom of the test tube and the axis of rotation is 20 cm. If the centrifuge is to be operated at 700 rpm, how long will it take to completely recover spherical particles of 0.5 mm diameter and $1.06 \, \text{g/cm}^3$ density? Assume that the slurry density and viscosity are close to those of water.

9.3 A tubular bowl centrifuge is used to concentrate a cell suspension prior to disruption. The cells are spherical with a $2 \, \mu\text{m}$ diameter and a density of $0.8 \, \text{g/cm}^3$. The centrifuge has a 14 cm diameter, is 30 cm long, and is operated at 5000 rpm.

(a) What is the maximum feed flow into the centrifuge?

(b) After disruption, the cell diameter is reduced by one-third and the viscosity is increased by a factor of 4. Calculate the new feed flow, assuming that the centrifuge operates at the same speed.

9.4 A tubular bowl centrifuge is used to clarify a liquid containing cells of 20 μm diameter and 1.01 g/cm³ density. The feed density and viscosity are close to those of water. The tubular centrifuge is 38 cm high and has a 20 cm diameter. When operated at 5000 rpm, it removes nearly all of the cells having a diameter greater than 20 μm. To further clarify the suspension by removing cells of 8 μm diameter, the discharge from the tubular bowl centrifuge is fed to a disc-type centrifuge. The centrifuge has 50 discs at an angle of 45°, an outer radius of 18 cm, and an inner radius of 8 cm, and it operates at 15,000 rpm.

(a) Calculate the maximum feed flow into the tubular bowl centrifuge when the distance to the liquid interface is 8 cm.

(b) If the maximum flow into the disc-type centrifuge is 37 L/s, what is the minimum diameter of the cells that can be recovered?

9.5 A tubular bowl centrifuge is being used to concentrate a suspension of *E. coli* at a feed flow rate of 30 L/s. The bowl of the unit has an inside diameter of 13 cm, is 75 cm long, and operates at 16,000 rpm. If a disc-type centrifuge with 32 discs at an angle of 45°, an outer radius of 12 cm, and an inner radius of 6 cm is to be operated at the same rotational speed as the tubular bowl centrifuge, estimate the Σ factors for both centrifuges.

(a) Which centrifuge should be used for scale-up purposes?

(b) If we double the Σ factor of the centrifuge of choice, how does it influence the feed flow rate?

9.8 REFERENCES

1.   Rickwood D, ed. Centrifugation: A practical approach, 2nd ed. Oxford: IRL Press, 1984.
2.   Blanch HW, Clark DS. Biochemical engineering. New York: Marcel Dekker, 1996.
3.   Belter PA, Cussler EL, Hu W-S. Bioseparations: downstream processing for biotechnology. New York: John Wiley and Sons, 1988.

10

Chromatography

Chromatography is a useful and popular technique that can provide a high degree of resolution for a wide range of bioseparations. A chromatographic system includes four basic elements: a stationary phase, a mobile phase, a pumping system, and an on-line detector. Using these core elements, many different physical, chemical, and biological interactions can be exploited to separate biochemicals and biopolymers. Before getting to the specifics of this operation, it is useful to focus on the general features of a chromatographic separation process.

Figure 10.1 shows a simple schematic of a chromatography system. The solution known as the mobile phase is pumped through a column containing the stationary phase. The stationary phase typically consists of a spherical insoluble polymer, ranging in size from 5 to 300 µm. While the mobile phase flows at a constant rate through the column packed with the stationary phase, the sample is introduced at a specified time. At the end of the column, a detector tracks the concentrations of the now-separated components of the sample. After detection, the column effluent can be collected as separate fractions for further analysis or processing.

Many variations on this basic chromatographic method exist. In this chapter, we will focus our attention on well-established methods and useful mathematical analyses that can guide chromatographic method operation, selection, and design. The latter part of this chapter will discuss useful variations on the standard chromatographic method (some of which have been developed relatively recently) and strategies for scaling up a chromatographic method to an industrial scale.

Our goal in this chapter is to quickly and efficiently introduce mathematical analyses that can facilitate the design of bioseparation processes. More rigorous and detailed mathematical analyses are available in the technical literature. Because such detailed analyses would detract from our focus here and do not significantly benefit the practicing scientist and engineer, we have deferred the discussion of these mathematical models to the references provided at the end of this chapter.

10.1 DETECTION METHODS

The detector, though it is placed at the end of the column, can be considered the beginning of the chromatographic process. This approach is only logical, because if we do not know what the chromatographic separation process accomplished, then in effect no separation has been accomplished. Detection has four basic objectives: sensitivity, linearity, reproducibility, and ease of use in a continuous system. The chromatographer must also know the limitations of the selected detection system so as to draw appropriate conclusions from experimental and process data.

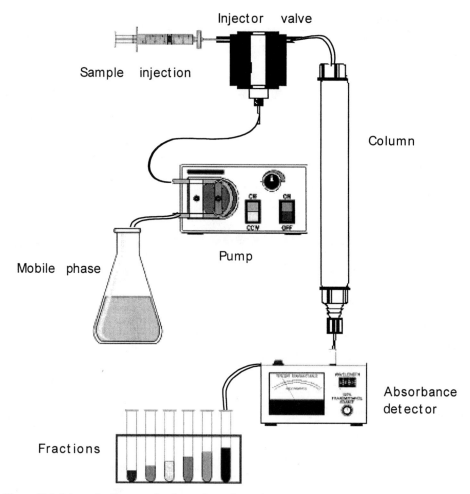

Figure 10.1 Schematic diagram of a chromatography system.

The preferred detection method will function in a continuous, or on-line, fashion. Table 10.1 lists some common on-line detection methods, including comments on how they satisfy the four previously mentioned criteria. By far, the most popular method is light absorbance—using this technique, a wide range of components can be tracked. The system is also easy to use, especially in protein separation, because proteins absorb light at wavelengths that buffer components, such as salts, do not.

With the advent of high-speed, economical microprocessors, detectors can now scan a wide range of wavelengths quickly. Such detectors, known as variable diode arrays, ensure that the user obtains a set of spectral "fingerprints" for the column effluent—a useful characterization tool when complex mixtures must be separated. Moreover, tracking multiple wavelengths gives the chromatographer greater flexibility in choosing wavelengths for different purposes; for example, the wavelength might be selected so as to improve the process' sensitivity or track impurities.

Using combinations of detection methods can be important or even critical to the chromatographic separation. For example, on-line detection is often complemented with off-line methods, such as electrophoresis, affinity chromatography,

Table 10.1 A Description of Selected Liquid Chromatography Detectors

	Principle	Sensitivity	Linearity	Stability	Ease of Use
Absorbance	Differences in light absorbance	Varies, 1 µg/mL	Beer's law for low concentrations	Stable	Inexpensive single wavelength, flexible photodiode array
Chemiluminescence	Light-emitting label	Pico- or femtomole range	Linear with concentration	Indirect techniques may be noisy	Lower detection limit than fluorescence
Electrochemical	Oxidation or reduction of eluted species	Nanogram range for phenols, organic acids	Linear with concentration	No oxygen, metals, or halides	Mobile phase must be conductive
Electrolytic conductivity	Ionic species-influenced changes in conductivity		Linear with concentration	Solvent gradients cause a baseline shift	Cell stability limits use in ion exchange
Fluorescence	Light-excitable functional groups	1000 times more sensitive than ultraviolet light	Emission linear with concentration	Specific, selective optical detectors	Requires natural fluorescence or derivatization
Fourier transform infrared (FTIR)	Nebulizer removes water, solutes deposited on ZnSe	On-line methods not as sensitive as off-line FTIR	Linear with absorbance	Repro-ducibility within 10%	Removal of water can allow for FTIR spectroscopy
Refractive index	Light ray changes direction based on ρ	Depends on refractive indexes	Linear at low concentrations	Rezeroing required	Cannot be used with gradient
Mass spectrometry: fast atom bombardment	Mobile phase in glycerol, bombarded and detected with mass spectrometry	2 ng/mL	Linear with abundance of specifc ions	Good for polar, labile solutes	Recent technique
Mass spectrometry: electrospray	Ionization through solvent evaporation	10 pg	Linear with abundance of specifc ions	Peptides, proteins, oligonucle-otides	May be used as a sequencing tool
Mass spectrometry: particle beam	Mobile phase aerosol hit with particle beam	0.1–1 ppb	Linear calibration with abundance verus concentration	Sophist-icated equipment	New developments have improved technique
Evaporative light scattering	Intensity of desolvated spray	100 ng/mL	Nonlinear calibration	Sensitivity depends on refractive index	Promising method

or a bioassay, so as to positively identify which biomolecule is exiting the column at a specific time. For example, affinity chromatography with protein A is used to pinpoint the time at which antibodies exit the column during the chromatographic procedure.

The second step in performing a chromatographic separation involves the choice of a stationary phase. To select the appropriate material, we must know what kind of physical, chemical, or biological interactions can be exploited to effect the separation. Use of a specific interaction can often influence the choice of the stationary phase.

10.2 SUMMARY OF THE TYPES OF CHROMATOGRAPHY

Table 10.2 illustrates some popular chromatographic techniques. These methods all deal with design of the stationary phase (and the complementary mobile phase), which exploits a property of the solute to effect a separation. These chromatographic techniques are briefly described in this section.

Table 10.2 Popular Chromatographic Methods for Bioseparations

Chromatographic Method	Types of Solutes Most Commonly Separated	Separation Principle	Possible Advantages	Possible Limitations
Adsorption (hydroxyapatite)	DNA, RNA, proteins	Nonspecific chemical interactions	Separation of double- from single-stranded DNA	Not specific for many protein mixtures
Affinity	Proteins, peptides	Complex, specific interactions	Wide range of specific ligand interactions	Expensive stationary phases
Gel permeation (GPC)	Removing buffer ions, proteins	Size	Simple, noninteracting mechanism	Soft stationary phases limit flow rates
Hydrophobic interaction (HIC)	Vitamins, drugs, proteins	Hydrophobicity	Wide pH range, high recovery	Need to increase ionic strength
Ion exchange (IEC)	Proteins, organic ions	Electrostatic interactions	Sensitive to even small changes in protein sequence, can alter surface charge	Limited pH range, need for high salt or displacer concentrations for full recovery
Ion exclusion	Carbohydrates, organic acids	Donnan effect: like charges repel one another	High recovery of carbohydrates	Not useful for protein separations
Perfusion	Proteins	Pores large enough to allow flow, other interactions possible	Extremely fast volumetric throughput useful for large-scale separations	Does not allow for separation by size
Reverse phase (RPC)	Vitamins, drugs, peptides, proteins	Hydrophobic interactions	Works well with low water solubility or aqueous labile solutes	Can denature solutes

Affinity chromatography is a broad classification of techniques whereby complex, specific interactions arise between the solute and the stationary phase because an affinity ligand is present in the mixture. Affinity ligands range from monoclonal antibodies, which can have a high and very specific binding capacity to a particular solute, to chemical dyes or metal ions, which interact most strongly with one or more types of amino acids on proteins. In the basic construct of an affinity media, an affinity ligand is covalently attached to a spacer arm, which is in turn immobilized on the stationary phase. During the chromatographic process, this ligand becomes preferentially separated from the bulk solution.

Gel permeation chromatography (GPC), also known as size-exclusion chromatography (SEC), controls the pore size of the stationary phase so as to separate the solution's components. Smaller solutes can freely pass in and out of its pores; larger solutes, on the other hand, have more restricted pore movement—some are even totally excluded from the fluid in its pores. Thus the molecular weight cut-off point, which provides a general guide as to the largest solute that can enter the pores, is an important design variable. Smaller solutes spend a longer time in the column because they can freely enter the pore volume, while larger solutes pass quickly through the column. Solutions may also be desalted during the GPC process, as stationary phases with low molecular-weight cut-offs will quickly separate biopolymers from any buffer salts present in the sample.

Ion-exchange chromatography (IEC) and a related technique called ion-exclusion chromatography—two techniques that are based on charge-charge interactions—are useful for separating ionogenic biopolymers and small ions. Proteins have a complex array of charges on their outer surface because of their constituent amino acids' Zwitterionic character. Nevertheless, proteins that have a preponderance of one charge on a section of their outer surfaces will show either a net attraction or repulsion to the fixed charges on the stationary phase. This attraction or repulsion, respectively, depends on whether the charge on one section of the protein is the opposite of or the same as the stationary phase charge. The ionic strength of the mobile phase plays a major role in the separation of both small molecules and biopolymers.

In ion-exclusion chromatography, the fixed charge on the surface of the pores hinders similarly charged ions from passing through the pores. Thus non-ionic solutes can enter freely, weakly ionogenic solutes such as carboxylic acids are partially excluded from the pores, and strong ions are totally excluded. The order of elution is salts, followed by weak acids or bases, and finally uncharged molecules.

Hydrophobic interaction chromatography (HIC) and reverse-phase chromatography (RPC) columns take advantage of the molecular composition and structure of the mobile phase, along with the surface chemistry of the stationary phase, to retain solutes based on their hydrophobicity (or, conversely, their hydrophilicity). The unique capability of water to form relatively ordered hydrogen-bonded structures in solution has important implications for biopolymer structure and solubility.

Perfusion chromatography is a relatively new concept, whose strength derives from its ability to process samples quickly. In this technique, the mobile phase actually flows through the large pores in the stationary phase. All of the stationary-phase

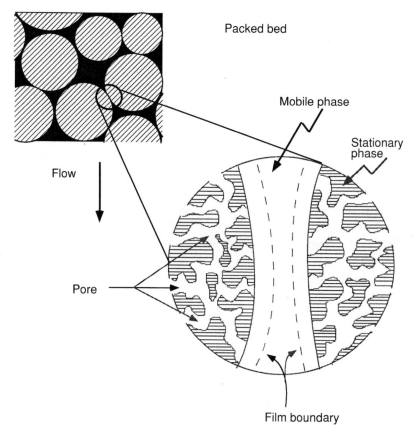

Packed bed

Mobile phase

Stationary
phase

Flow

Pore

Film boundary

Figure 10.2 Perfusion chromatography.

chemistries available for standard chromatography can be applied to perfusion phases as well, with the natural exception of GPC. Figure 10.2 shows the difference in mobile-phase flow in standard phases (where movement through pores is based on diffusion) versus the flow with perfusion stationary phases.

Hydroxyapatite chromatography capitalizes on the important biological role that bone mineral matter plays in higher organisms. Nucleic acids interact with calcium or phosphate ions, while proteins interact through specific groups as well as with the charged surface resulting in less specific interactions. Double-stranded DNA binds more strongly; it can be separated from single-stranded DNA because of the higher surface exposure of the sugar backbone per linear length of DNA.

10.3 STATIONARY PHASES

The heart of a chromatographic separation process is the stationary phase. There are myriad polymer matrices and modification chemistries from which to choose (with more being developed every day). The separation scientist or engineer welcomes this diversity, as each new stationary phase represents a potentially useful tool for a specific separation need.

Most stationary phases include two major components: 1) the composition of the polymer matrix, which serves as the structural component, dictating the stationary phase's size and porosity, and 2) the chemical or biopolymer functional component, which confers a specific ability for interactions with the particular solute. The best design matches these properties to the other needs for the separation, such as sample size, cost considerations, and processing speed.

Table 10.3 lists polymer matrices that are often used in the chromatography of biological molecules, along with some of their applications and specific examples of separations performed with these stationary phases. The major classification scheme for stationary phases is matrices consisting of naturally derived polymers versus wholly synthetic matrices. Of paramount importance in choosing suitable polymers, both synthetic and naturally derived, is "biocompatibility"—that is, whether the materials fail to exhibit detectable, irreversible reactivity with biological molecules. Generally, any nonspecific interactions (interactions that occur with a wide range of biomolecules) are considered undesirable, and stationary phases are synthesized and tested with this design parameter in mind.

Stationary-phase porosity is controlled in different ways depending on the polymerization method (Figure 10.3). Gel structures, where "gel" denotes a random structure that differs based on the type of chromatography employed, are generated by covalently linking polymer strands so as to form a spherical mass containing a labyrinthine network of randomly shaped openings. The pore size and

Table 10.3 Selected Commercially Available Polymer Matrices

Stationary-Phase Matrix	Principal Use	Possible Advantages	Possible Limitations
Silica	HIC, RPC	Withstands high hydrostatic pressures	High density
Agarose	Affinity	High biocompatibility	Limited to low flow rates, temperature labile
Polystyrene-divinylbenzene	IEC	Very high surface areas, stable in organic solvents	Biocompatibility limitations
Polyacrylamide	GPC	Inexpensive, versatile	Limited to low flow rates
Dextran	GPC	Stronger than agarose due to cross-linking	Unstable in organic solvents
Hydroxyapatite	Adsorption chromatography	Withstands high hydrostatic pressures	Limited to smaller particle sizes for high performance
Cross-linked polysaccharides, cellulose	HIC, affinity chromatography	High biocompatibility	Low flow rates
Agarose/dextran, dextran/ acrylamide	GPC	Stronger particles, high flow rates	Unwanted adsorption of proteins
Polymethacrylate	IEC, affinity chromatography	Easily modified	Generally limited to low flow rates

distribution can be controlled by manipulating the degree of cross-linking of individual polymer strands and by using strands of different lengths. Macroreticular phases can be generated through careful manipulation of the polymer's solubility as the sphere is created during polymerization. A macroreticular stationary phase comprises a network of tiny, covalently linked spheres, typically 50 to 100nm in diameter. Pore size and distribution are also controlled by manipulating the polymer's solubility during polymerization, in an effort to dictate the size of the spheres.

Chemical functional groups, such as ionogenic groups, can be carried by one of the polymers used to form the stationary phase; alternatively, they can be created by modifying the stationary phase after polymerization is complete. Table 10.4 summarizes the types of chemical groups that are commercially available today. Some of these groups are used in affinity chromatography as group-specific ligands; others are employed in ion-exchange or hydrophobic interaction chromatography.

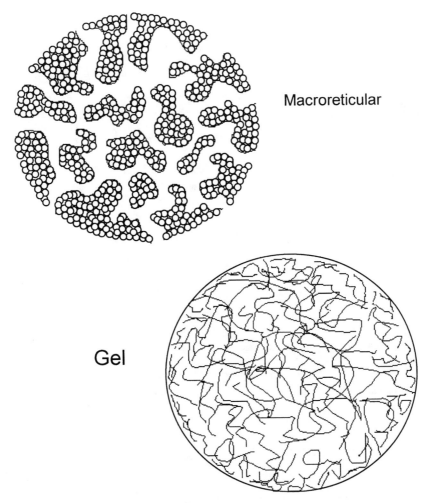

Macroreticular

Gel

Figure 10.3 Types of stationary-phase porosity.

Table 10.4 Popular Chemical Functional Groups on Stationary Phases

Functional Group	Principal Use	Chemical Structure, Ligand, or Binding Site Schematic
Avidin, streptavidin	Purification of biotin-labeled biomolecules	Biotin
Cebacron blue	A group-specific dye ligand— mimics biological substances	
Glutathione	Purification of detoxification enzymes	
Auaternary amines	Anion exchange	
Sulfonic acid	Cation exchange	
Protein A or G	Purification of IgG	Protein A IgG
Heparin	Separation of lipoproteins	
Lysine	Polysaccharides	
Phenyl	Hydrophobic interaction chromatography	
Carboxymethyl	Cation exchange	
Imminodiacetate -Cu(II)	Immobilized metal affinity chromatography	
5'-Adenosine monophosphate	Purification of NAD and ATP enzymes	

10.4 SIX WAYS TO ANALYZE CHROMATOGRAPHIC PROCESSES

Chromatographic processes are critically important in both analytical laboratories and commercial-scale operations. Hence many mathematical treatments have been developed to aid in method design and to analyze results. This section introduces six of the most frequently used mathematical approaches. The scientific and engineering literature are replete with many variations on these equations, which were developed to increase the utility and precision of the basic equations provided here. The equations given here will suffice for most purposes, however, and the reader can review the references at the end of this chapter to obtain further information.

The six useful mathematical treatments are as follows:

- Gaussian solution
- Van Deemter equation
- Newtonian continuum mechanics and linear equilibria
- Constant pattern and saturation equilibria
- Staged model
- Gel partition model

These analyses are not equivalent in terms of their scope or the results that they provide. Instead, they are generally complementary mathematical developments whose use depends on the available experimental information and the user's preference. The gel partition model is an exception because of its specific application—in this technique, the stationary phase is employed as a separate phase to hold back solutes that can permeate the gel through its pores without the complementary use of any chemical or biological interactions.

It is always important to determine what is needed to accomplish the separation task and to set out in a focused way to reach a goal. The reader is encouraged to critically review the equations and strategies provided here, then apply this knowledge to make appropriate choices of experiments and mathematical tools in a particular situation.

10.4.1 Gaussian solution

In a straightforward, phenomenological analysis of chromatography data, one can fit the data to an equation and use the resulting curve to extract as much information as possible. Giddings (1) suggested using a form of the Gaussian (or normal) probability distribution equation to describe chromatography peaks. A Gaussian distribution is appropriate because, in a properly designed chromatographic system, the distribution of molecules' residence times in the column most likely reflects differences in the paths that the molecules take as they travel through the column.

Figure 10.3 illustrates a simple example in which a small population of molecules travels through and around a group of stationary phase particles in GPC. The equation used to describe chromatography peaks, such as those shown in Figure 10.4, differs from a Gaussian distribution in that we have replaced the term $\dfrac{1}{\sqrt{2\pi}\sigma}$ with 1 and used S_{max}, the maximum signal provided by the detector as that peak leaves the column:

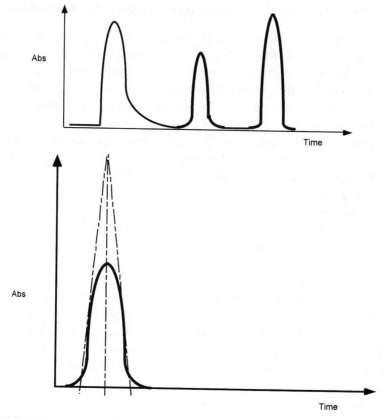

Figure 10.4 Typical chromatography peaks and Gaussian curve fits.

$$S = S_{max} \exp\left(-\frac{\left(\frac{t}{t_{max}} - 1\right)^2}{2\sigma^2} \right) \tag{10.1}$$

or

$$S = S_{max} \exp\left(-\frac{\left(\frac{V}{V_{max}} - 1\right)^2}{2\sigma^2} \right) \tag{10.2}$$

Alternatively, rather than use time as the independent variable, we can track the total volume collected at the column exit (as given in Equation 10.2).

A very slight error occurs when we modify the Gaussian distribution equation. This error, however, is usually much less than the experimental error associated with detection.

Using calculus principles for determining maxima, minima, and points of inflection, we can define some useful terms in this equation. By inspection, or by taking the first and second derivatives of Equations 10.1 and 10.2, we can show that the parameters t_{max} and V_{max} are the time and volume at the maximum detection signal, $S = S_{max}$ (i.e., maximum concentration). Setting the second derivative of Equation 10.1 equal to zero, the width of the peak at the inflection point (see Figure 10.4), occurs at $t/t_{max} = 2(\sigma + 1)$ and is equal to 2σ. By fitting the points of the peak data, we can obtain the standard deviation σ. Linearizing Equation 10.1 or 10.2 gives

$$\ln\left(\frac{S}{S_{max}}\right) = \left(\frac{1}{2\sigma^2}\right)\left(\frac{t}{t_{max}} - 1\right)^2 \tag{10.3}$$

Plotting $\ln(S/S_{max})$ versus $(t/t_{max} - 1)^2$ and drawing a line through the origin gives a data fit where the slope $(-1/2\sigma^2)$ provides the standard deviation.

Figure 10.4 illustrates the importance of graphically verifying that the computer curve fits are accurate. At times, we must apply our judgment to estimate the parameters, as deviations from a normal distribution are possible because of operational problems.

Armed with the Gaussian solution, we can determine the amount of solute that exited the column in a peak by calculating the area under the peak. The amount of solute exiting the column at some time t' is calculated as follows:

$$\text{Amount} = \int_0^{t'} QS_{max} \exp\left(-\frac{\left(\frac{t}{t_{max}} - 1\right)^2}{2\sigma^2}\right) dt \tag{10.4}$$

This amount is simply the integral of S times Q (the flow rate) from $t = 0$ to some time t'. If, as is always the case in chromatography, the flow rate is not a function of time, then the integration becomes

$$\text{Amount} = QS_{max} \int_0^{t'} \exp\left(-\frac{\left(\frac{t}{t_{max}} - 1\right)^2}{2\sigma^2}\right) dt \tag{10.5}$$

The exact solution is provided by the following series:

$$\text{Amount} = \sqrt{2}\sigma t_{max} QS_{max} \sum_0^{\infty} (-1)^{k+1} \frac{x^{2k-1}}{(2k-1)(k-1)!} \tag{10.6}$$

In this series,

$$x = \frac{1}{\sqrt{2}\sigma}\left(\frac{t}{t_{max}} - 1\right) \tag{10.7}$$

Other references provide the series solution in terms of the error function $\left[\text{erf}(x) = \frac{\sqrt{2}}{\pi}\int_0^x \exp(-u^2)du\right]$, but tables are needed to use this solution. With an electronic spreadsheet, the series solution can be calculated easily to 100 or more terms. This solution provides a highly accurate estimate of the answer in practically all cases.

An even simpler answer appears when the peaks do not overlap. In this case, the integral can be extended to infinity, giving the following result:

$$\text{Amount} = \sqrt{\pi}\sigma^2 t_{max} Q S_{max} \tag{10.8}$$

As long as we integrate the entire peak (essentially integrating past the time of the peak maximum by four standard deviations), the error is negligible. Thus most of the work in determining the peak area of nonoverlapping peaks involves fitting the data to Equation 10.1 or 10.2.

Before leaving this discussion on the Gaussian shape of peaks, we should define several other terms that are helpful in designing chromatographic methods. First, for a Gaussian peak, the width of the peak at the baseline is a more useful value than the width at the inflection point. This width w_b turns out to be equal to $2w_i$ and thus

$$w_b = 2w_i = 4t_{max}\sigma \tag{10.9}$$

Another important term to define is R (resolution) for two peaks:

$$R = \frac{t_{max,2} - t_{max,1}}{\dfrac{(w_{b,2} + w_{b,1})}{2}} = \frac{t_{max,2} - t_{max,1}}{2(\sigma_2 + \sigma_1)} \tag{10.10}$$

In this equation, the subscripts 1 and 2 refer to peaks 1 and 2, respectively. To simplify the discussion, we assume that the peak widths are essentially equal, as peak shape is primarily influenced by column properties and conditions, not by the solute's inherent properties. Thus R can be defined as follows:

$$R = \frac{t_{max,2} - t_{max,1}}{4\sigma} \tag{10.11}$$

When $R = 1$, the two peaks have an overlap of only about 2%. When R exceeds 1.5, peaks are said to have "baseline separation" and do not contain appreciable amounts of the other solute.

The Gaussian solution provides a way to calibrate data and characterize important features of the peak data. Unfortunately, this method provides no information as to why the peak has a particular retention time or volume. It also does not predict retention times and peak widths. This method is not very useful if you already have software and hardware that can perform peak integrations. In contrast, the next two methods discussed enable us to build predictive tools for these parameters.

10.4.2 Staged models

A useful, but somewhat simplistic kinetic model for viewing how solutes interact with the stationary and mobile phases as they pass through the column is to imagine that the column actually consists of a number of individual stages. In each of these stages, the stationary and mobile phases are in equilibrium. Somehow, these phases become well mixed. Using the law of conservation of matter, we can develop an equation for each stage to keep track of changes in the concentration of solute in that stage. In words, this equation is described as follows:

$$\begin{bmatrix} \text{Amount of} \\ \text{solute A} \\ \text{entering stage} \end{bmatrix} - \begin{bmatrix} \text{amount of} \\ \text{solute A} \\ \text{exiting stage} \end{bmatrix} = \begin{bmatrix} \text{rate of} \\ \text{accumulation of} \\ \text{solute A in} \\ \text{stationary phase} \end{bmatrix} + \begin{bmatrix} \text{rate of} \\ \text{accumulation of} \\ \text{solute A in} \\ \text{liquid phase} \end{bmatrix}$$

$$(10.12)$$

The staged model assumes that the mobile phase leaving the stage is in equilibrium with the stationary phase in that stage. Figure 10.5 summarizes the equations and curves that typify these equilibria for a solute that distributes between a solid and liquid phase. The concentration of the solute in the liquid phase is denoted by C, while the solute concentration within the boundaries of the stationary phase is q.

We can use calculus to describe the rate of change in the solid- and liquid-phase concentrations as the derivative of the concentration of solute A in the stationary or mobile phases with respect to time:

$$QC_{n-1} - QC_n = \varepsilon V_n \frac{dC_n}{dt} + (1-\varepsilon)V_n \frac{dq_n}{dt} \tag{10.13}$$

The void fraction ε, which usually ranges between 0.3 and 0.38, is the fraction of the total stage volume that is occupied by the mobile phase. By assuming equilibrium and using the simplest equilibrium expression in Figure 10.5 ($q_n = KC_n$), we can relate the derivative for the stationary phase to the derivative for the mobile phase as $\frac{dq_n}{dt} = K\frac{dC_n}{dt}$. Thus the equation can be rearranged and simplified to give

$$\frac{dC_n}{d\tau} = (C_{n-1} - C_n) \tag{10.14}$$

To simplify Equation 10.14, we define a new time variable as the dimensionless quantity $\tau = \dfrac{tQ}{V_n(\varepsilon + (1-\varepsilon)K)}$. This equation can be solved using the initial condition

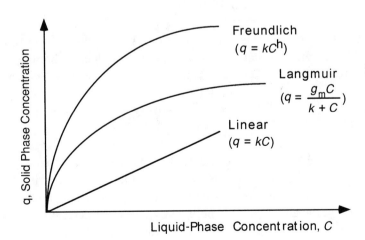

Figure 10.5 Possible equilibrium equations for a solute.

that for $t < 0$, $C_n = 0$; at $t = 0$, an instantaneous pulse, or injection, occurs with concentration C_f.

To solve the set of equations that result from the staged model, we can employ Laplace transforms (see Appendix A for a review of this method). Thus the most useful way of describing the initial and boundary conditions is as follows:

$$\text{at } \tau = 0, C_n = 0 \tag{10.15}$$

$$C_0 = C_f \delta(\tau) \tag{10.16}$$

In Equation 10.16, $\delta(t)$ is the impulse function that describes a perfect pulse at $t = 0$. After applying Laplace transforms, we can obtain

$$sY_n = Y_{n-1} - Y_n \tag{10.17}$$

Starting at stage 1, this equation can be solved for the Laplacian variable Y_n so that

$$sY_1 = C_f - Y_1 \tag{10.18}$$

and

$$Y_1 = \frac{C_f}{s+1} \tag{10.19}$$

as the Laplace transform of the impulse function is 1. For the second and nth stage, we obtain the following transforms:

$$Y_2 = \frac{C_f}{(s+1)^2} \tag{10.20}$$

$$Y_n = \frac{C_f}{(s+1)^n} \tag{10.21}$$

The inverse transform of the nth-stage equation can be obtained using tables provided in various mathematics handbooks, giving the final form:

$$C_n = C_f \frac{\tau^{n-1} \exp(-\tau)}{(n-1)!} \tag{10.22}$$

Equation 10.22 describes a Poisson distribution. For the next step in this analysis, we also need to define the total volume of the column, which is the sum of N stages as $V = NV_n$. We then substitute the definition of t back into the equation:

$$C_n = C_f \frac{\left(\dfrac{tQN}{V(\varepsilon + (1-\varepsilon)K)}\right)^{n-1} \exp\left(-\dfrac{tQN}{V(\varepsilon + (1-\varepsilon)K)}\right)}{(n-1)!} \tag{10.23}$$

For the sake of convenience and to relate this analysis to the Gaussian model, we note that if the number of stages is large, then the Poisson distribution can be approximated by the Gaussian distribution. To put this result in the equivalent Gaussian form at the column exit (where $n = N$), we must determine the mean and the variance of the Poisson distribution. Although we can use a standard textbook on probability and statistics to make these calculations, it is more instructive to solve the equation for Y_n and obtain the first and second moments using van der Laan's theorem (see Appendix B). The first moment is the mean, which represents the peak residence time t_{max}. The second moment is the variance, σ^2. These terms can be then substituted into the general equation for the Gaussian distribution:

$$\frac{C}{C_f} = \frac{1}{\sqrt{2\pi}\sigma} \exp\left(-\frac{1}{2}\frac{(t-\mu)^2}{\sigma^2}\right) = \frac{1}{\sqrt{2\pi}\sigma} \exp\left(-\frac{\left(\frac{t}{\mu}-1\right)^2}{\frac{2\sigma^2}{\mu^2}}\right) \tag{10.24}$$

Using van der Laan's theorem, we find that

$$\hat{\mu} = \frac{1}{C_f}\left(\lim_{s\to 0}\frac{\partial Y_n}{\partial s}\right) \tag{10.25}$$

$$\hat{\mu} = \frac{1}{C_f}\left(\frac{NC_f}{1^{N+1}}\right) = N \tag{10.26}$$

$$\mu = \frac{V(\varepsilon + (1-\varepsilon)K)}{Q} \tag{10.27}$$

The variable $\hat{\mu}$ is the dimensionless first moment, which is equal to $\hat{\mu} = \frac{\mu Q}{V_n(\varepsilon + (1-\varepsilon)K)}$.

We then apply van der Laan's theorem to find the second moment:

$$\hat{\sigma}^2 = \frac{1}{C_f}\left(\lim_{s\to 0}\frac{\partial Y_n}{\partial s}\right) - \hat{\mu}^2 \tag{10.28}$$

$$\hat{\sigma}^2 = (N)(N+1) - N^2 = N \tag{10.29}$$

$$\sigma^2 = \frac{V(\varepsilon + (1-\varepsilon)K)}{Q} \tag{10.30}$$

Substituting these results into the Gaussian equation form, we obtain a useful equation:

$$C_n = C_f\sqrt{\frac{Q}{2\pi V(\varepsilon + (1-\varepsilon)K)}} \exp\left[-\frac{\left(\frac{Qt}{V(\varepsilon + (1-\varepsilon)K)}-1\right)^2}{\frac{2}{N}}\right] \tag{10.31}$$

In this equation, the important properties—such as the time for the peak maximum and the standard deviation of the peak (i.e., the width of the peak at the point of inflection)—are defined as follows:

$$t_{max} = \frac{V(\varepsilon + (1-\varepsilon)K)}{Q} \tag{10.32}$$

$$w_i = \frac{2V(\varepsilon + (1-\varepsilon)K)}{Q\sqrt{N}} \tag{10.33}$$

Table 10.5 Summary of Results for the Staged Model

Chromatography Variable	Effect of Increasing Value
Column length, L	Increases residence time and peak width
Flow rate, Q	Decreases residence time and peak width
Equilibrium constant, K	Increases residence time and peak width
Void fraction, ε	Not much variation possible; for high K decreases residence time and peak width; for low K increases residence time and peak width
Number of stages, N	Decreases peak width
Column diameter, D	Increases residence time and peak width

The resultant model now provides insight into some mechanisms that dictate peak shape. For example, if K is large, then the solute is attracted to the stationary phase because the concentration in the stationary phase greatly exceeds the concentration in the mobile phase. Thus the time at which the midpoint of the peak exits the column will increase—an important result that makes intuitive sense. Similarly, as the flow rate or column length increases, the time at which the midpoint of the peak exits the column will decrease. In addition, as the equilibrium constant increases, the peak width will increase as well. A column that contains more stages but holds the same volume as another column will have a less wide peak than the second column. Similarly, using the definition of the height equivalent of a theoretical plate (H.E.T.P., or H) where $N = L/H$, we see that for a given column length, columns with a lower value of H.E.T.P. possess narrower peaks.

These results (summarized in Table 10.5) illustrate the inevitable trade-offs that must be made when designing a chromatographic process. It is clearly desirable to have large differences in solute peak residence times when separating two or more components; unfortunately, the longer residence times achieved by increasing the column length or K are associated with broader peaks. Broader peaks are undesirable, because the solutes are collected at the end of the column at a lower concentration and take longer to elute.

To obtain a better understanding of how solutes exit a column, we can extend the staged model to deal with a more realistic input, such as an injection of a solution of concentration C_f over a finite period of time (Figure 10.6).

We can analyze the square wave input given by Equation 10.34 for stage C_o (the column input) as a function of time and as a Laplacian transform variable Y_o:

$$C_o = \frac{C_f}{k}[u(t) - u(t - k)] \tag{10.34}$$

$$Y_o = \frac{1 - e^{-ks}}{s} \tag{10.35}$$

The expression for plate n then becomes

$$Y_n = \frac{C_f}{k} \frac{1 - e^{-ks}}{s(s + 1)^n} \tag{10.36}$$

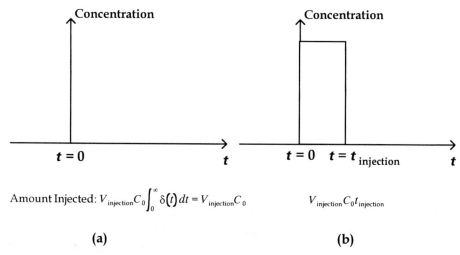

Amount Injected: $V_{\text{injection}}C_0\int_0^\infty \delta(t)\,dt = V_{\text{injection}}C_0$ $V_{\text{injection}}C_0 t_{\text{injection}}$

(a)   **(b)**

Figure 10.6 (a) Perfect pulse and (b) square wave input to a column.

We have two ways to solve this expression to obtain the peak residence time. The easier way of obtaining the residence time is to realize that a perfect pulse with no width differs from a real pulse of width k by only the addition of half the pulse width to the perfect-pulse answer. A rational analysis of this situation suggests that the answer is

$$t_{\max} = \frac{V_n\left(N+\dfrac{k}{2}\right)(\varepsilon + (1-\varepsilon)K)}{Q} \tag{10.37}$$

When $k = 0$ (that is, as we approach a pulse of infinitesimal width), the answer reverts to that for the perfect-pulse input because $V_n N = V$. To verify this answer and to obtain the peak width, we can apply van der Laan's theorem to obtain the first and second moments in Laplacian space. We also need to use L'Hopital's rule to apply the limit as $s \to 0$. Because the repetitive differentiations needed to apply van der Laan's theorem and L'Hopital's rule are quite tedious, we can preserve our sanity by using computer-aided symbolic calculus programs to find the first moment (the peak residence time) and the second moment (the variance σ^2). Applying van der Laan's theorem and L'Hopital's rule twice gives

$$\hat{\mu} = -\frac{1}{C_f}\left(\lim_{s\to 0}\frac{\partial Y_n}{\partial s}\right) \tag{10.38}$$

$$\hat{\mu} = \frac{1}{C_f}\left(\frac{C_f}{k}\left(\frac{-k^2-2kn}{2}\right)\right) \tag{10.39}$$

$$\hat{\mu} = n + \frac{k}{2} \tag{10.40}$$

This answer is the same as the one obtained earlier.

The more complex determination of the variance, $\hat{\sigma}^2$, can be carried out by using symbolic math computer programs accompanied by van der Laan's theorem. This calculation will provide a function that includes both n and k. Using a symbolic math computer program, the variance can be determined as

$$\hat{\sigma}^2 = \frac{1}{C_f}\left(\lim_{s \to 0} \frac{\partial Y_n}{\partial s}\right) - \mu^2 \tag{10.41}$$

$$\hat{\sigma}^2 = n + \frac{k^2}{12} \tag{10.42}$$

Thus the input width has a more significant effect on the peak width if k^2 is close to the magnitude of n. Note that k is a dimensionless term such that $k = \dfrac{aQ}{V_n(\varepsilon + (1 - \varepsilon)K)}$, where a is the width of the square wave input (in units of time).

This model can be applied to analyze the effects of many different types of inputs on the first and second moments, though we will not develop these results here. In Section 10.4.3, however, we will see how relaxing the assumption of equilibrium between the stationary and mobile phases in a specific section of the column affects the peak shape. To carry out this analysis and to elucidate how the mobile-phase flow pattern can affect peak shape, we must regard the column as a continuous mixture of stationary and liquid phases. Altering our assumptions in this way does not imply that the staged model has no utility. In fact, the simplicity and power of the results developed with this model yield many useful equations for modeling column behavior. The continuum approach developed in the next section may afford greater flexibility in studying important column phenomena, but this tactic comes at a price of increased complexity, requiring experimental determinations of new parameters.

10.4.3 Newtonian continuum mechanics and linear equilibria

To apply Newton's laws of motion to carry out a "microscopic" analysis of how solutes move through a column, we must make two assumptions: 1) both the mobile and stationary phases are distributed throughout the column in the same manner, and 2) an infinitesimally small control volume anywhere in the column will typify the functioning of the entire column. Contrary to the staged model's view that the column can be broken down into a series of well-mixed tanks, the continuum model states that the mobile phase enters and leaves the control volume based on microscopic fluid phenomena such as diffusion and forced convection (Figure 10.7).

The transport of solutes within the confines of the stationary phase in the control volume is similarly governed by diffusion or, if the pores are relatively large, by restricted convection (as in perfusion chromatography). Accounting for all of the phenomena when performing the accounting of solute, as stated from the staged model, we have

$$\begin{bmatrix} \text{Rate of solute} \\ \text{A entering} \\ \text{control volume} \end{bmatrix} - \begin{bmatrix} \text{rate of solute} \\ \text{A exiting} \\ \text{control volume} \end{bmatrix} = \begin{bmatrix} \text{rate of} \\ \text{accumulation of} \\ \text{solute A in} \\ \text{stationary phase} \end{bmatrix} + \begin{bmatrix} \text{rate of} \\ \text{accumulation of} \\ \text{solute A in} \\ \text{liquid phase} \end{bmatrix}$$

$$\tag{10.43}$$

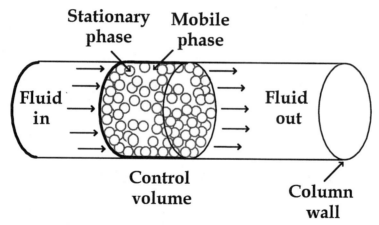

Figure 10.7 Continuum model approach.

We then face the daunting task of solving several complex equations. Rather than tackle this job, we will use several standard simplifying assumptions.

The first assumption is that the flow through a packed column is a steady, plug flow. With this assumption, the mobile-phase velocity is taken to be a constant that does not depend on time or radial position. The second assumption is that properties such as solute diffusion coefficients are also constants throughout the column. Although more assumptions are needed, these two apply specifically to the two models that will be developed in this section—the local equilibrium theory (LET) and the dispersion and linear rate model (DLRM). The LET model, the simpler of the two, illustrates how the microscopic analysis relates to the macroscopic analysis of the staged model.

10.4.3.1 Local equilibrium theory

The following equation accounts for solute A in the control volume of ΔV in Figure 10.7:

$$\Delta x \Delta y (\varepsilon U C_A|_z - \varepsilon U C_A|_{z+\Delta z}) = \Delta x \Delta y \Delta z \left[(1-\varepsilon) \frac{\Delta q_A}{\Delta t} + \varepsilon \frac{\Delta C_A}{\Delta t} \right] \tag{10.44}$$

To keep the mathematics simple, this equation does not include cylindrical coordinates. Instead, we will assume that no radial dependencies exist, because each cross section has a uniform flow rate and the stationary and mobile phases are at equilibrium. After dividing by $\Delta x \Delta y \Delta z$ and taking the limits as t and z approach zero, the following differential equation is found:

$$U \frac{\partial C_A}{\partial z} + \frac{\partial C_A}{\partial t} + \left(\frac{1-\varepsilon}{\varepsilon} \right) \frac{\partial q_A}{\partial t} = 0 \tag{10.45}$$

The assumption of linear equilibrium between the stationary and mobile phases at each specific cross section is manifested in

$$\frac{\partial q_A}{\partial t} = K \frac{\partial C_A}{\partial t} \tag{10.46}$$

which permits us to write the following simplified form:

$$U\frac{\partial C_A}{\partial z} + \left[1 + \left(\frac{1-\varepsilon}{\varepsilon}\right)K\right]\frac{\partial C_A}{\partial t} = 0 \qquad (10.47)$$

Using the impulse function introduced in our discussion of the staged model, and with the initial condition that the column contains no solute A, we have

at $t = 0$, $C_A = 0$ $\qquad\qquad\qquad\qquad\qquad\qquad\qquad\qquad\qquad\qquad (10.48)$

$C_A(z = 0) = C_f\delta(t)$ $\qquad\qquad\qquad\qquad\qquad\qquad\qquad\qquad\qquad (10.49)$

We can apply van der Laan's theorem to obtain the following results for the first and second moments:

$$\mu = \frac{L}{U}\left[1 + \left(\frac{1-\varepsilon}{\varepsilon}\right)K\right] \qquad (10.50)$$

$$\sigma^2 = 0 \qquad (10.51)$$

The first moment is identical to the result obtained with the staged model, which used the impulse function as the column feed. The continuum model, however, states that if local equilibrium holds, the peak does not widen as it goes through the column. In other words, the peak shape is not affected by its passage through the column and does not incur the dispersion, or Gaussian shape, normally attributed to chromatography.

Obviously, this model is not realistic. We can include a measure of realism by recognizing ways in which the peak shape can change while passing through the column. For example, fluid mixing may occur along the axis of the column. We can also account for the kinetics of the solute transfer from the mobile to the stationary phase. These two mechanisms are explored in the DLRM model.

10.4.3.2 Dispersion and linear rate model

We can include a term in the LET model that accounts for both axial molecular diffusion and turbulent mixing around the stationary-phase particles. This strategy is a reasonable way to acknowledge peak-broadening processes in chromatography:

$$D_L\frac{\partial^2 C_A}{\partial z^2} \qquad (10.52)$$

Moreover, by including a simple kinetic model for the transfer of solute between mobile and stationary phases, we can use another parameter to account for peak shape effects:

$$\frac{\partial q_A}{\partial t} = b(C_A - C_A^*) \qquad (10.53)$$

In Equation 10.53, the parameter C_A' is the concentration of solute A in equilibrium with the solid-phase concentration q_A. In words, this first-order kinetic model states that the rate of change in the stationary-phase concentration of solute A is proportional to the difference between the mobile-phase concentration near the stationary phase and the mobile-phase concentration that would be in equilibrium with the stationary-phase concentration. This simple kinetic model, which is used in many engineering applications in heat and mass transfer, permits the measurement of the parameter b. Experimentally obtained data can then be fitted to this calculation, enabling the engineer to design process equipment during scale-up.

Combining these two terms into the control volume expression derived for LET gives the following DLRM expression:

$$-D_L \frac{\partial^2 C_A}{\partial z^2} + U \frac{\partial C_A}{\partial z} + \frac{\partial C_A}{\partial t} + \left(\frac{1-\varepsilon}{\varepsilon}\right) b(C_A - C_A^*) = 0 \tag{10.54}$$

Using Laplace transforms and van der Laan's theorem for the initial and boundary conditions,

at $t = 0$, $C_A = 0$ $\tag{10.55}$

$C_A(z = 0) = C_f \delta(t)$ $\tag{10.56}$

as $z \to \infty$, C_A is finite $\tag{10.57}$

gives the following values for the first and second moments:

$$\mu = \frac{L}{U}\left[1 + \left(\frac{1-\varepsilon}{\varepsilon}\right)K\right] \tag{10.58}$$

$$\sigma^2 = 2\left[\frac{\mu^2}{Pe} + \frac{L}{U}(1-\varepsilon)\frac{K^2}{b}\right] \tag{10.59}$$

In these equations,

$$Pe = \frac{UL}{D_L} \tag{10.60}$$

$$D_L = \frac{U^2 d_p^2}{192 D_{\text{mobile phase}}} \tag{10.61}$$

When comparing the DLRM result with the LET result, the only difference arises from the additional terms associated with the variance, which dictates peak shape. The LET results indicates the peak residence time in chromatography, even though it does not account for many fluid and mass transport phenomena. As noted earlier, the conditions that are applicable in both LET and DLRM are as follows:
- The isotherm is linear ($q_A = KC_A$).
- The temperature and flow rates are maintained at constant levels.
- The solutes interact independently with the stationary phase, as is the case in a trace system where the total solute mass injected into the column is much smaller than the column's theoretical capacity.

An additional constraint in LET is that the pulse must be introduced in a much shorter time interval than the peak residence time.

The DLRM model provides a rationale for fitting experimentally obtained peak shape data so as to guide column design. In particular, it influences the column length, the column superficial velocity $\left(U = \dfrac{Q}{\varepsilon \pi r^2}\right)$, the particle size ($d_p$), and the equilibrium constant. Peaks can be graphed by taking the first and second moments and then inserting those values in the Gaussian equation. Note, however, that the second moment is given as two terms: 1) the contribution by axial dispersion and diffusion, and 2) the contribution from the kinetics of stationary phase–mobile phase solute transport.

Many other proposed models incorporate more details regarding pore diffusion, stationary-phase film transport, radial diffusion, and other effects. After considering the staged and continuum models, however, it is more helpful to discuss the contributions of van Deemter, who developed a generalized model for interpreting the effects of flow rate, particle size, and column length. This equation builds upon the principle illustrated in this section—namely, that the variances are additive for trace systems with linear isotherms.

First, we will state some ways in which the kinetic constant b can be related to column conditions—most notably, the equilibrium constant and particle size. Many analyses have empirically derived equations for column parameters (2). For our purposes here, we will outline the two contributions to the b term as follows:

- The sum of resistances for film diffusion (molecules diffusing through a stagnant layer around a stationary-phase particle)
- Pore diffusion (molecules entering the gel matrix or pores by diffusion)

These contributions can be used to develop the following equation:

$$b = \cfrac{1}{\left(\cfrac{d_p^{1.415}}{\alpha U^{0.585}} + \cfrac{d_p^2}{\beta \cfrac{\partial q_A}{\partial C_A}} \right)} \qquad (10.62)$$

In Equation 10.62, note that α and β are experimental constants and that the derivative is the slope of the equilibrium isotherm (K in our case, which assumes linear equilibrium). In the denominator, the first term on the left is the film "resistance," while the second term represents the resistance caused by pore diffusion. Thus, if the slow step is film transport, then the second term in the denominator can be ignored, or vice versa. Practically speaking, we can determine whether film transport is the bottleneck that dictates the rate at which molecules enter the stationary phase by seeing whether the velocity and the equilibrium constants are large. In any event, experiments are needed to find how b changes with U and d_p, as this consideration is important during scale-up.

10.4.4 Constant pattern and saturation equilibria

In the Newtonian continuum mechanics approach, the column was operated by injecting a pulse and measuring the resultant peak exiting the column. In affinity and ion-exchange chromatography, adsorption columns can be operated in binding-elution cycles. We will model what occurs inside the column and at the column exit by breaking down the cycles into separate steps. The resultant model is, of course, greatly simplified—but it nevertheless describes the basic phenomena well enough to guide process design. The constant pattern model is the context for this description of the binding and elution cycles.

Figure 10.8 schematically describes the binding cycle. Rather than injecting a sample, the mobile phase contains the species that is removed from solution by binding to the stationary phase. The concentration bound to the stationary phase is denoted by q_A. Shortly after starting the binding cycle (denoted by Step I), q_A reaches its maximum value in the zone near the column entrance. Thus the entrance of the column contains a stationary phase that is saturated with species A; it can no longer remove solute A from the mobile phase. The mobile phase near the column entrance is at its maximum concentration $C_{A,f}$ as well. During this part of the binding cycle, the concentration of solute A exiting the column ($C_{A,e}$) is very low, essentially zero.

In Step II, the profiles of q_A and C_A (i.e., the curve describing how q_A varies with column length) develop into consistent patterns that move down the column with time. In contrast, in Step I these patterns were not well defined, because they were evolving into the patterns that became evident in Step II. During Step II, the column exit concentration is still essentially zero.

The most important development in the binding cycle comes in Step III, when the leading edge of the q_A and C_A profiles reach the column exit. At this point, a breakthrough of species A occurs—the mobile phase concentration of solute A starts to climb, which usually signifies the end of the binding cycle in real-world production (a significant amount of product is being lost from the column).

For analytical purposes, it is useful to continue to operate the column though the next step (Step IV), where we record the change in the column exit concentration $C_{A,e}$ with time. These data are known as the breakthrough curve.

Finally, in Step V the column is fully saturated. Thus $C_{A,e}$ equals the feed concentration $C_{A,f}$.

Constant pattern analysis addresses a major design issue—the width of the mass transfer zone (also known as the width of the breakthrough curve), shown in Figure 10.8. A perfect column would have zero width, meaning that at the breakthrough time, t_b, the entire column has become saturated with species A. Unfortunately, in real-world column operation, a portion of the column near the column exit is not used for binding species A. Thus knowing only the binding capacity of the stationary phase will cause us to underestimate the amount of stationary phase needed for a separation cycle. The mass transfer zone also has implications for the elution cycle.

After the binding cycle, the column can be washed with a simple buffer to remove the binding-cycle mobile phase, which may contain unwanted impurities. The elution cycle then begins. An eluent solution is administered as the mobile phase. Because the mass transfer zone exists at the binding cycle column exit, it is usually better to flow the eluent mobile phase in the reverse direction; thus the mobile phase exiting at the top of the column will contain the highest concentration of species A. A mobile phase containing species A is usually collected after this point, so that the mobile phase exiting the top of the column contains only the wash buffer and very little of species A.

In the next step, species A is collected in the eluent mobile phase at essentially the maximum concentration. Just as in Step II of the binding cycle, the bottom of the column will show a constant profile of q_A that travels up the column. The elution profile may not be the same as the binding cycle, however, because the driving force for elution may not simply be the reverse of the driving force for binding. To ensure

Step I: The Binding Cycle Begins

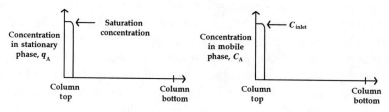

Step II: Development of Constant Patterns

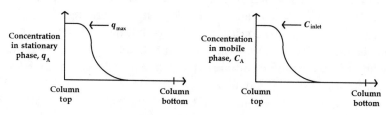

Step III: Breakthrough

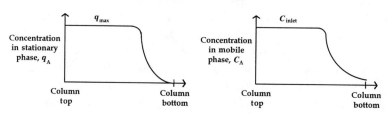

Step IV: Further Loss of Species A in Mobile Phase

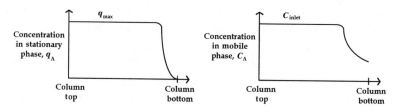

Step V: Column Exhaustion

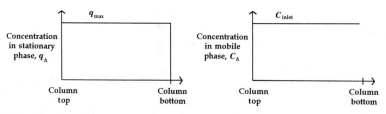

Figure 10.8 Binding of a solute as a series of steps during on-off cycling.

that the eluent does not become significantly diluted, the collection of species A for further processing may be terminated before all of this solute is removed.

Finally, the column is prepared once again for the binding cycle—that is, it is washed thoroughly with a wash buffer solution or the buffer solution used for preparing the sample. Sometimes a regeneration cycle may be needed before revisiting the binding cycle so as to restore column capacity.

During this idealized walkthrough of a binding-elution cycle, two recurrent events are worth noting. First, the concentration profiles in the stationary and mobile phases move through the column as a constant pattern or wave. Second, the mass transfer zone is measured by graphing the breakthrough curve in both the binding and elution cycles. Using this information, we can develop a model for predicting the breakthrough curves based on the constant pattern assumption.

Constant patterns in both q_A and C_A suggest that directly behind the wave

$$\frac{q_A}{C_A} = \left[\frac{q}{C}\right]_{\text{equilibrium}}$$

(10.63)

The ratio on the right side of the equation involves the equilibrium concentrations. Equation 10.63 does not indicate that local equilibrium occurs at the leading fronts of the waves. Instead, the assumption of constant patterns leads to a simple equation relating the stationary- and mobile-phase concentrations, similar to the equation used in the DLRM model:

$$\frac{(1-\varepsilon)}{\varepsilon}\frac{dq_A}{dt'} = k_{\text{cp}}(C_A - C_A^*)$$

(10.64)

Rather than relating the driving force of equilibrium to a rate expression, the constant pattern suggests that the equation

$$t' = t\frac{z}{U}$$

(10.65)

can be used. The Langmuir isotherm describes the equilibrium concentration C' for the case of saturation of a stationary phase:

$$q_A^* = \frac{q_m C_A}{K + C_A}$$

(10.66)

or

$$C_A^* = \frac{q_A K}{q_m - q_A}$$

(10.67)

Our next move in solving the rate equation is to obtain an expression in terms of the mobile-phase concentration. We will then use the constant pattern ratio for relating q_A to C_A. Although this approach may seem to rely on circular logic, the introduction of the Langmuir expression into the constant pattern expression shows us that

$$\frac{q_A}{C_A} = \left[\frac{q}{C}\right]_{\text{equilibrium}} = \left[\frac{q}{C}\right]_{\text{feed}} = \frac{q_m}{K + C_f}$$

(10.68)

which allows for the following result:

$$\frac{(1-\varepsilon)q_m}{\varepsilon(K+C_f)}\frac{dC_A}{dt'} = k_{\text{cp}}(C_A - C_A^*) = k_{\text{cp}}C_A\left(\frac{C_f - C_A}{K + C_f - C_A}\right)$$

(10.69)

Equation 10.69 can be rearranged to obtain an expression for the width of the break-through curve:

$$t'_{0.99C_f} - t'_{0.01C_f} = \frac{(1-\varepsilon)q_m}{\varepsilon k_{cp}(K+C_f)} \int_{0.01C_f}^{0.99C_f} \left(\frac{K+C_f-C_A}{C_fC_A-C_A^2} \right) dC_A \tag{10.70}$$

This value has been arbitrarily chosen to be the width for exit concentrations between 0.01 and 0.99 of the feed concentration of species A:

$$t'_{0.99C_f} - t'_{0.01C_f} = \frac{(1-\varepsilon)q_m}{\varepsilon k_{cp}(K+C_f)} \left[\left(\frac{K+C_f}{C_f} \right) \ln \left(\frac{C_A}{C_f-C_A} \right) + \ln(C_f-C_A) \right]_{0.01C_f}^{0.99C_f} \tag{10.71}$$

Using Equation 10.71, we can obtain the width of the breakthrough curve for other measurable concentration ranges, such as from $0.05C_f$ to $0.95C_f$. We can also plot this equation by setting the lower limit at some easily measured value, such as $0.01C_f$, and then varying the upper limit to identify discrete points on the breakthrough curve. Although the width of the elution breakthrough curve can be obtained as well, plotting the data points requires us to use the reverse limits (i.e., from $0.99C_f$ to some lower value). In practice, this equation is plotted for values of q_m, K, and k_{cp} that will best fit the data. Usually q_m and K are obtained by batch equilibrium experiments first; only k_{cp} is determined from the column experiment.

The value of this derivation derives from its ability to give us a better under-standing of how equilibria and mass transfer kinetics affect the mass transfer zone. From the Langmuir isotherm, we know that as K decreases so that $q = q_m$, the width of the mass transfer zone shrinks to its smallest value. We can prove this point by working with Equation 10.71. First, we multiply each term on the right-hand side by $1/(K + C_f)$. Then we set $K = 0$. As $K \to 0$,

$$t'_{0.99C_f} - t'_{0.01C_f} = \frac{(1-\varepsilon)q_m}{\varepsilon k_{cp}C_f} [\ln(C_A)]_{0.01C_f}^{0.99C_f} = \frac{(1-\varepsilon)q_m}{\varepsilon k_{cp}C_f} [\ln(99)] \tag{10.72}$$

Similarly, as $K \to \infty$,

$$t'_{0.99C_f} - t'_{0.01C_f} = \frac{(1-\varepsilon)q_m}{\varepsilon k_{cp}C_f} \left[\ln \left(\frac{C_A}{C_f-C_A} \right) \right]_{0.01C_f}^{0.99C_f} = \frac{(1-\varepsilon)q_m}{\varepsilon k_{cp}C_f} [2\ln(99)] \tag{10.73}$$

The results are the same for other limits of the breakthrough curve—namely, the width of the breakthrough curve doubles as K goes from zero to infinity. Thus the minimum width occurs where $q = q_m$.

Because the width of the breakthrough curve is inversely proportional to the rate constant k_{cp}, it is important to understand how this parameter can vary with column conditions and properties. These issues are discussed in more depth in the references cited at the end of this chapter. In general, the more important relation-ships are that the rate constant increases with decreasing particle size, increasing column superficial velocity ($4Q/d_p^2$), and increasing solute diffusivity.

Binding-elution cycles are used in affinity chromatography and sometimes in the recovery of small biomolecules in ion-exchange chromatography. This approach is also used in adsorption chromatography, where biomolecules become bound through nonspecific interactions.

10.4.5 Van Deemter equation

After reconciling the results from the staged and Newtonian continuum mechanics models, van Deemter proposed that H.E.T.P. (or H) is related to the superficial velocity by a simple expression:

$$H.E.T.P. = \frac{A}{U} + BU + C \tag{10.74}$$

$$A \propto D_{\text{mobile phase}} \tag{10.75}$$

$$B \propto k_{\text{mt}} \tag{10.76}$$

$$C \propto d_{\text{p}} \tag{10.77}$$

In these equations, A takes into account the mobile-phase solute's ordinary diffusivity (usually a negligible value in liquid phases), B potentially involves many terms dealing with solute transport between the stationary and mobile phases, and C deals with the effect of the particle size (which can cause eddies and vortices in the mobile phase, leading to axial dispersion).

One form of the van Deemter equation that can be used to correlate experimental data is provided by Wankat (2):

$$H \propto \frac{U d_{\text{p}}^2}{D_{\text{mobile phase}}} + \text{constant} \tag{10.78}$$

As in the DLRM equation, this form of the van Deemter equation incorporates the effects of stationary particle size as well as mobile phase velocity and solute diffusivity on the column efficiency. Many other researchers have built upon this model, proposing a variety of different equations intended for many applications (Table 10.6). For column design, the van Deemter equation can be effectively used with the staged model by defining how N varies, because $H = L/N$.

10.4.6 Gel partitioning model

Before discussing specific separations and operational issues of gel-permeation chromatography, we should consider the gel partitioning model (GPM) in the context of the previous analyses. This model offers a more limited, but still useful

Table 10.6 Variations on the van Deemter Model (2)

Alternate Equation	Utility
$H = \dfrac{B}{U} + C_{\text{sm}}U + \dfrac{1}{\left(\dfrac{1}{A} + \dfrac{1}{C_{\text{m}}U}\right)}$	Accurate
$H = kU^n$	Simple, for velocity effect
$H_{\text{min}} = 2d_{\text{p}}$	Theoretical minimum of H
$H_{\text{min}} = 3d_{\text{p}}$ to $3.5d_{\text{p}}$	Minimum of H in "real" system

way of defining the key variables for chromatography. It uses the stationary phase as a porous barrier. Because it is more useful to deal with volume quantities, this model first defines the different volume elements:

$$V_T = V_o + V_i + V_s \tag{10.79}$$

In Equation 10.79, V_T is the total column volume, V_o is the column void volume, V_i is the volume in gel stationary-phase pores, and V_s is the solid volume of gel in the gel stationary phase. The maximum volume that mobile-phase solutes can occupy is $(V_T - V_s)$. Solutes that do not fully permeate the gel can be characterized through the gel partition coefficient and the elution volume:

$$K_p = \frac{(V_e - V_o)}{V_i} \tag{10.80}$$

In this equation, V_e is the elution volume of a solute. When $V_e = V_i$, K_p is a maximum and the product of K_i and V_i is equal to $V_T - V_s$. When $V_e = V_o$, $K_p = 0$. Thus the solute is considered totally excluded from the stationary-phase pores.

This model can be applied to predict the elution volume based on experimental determination of K_p and the calculation of V_i by the user or the gel manufacturer:

$$V_e = V_o + K_p \left[\frac{W_t \rho_g (V_T - V_o)}{(1 + W_t \rho_w)} \right] \tag{10.81}$$

Its strengths derive from the fact that the terms in the equation do not depend on column size and can be readily determined through experimentation and manufacturer-supplied information. If information about the resolution or peak shape is needed, however, the previously described models must be invoked.

Armed with the mathematical analyses discussed in this chapter, we now examine specific applications of chromatography from the scientific literature and practical uses of these analytical methods.

10.5 GEL-PERMEATION CHROMATOGRAPHY

Most gel-permeation chromatography (GPC) users first calibrate their columns for a specific set of conditions using known molecular-weight standards. The unknown sample is then injected, and the peaks are interpreted through a comparison with the calibration standards. The following sample problems illustrate this common procedure.

Example 10.1

Warzecha and colleagues (3) performed GPC on several industrial proteins using a 1.5 cm inner diameter by 55 cm column packed with Sephadex G-100. Figure 10.9 reproduces some of their chromatography data and indicates the buffer and flow rate used. Plot the log of the molecular weight versus V_e/V_o, where V_o is obtained from the elution volume of blue dextran, which is totally excluded from the gel.

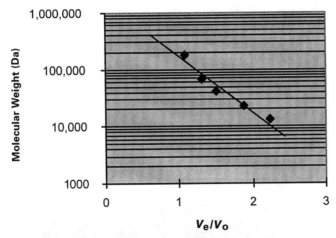

Figure 10.9 Retention volumes of selected proteins using Sephadex G-100, a dextran-based gel. (Adapted from Warzecha L, Piwowar Z, Bodzek D. Gel permeation chromatography for the determination of the molecular mass distribution of some industrial proteins. J Chrom 1990;509:227–231.)

Solution

Plotting $\ln(M)$ versus V_e/V_e and regressing the data to obtain a linear fit results in the following equation:

$$\frac{V_e}{V_o} = -0.44\ln(MW) + 14.9 \tag{10.82}$$

The least squares fit residual is $r^2 = 0.9$. This calibration permits us to predict the retention time of other proteins with a molecular weight less than approximately 200,000 Da, such as enzymes, and will allow the determination of protein purity from industrial separation processes such as ultrafiltration. This calibration has some limitations, however: It is specific for the column dimensions, the flow rate, the gel pore size distribution, and, to a lesser extent, the mobile-phase composition for biopolymers that undergo significant conformational changes in various buffers. To better address these issues, Flapper and colleagues (4) have suggested that the data could be calibrated based on the partition parameter K_D, and that this term be related to protein hydrodynamic radius.

Example 10.2

Flapper and colleagues (4) used the Stokes-Einstein equation $R = kT/6\pi\eta D$ to calculate the hydrodynamic radii of proteins. They argue that this parameter, which accounts for the equivalent spherical space that the protein occupies as it moves in a fluid, has been demonstrated by Ackers (4) to correlate well with K_D (which is equivalent to K_p) via the following equation:

$$R = A + \mathrm{erf}_c^{-1}(K_D) \tag{10.83}$$

Equation 10.83 assumes that the pore size distribution of the gel follows a Gaussian distribution. Given the relationship between the molecular weight and R (which can be obtained using the Stokes-Einstein equation or Equation 10.83)

Table 10.7 Selected R Values for Proteins

Protein	Molecular Weight (Da)	R
Bovine serum albumin	60,000	36.3
IgG	150,000	53.6
IgA	160,000	65
IgM	900,000	123.9

Source: Adapted from Flapper W, Theeuwes AGM, Kierkels J, Steenbergen J, Hoenders HJ. Separation of serum proteins by high-performance gel-permeation column systems. J Chrom 1990;533:49.

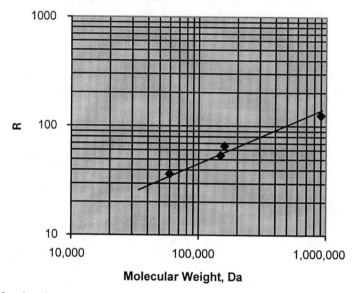

Figure 10.10 Log-log plot.

provided in Table 10.7, find K_D for a protein of molecular weight 100,000 Da. Use the equation for Superose 6 where $A = 3.7$ and $B = 175.6$.

Solution

First, we need to plot R versus the molecular weight to obtain the proper value of R for a protein of 100,000 Da. We will use the equation derived by Flapper and colleagues, in which R is proportional to the molecular weight raised to the 1/3 power. Using the log-log plot shown in Figure 10.10, we estimate that an R value of 45 is justifiable for a protein of 100,000 Da. Thus we find that $erf_c^{-1}(K_D) = 0.235$ and that $K_D = 0.21$.

10.6 ION-EXCHANGE CHROMATOGRAPHY

Both proteins and small charged biochemicals can be separated based on their charge. Also, because proteins are Zwitterions, they can be separated by isoelectric point (pI). To understand how ion exchange can be used in chromatography, we can explore the mass action that results from this process. The exchange can be represented as a displacement reaction:

$$n\mathrm{R}^-\mathrm{Na}^+ + \mathrm{P}^+ \rightleftharpoons \mathrm{R}_n^-\mathrm{P}^+ + n\mathrm{Na}^+ \tag{10.84}$$

$$K_{\mathrm{IE}} = \frac{[\mathrm{R}_n^-\mathrm{P}^+][\mathrm{Na}^+]^n}{[\mathrm{R}^-\mathrm{Na}^+]^n[\mathrm{P}^+]} \tag{10.85}$$

This reaction involves the cation exchange of a protein, where the stationary phase has ionogenic groups denoted as R^-. There are n molecules of sodium ions and ionogenic sites on the stationary phase in the equilibrium expression, as each protein molecule has many ionogenic species and can displace multiple sodium ions.

Because proteins are biological polymers of Zwitterion amino acids, some amino acid residues can have a net positive, net negative, or neutral charge depending on the pH. Proteins can have clusters of amino acids of similar charge where—unlike with simpler biochemicals—charge heterogeneity can occur on the protein surface. Thus, a mostly positively charged section of the protein surface can interact with the stationary phase, displacing the small cations.

The equilibrium expression defined in Equations 10.84 and 10.85 explains several important points. In ion-exchange chromatography, the total number of moles of ionogenic sites on the stationary phase is much greater than the total protein applied to the column. The mobile-phase concentration of the counter-ion sodium also remains relatively constant. Thus the equilibrium expression can be simplified to

$$K_{\mathrm{II}} = \frac{\dfrac{q}{q_{\max}}}{[\mathrm{P}^+]\left(1 - \dfrac{q}{q_{\max}}\right)} \tag{10.86}$$

where q is obtained by dividing by the total concentration of ionogenic sites:

$$R_{\mathrm{total}} = q_{\max} \cong [\mathrm{R}_n^-\mathrm{P}] + [\mathrm{R}^-\mathrm{Na}^+] = q + (q_{\max} - q) \tag{10.87}$$

Because $q \ll 1$, the ion-exchange expression can be further simplified to a linear relationship:

$$q = q_{\max}K_{\mathrm{II}}[\mathrm{P}^+] = K'[\mathrm{P}^+] \tag{10.88}$$

Consequently, the LET and DLRM equations can be used to predict column behavior. A protein mixture can be resolved into separate peaks, because K_{IE} accounts for binding affinities that increase as the net positive charge on the protein increases. At a given pH, proteins with higher isoelectric points are retained longer on cation-exchange columns.

Example 10.3

Fuchs and Keim (5) separated rat pancreatic secretory proteins by cation exchange. Table 10.8 lists the retention volumes and isoelectric points for this separation. Determine the linear distribution coefficient K' for the data given. What relationship can you find between the isoelectric point and protein retention time for this separation?

Table 10.8 Cation Exchange of Rat Secretory Protein

Protein	Retention Volume* (mL)	Molecular Weight (Da)	pI
Procarboxypeptidase A	2.3	47,000–52,000	4.3–5.3
Procarboxypeptidase B	8.2	50,000	4.5–5.1
Amylase 1	6.06	50,000	8.5–9.0
Chymotrypsin(b)	11.88	25,000	9.0–9.3
Lipase	14.95	50,000	6.5
Proelastase	12.4	26,000	9.0

* Flow rate: 0.5 mL/min; column: 5 cm × 5 mm; injection: 320 mg of protein.
Source: Adapted from Fuchs MJ, Keim V. Separation of rat pancreatic secretory proteins by cation-exchange fast protein liquid chromatography. J Chrom 1992;576:291.

Table 10.9 Cation Exchange of Rat Secretory Protein

Protein	Retention Volume* (mL)	K'
Procarboxypeptidase A	2.3	0.58
Procarboxypeptidase B	7.1	2.7
Amylase 1	6.06	2.2
Chymotrypsin(b)	11.88	4.8
Lipase	14.95	6.1
Proelastase	12.4	5.0

* Flow rate: 0.5 mL/min; column: 5 cm × 5 mm; injection: 320 mg of protein.

Solution

Taking the data for retention volume, flow rate, and column dimensions, we can calculate K' assuming a void fraction of 0.3 (Table 10.9). Upon graphing the values of either m or K' to determine a correlation between either of these values and the pI of the protein, however, we find a very low correlation coefficient (Figure 10.11). Other researchers have made similar observations. The best explanation for this disparity is that proteins exhibit charge asymmetry; this property enables ion interactions to occur even when the net charge of the molecule is zero.

The equilibrium expression also yields another important point: Increasing the salt concentration increases the concentration of ionogenic sites on the stationary phase that are associated with sodium ions and decreases the concentration of bound protein. As a result, a mobile-phase salt gradient can be used to separate and sharpen peaks by speeding the elution of protein from the column.

One important procedure in ion-exchange chromatography that follows from changes in the mobile phase is chromatofocusing. If we start a chromatographic method at one pH and then introduce a new mobile phase that has a lower pH, we can separate proteins based on their isoelectric points (if the pH range encompasses their pIs, that is). Chromatofocusing is best visualized as a moving titration. As the pH declines, proteins move through the column in conjunction with the pH microenvironment that matches their pIs. This method can resolve proteins that differ in isoelectric points by as little as 0.02 pH units.

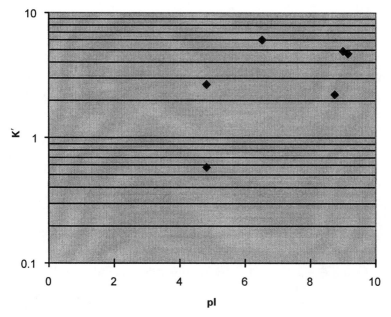

Figure 10.11 Examination of the correlation coefficient.

The mobile phase can also promote water structures that either help or hinder solubilization of hydrophobic biological molecules. Chaotropic buffers, for example, promote water solubilization of hydrophobic molecules in the presence of salts. In contrast, antichaotropic buffers tend to favor salting out or destabilization of hydrophobic biomolecules upon salt addition.

Common mobile-phase buffer salts can be ordered based on their chaotropic nature, from highest to lowest:

For anions: $PO_4^{3-} < SO_4^{2-} < CH_3COO^- < Br^- < NO_3^- < ClO_4^- < I- < CF_3COO^-$
$< SCN < CCl_3COO^-$

For cations: $NH^{4+} < Rb^+ < K^+ < Na^+ < Cs^+ < Li^+ < Mg^{2+} < Ca^{2+} < Ba^{2+}$

Failure to include a solubilizing buffer in tandem with a salt gradient can lead to peak tailing and loss of resolution. Salting out can enhance protein precipitation so as to purify and help recover products.

Ion-exclusion chromatography represents a useful tactic for operating ion-exchange columns when separating ionic solutes from weakly dissociating or non-ionic species. Inside the pores of the stationary phase, ion repulsion prevents like-charged mobile-phase ionic species from entering the stationary phase. Likewise, because electroneutrality must be maintained, the mobile-phase counter-ions do not enter. In contrast, uncharged species such as sugars or alcohols enter the stationary-phase pores and are therefore retained longer. The residence times of weakly charged species depend upon pK_a and pH. Thus, when separating weak acids such as carboxylic acids, the more acidic species spend less time inside the stationary phase and have shorter residence times.

Ion exclusion can be modeled as a phase partition:

$$K_{\text{ion exclusion}} = \frac{C_{\text{stationary}}}{C_{\text{mobile}}} \tag{10.89}$$

For weakly dissociating acids, we can derive the partition ratio by assuming that only uncharged species enter the stationary phase. The fraction of uncharged species in aqueous solution can be determined from the acid dissociation constant:

$$K_a = \frac{C_A^{\text{aq}}(10^{-\text{pH}})}{C_{\text{HA}}^{\text{aq}}} = 10^{-pK_a} \tag{10.90}$$

Also, the distribution equation must be rewritten to account for having both dissociated and undissociated species in solution, but only nonionic species in the stationary phase:

$$K_{\text{HA}} = \frac{C_{\text{HA(org)}}}{C_{\text{HA(aq)}} + C_{A^-(\text{aq})}} \tag{10.91}$$

Substituting into Equation 10.91 and assuming that only the undissociated species enters the stationary phase gives

$$K_{\text{HA}} = \frac{C_{\text{HA(org)}}}{C_{\text{HA(aq)}}(1 + 10^{\text{pH}-pK_a})} \tag{10.92}$$

and

$$K_{\text{HA}} = \frac{K_0}{1 + 10^{\text{pH}-pK_a}} \tag{10.93}$$

Combining these equations with the expression for residence time gives the final result:

$$\mu = \frac{L}{U}\left[1 + \left(\frac{1-\varepsilon}{\varepsilon}\right)\frac{K_0}{1 + 10^{\text{pH}-pK_a}}\right] \tag{10.94}$$

Example 10.4

Bio-Rad Laboratories has provided a chromatogram of organic acid standards separated on its Aminex HPX-87H column (6). The company used ion exclusion to separate oxalic, citric, malic, succinic, formic, and acetic acid, resulting in the retention time data given in Table 10.10. If the pK_a values of fumaric acid are 3.02 and 4.38 at 25 °C, where would fumaric acid elute?

Solution

Based on the data given and our earlier mathematical analysis, the order of elution appears to follow the expected trend, with compounds having lower values for the first pK_a eluting first. A discontinuity becomes apparent, however, when we compare monocarboxylic acids with di- and tricarboxylic acids. These acids have a higher degree of exclusion because they contain more ionogenic groups than monocarboxylic acids. Thus we can concentrate on the dicarboxylic acid data. Because oxalic acid is fully dissociated at the mobile-phase pH, we must also remove this data point from the analysis.

Table 10.10 Ion Exclusion

Acid	Retention Time* (min)	pK_a (25 °C)
Oxalic acid	6.4	1.25, 3.67
Citric acid	8.0	3.13, 4.76, 6.40
Malic acid	9.7	3.46, 5.10
Succinic acid	12.1	4.21, 5.72
Formic acid	13.8	3.74
Acetic acid	15.3	4.76

*Flow rate: 0.6 mL/min; column: 300×7.8 mm; mobile phase: 0.004 N H_2SO_4.
Source: Adapted from Bio-Rad Laboratories. Life science research products 1997. Bio-Rad Laboratories 1997:46.

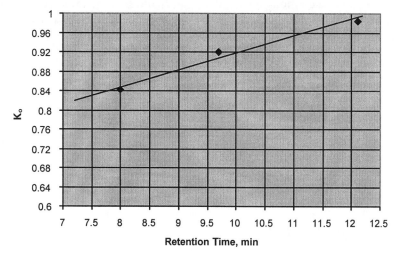

Figure 10.12 Retention time of fumaric acid.

We can therefore plot the data as shown in Figure 10.12. We find that, for fumaric acid,

$$\frac{1}{1 + 10^{(2.4-3.02)}} = 0.807 \tag{10.95}$$

This equation yields a retention time estimate of 7.0 minutes based on the data given in the problem.

10.7 AFFINITY CHROMATOGRAPHY

The specific affinity interactions of biological molecules can be exploited to select a particular molecule or class of molecules from a complex mixture. As the column goes through a binding-elution cycle in this type of chromatography, its operation can be analyzed with the constant pattern solution discussed in Section 10.4.4.

During the binding step, the sample is introduced continuously. Ideally, its introduction stops just prior to the detection of the biomolecule of interest at the column exit. During this process, the undesired solutes pass through the column. If the binding affinity is high, much of the column becomes saturated with the bound bio-molecule. The elution mobile buffer phase is then introduced into the column, thereby enabling the removal of the now highly pure biomolecule.

Before examining the column dynamics, we must first understand the affinity interaction. The most straightforward representation of this interaction during the binding phase is a mass-action equilibrium expression:

$$B^{aq} + L^{stat} \rightleftharpoons BL^{stat} \tag{10.96}$$

$$K_{affinity} = \frac{[BL]}{[B][L]} \tag{10.97}$$

Here L represents the affinity agent (the ligand) and B is the targeted biomolecule. Defining the total number of sites on the stationary phase and the fraction of occupied sites, we have

$$L_{total} = q_{max} = [L] + [BL] = q + (q_{max} - q) \tag{10.98}$$

The affinity equilibrium expression,

$$K_{affinity} = \frac{\dfrac{q}{q_{max}}}{[B]\left(1 - \dfrac{q}{q_{max}}\right)} \tag{10.99}$$

can be rearranged into the familiar Langmuir isotherm:

$$q = \frac{q_{max}K_{affinity}[B]}{1 + K_{affinity}[B]} = \frac{q_{max}[B]}{K_{diss} + [B]} \tag{10.100}$$

In this equation, K_{diss} is the equilibrium expression for the dissociation of the ligand-biomolecule complex.

Referring back to the derivation of the constant pattern solution for binding-elution cycles, we see that affinity complexation is favorable when the dissociation constant is small; this condition means that the width of the mass transfer zone is at a minimum when $K_{diss} = 0$. For this reason, affinity columns with low dissociation constants offer highly efficient operation. For example, the highly specific biotin-streptavidin interaction has a K_{diss} on the order 10^{-15}.

Example 10.5

Kang and Ryu (7) have studied the affinity chromatography of IgG using protein A. Figure 10.13 plots one of their breakthrough curves for a protein A ligand concentration of 1.22 mg/mL of gel. Using the information shown in Table 10.11, determine k_{cp}.

Solution

Because K_{diss} is so small, we will assume that it is essentially zero. We use the equation derived for the constant pattern solution:

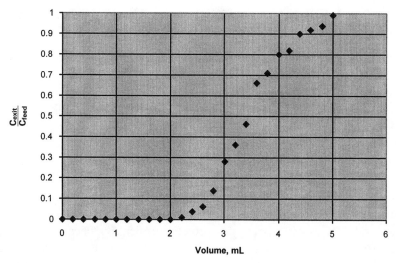

Figure 10.13 Breakthrough curve.

Table 10.11 Affinity Chromatography of IgG Using Protein A

Parameter	Value
Column diameter	0.7 cm
Superficial velocity	16 cm/h
Column length	1.5 cm
IgG in buffer	0.092 mg/mL
K_{diss}	10^{-8} M
q_m	0.12 mg/mL
Mean particle diameter	100 mm

Source: Adapted from Kang and Ryu, 1991.

$$[t'_{0.99C_f} - t'_{0.01C_f}]_{for\,K_{diss}=0} = \frac{(1-\varepsilon)q_m}{\varepsilon k_{cp}C_f}[\ln(99)] \tag{10.101}$$

Substituting the data given in the problem specification and calculating the flow rate to obtain Δt, we have

$$Q = 4\pi U\left(\frac{D^2}{4}\right) = 4\pi\left(16\,\frac{cm}{h}\right)\left(\frac{(0.7\,cm)^2}{4}\right) = 24.6\,\frac{cm^3}{h} \tag{10.102}$$

$$k_{cp} = \frac{(1-\varepsilon)q_m}{\varepsilon C_f(t'_{0.99C_f} - t'_{0.01C_f})}[\ln(99)] \tag{10.103}$$

$$k_{cp} = \frac{(1-0.3)\left(0.12\,\frac{mg}{mL}\right)}{(0.3)\left(0.092\,\frac{mg}{mL}\right)(0.205\,h - 0.086\,h)}[\ln(99)] = 117.5\,h^{-1} \tag{10.104}$$

10.8 HYDROPHOBIC INTERACTION AND REVERSE-PHASE CHROMATOGRAPHY

Hydrophobic interaction chromatography (HIC) and reverse-phase chromatography (RPC) capitalize on regions of proteins or low-molecular-weight biochemicals that are not well solvated by water and are therefore hydrophobic. HIC relies on conditions that provide mild hydrophobic interactions (actually nonwater solvating effects), with an aqueous mobile phase at high salt concentrations. This chromatography method usually employs phenyl or butyl functional groups attached by an ether linkage to the stationary phase. In contrast, RPC uses a polar organic solvent as the mobile phase, with a stationary phase that exhibits highly nonpolar functional groups such as alkanes (those having 8 or 18 carbon atoms are the most common).

In general, HIC is used mainly for protein purification, because the conditions used in this method are relatively mild. RPC, on the other hand, is used as an analytical and preparatory procedure for a vast array of low-molecular-weight biomolecules. With both techniques, equilibrium is characterized by linear isotherms because the coverage of hydrophobic sites is low.

In a useful variation of RPC for charged molecules, the chromatographer can add a chemical species to the buffer that will form an ion pair with the solute of interest. Ion-pair chromatography is popular for the analysis of oligonucleotides and carboxylic acids. It can deliver a high degree of purification because the ion pair exhibits properties differing greatly from those of the original, charged biosolute and those of the other solutes. With a judicious choice of the ion-pairing agent, the ion pair can prove quite hydrophobic and demonstrate a much longer retention time than many other species normally present in a biological sample. It also represents an alternative to ion-exchange chromatography for resolving solutes such as oligonucleotides, which carry a large negative charge and can be eluted on this basis from the stationary phase during anion exchange.

10.9 PERFUSION CHROMATOGRAPHY

In a relatively new development, chromatographers have used stationary phases containing such wide channels that fluid can actually flow through the particle, rather than merely diffusing through the gel or pore network. This membrane-like operation combines rapid separation with a capacity similar to that noted with high-performance and gel-permeation chromatography.

One model of perfusion chromatography suggests that the mobile-phase velocity through the pore is proportional to the column interstitial velocity:

$$U_{pore} = \alpha U_{column} \tag{10.105}$$

Using the definition of the Peclet number (Pe) given in DLRM, we find that the plate height from a van Deemter analysis is

$$H \propto \frac{U d_p^2}{D_{mobile\ phase}} + constant \tag{10.106}$$

Adding this relationship to the standard definition of the diffusion coefficient in the pores of a particle yields an effective diffusivity,

$$D_{eff} = D_{pore} + \frac{U_{pore}d_p}{2} \tag{10.107}$$

that can be used in the van Deemter equation and the binding-elution constant pattern expression for k_c. When the pore size is large enough to allow significant pore velocities, the second term gives the effective diffusivity. Substituting the equations for U_{pore} and D_{eff} into the van Deemter equation, we find that the plate height is independent of mobile-phase flow rate and proportional to the particle size:

$$H \propto \frac{2d_p}{\alpha} + \text{constant} \tag{10.108}$$

The constant pattern analysis shows that, when separating large molecules such as proteins, pore diffusion is the slowest step in the mass transfer. Based on the pore diffusion equation for k_c,

$$k_c^{\text{perfusion}} = \frac{60D_{eff}(1-e)L}{d_p^2 U_{column}} = \frac{30\alpha(1-e)L}{d_p} \tag{10.109}$$

we can obtain a result similar to the van Deemter equation—that is, the width of the breakthrough curve does not depend on mobile-phase flow rate and is proportional to the particle size.

10.10 OTHER CHROMATOGRAPHIC METHODS

Adsorption and other nonspecific interactions are used for many types of recovery operations. Suitable adsorbents have a high surface area and bind biomolecules partly through hydrophobic interactions (such as the binding of carboxylic acids using activated carbons) or by generalized affinities (such as ionic interactions and steric effects—for example, protein and nucleic acid purification using hydroxyapatite). Adsorption chromatography is usually performed with binding-elution cycles. The resulting equilibrium can be modeled as a Langmuir or Freundlich isotherm.

Chromatographers have realized several important technological advances by making simple changes in the elution cycle or the way in which the sample is applied. Some of these modifications are discussed here; others are described in the problems at the end of the chapter.

10.10.1 Gradient methods

One useful strategy in chromatography is to control the composition of the mobile phase. In ion-exchange chromatography, for example, altering the salt concentration of the mobile phase in a linear fashion may sharpen peaks. As the salt concentration increases in the column, the linear equilibrium constant decreases because of the greater competition, which in turn reduces the peak width.

RPC also employs gradients, especially to speed the residence time of strongly bound solutes. By waiting until the weakly held species elutes and then adding a strongly interacting solute to the mobile phase, a tightly bound solute can be forced to elute more rapidly (2).

Another strategy involves the manipulation of the pH in affinity chromatography to elute species. In IgG purification using protein A, for example, pH manipulation can separate subtypes IgG1, IgG2, and IgG3 (7).

10.10.2 Displacement chromatography

Another enhancement of chromatography involves the addition of a displacer molecule after the feed pulse. Of course, the displacer molecule, which binds to the stationary phase more strongly than any compound in the feed, must be removed from the column before a new separation can be performed. In the case of protein separation via ion exchange, small molecules can function as displacers. This technique can prove both useful and economical if these displacers can be tolerated as a buffer component or removed during desalting using gel-permeation chromatography (8).

10.10.3 Radial-flow chromatography

To circumvent the problems associated with increasing the length of columns during scale-up (see Section 10.11), columns can be piped so that the mobile phase flows radially into the column (Figure 10.14). This geometry allows samples to be processed rapidly, as the species have a much shorter residence time when they flow across—rather than down—a column. This configuration is especially useful for soft stationary phases, which are more prone to collapsing at the higher hydrostatic pressures needed to flow mobile phases axially in long columns. Scale-up is also facilitated, because the column size and flow rate are increased proportionately without increasing the time for separation (9). Short processing times are especially desirable when dealing with unstable products or protein separations involving proteases.

10.10.4 Membrane chromatography

Combining the strengths of two useful bioseparation processes can yield an even more valuable process. For example, membrane chromatography—like perfusion chromatography—can rapidly resolve solutes while retaining the power of interaction with a solid phase. In this technique, interaction with the solid phase takes place inside the pores of the membrane (10). Thus large molecules are retained, small molecules that do not interact with the solid phase permeate the membrane rapidly, and small molecules that interact with the solid phase permeate more slowly. To date, most of the applications of membrane chromatography have involved affinity chromatography employing "on-off" cycles.

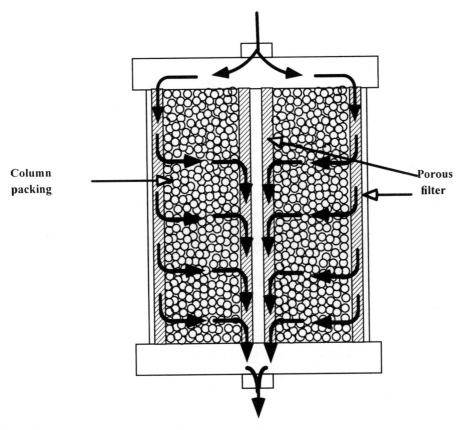

Figure 10.14 Radial-flow chromatography.

10.11 SCALE-UP STRATEGIES AND CONSIDERATIONS

To scale up a chromatographic method from the laboratory to commercial operation, the chromatographer must satisfy three needs:

- Planning for testing hypotheses and mathematical models at small and intermediate scales
- Consideration of practical issues and allowances (usually called safety factors) for less-than-expected capacities
- Development of an experiential database through actual testing or by drawing upon the expertise of vendors and industry practitioners that have dealt with the system of interest

Regardless of whether the reader has prior experience in scale-up, the following discussion will provide a brief introduction to several methods that can be used to develop a framework for effective scale-up.

Three issues of importance in chromatography scale-up are column volume, volumetric throughput (flow rate), and resolution.

A key factor in large-column operation is particle size. The pressure drop across a column packed with rigid particles of a uniform size can be predicted using a modified Ergun equation:

$$\frac{\Delta P}{L} = \frac{150U(1-\varepsilon)^2\eta}{d_p^2\varepsilon} + \frac{1.75\rho U(1-\varepsilon)}{d_p\varepsilon^3} \tag{10.110}$$

This equation is not particularly well suited to dealing with very compressible stationary phases, such as some GPC packings (11). For our purposes, however, it serves as an illustrative starting point.

Particles used in bioseparations can generally withstand hydraulic pressures between 3 and 5 MPa; in contrast, most gels used in GPC are weaker and can withstand a maximum of only 1.7 MPa. This practical limitation in the applied pressure needed to force an aqueous solution through a packed column leads to an important scale-up consideration—namely, a larger particle size is needed if the pressure applied to the column to achieve a given flow rate can destroy the stationary phase. For practical reasons it is usually desirable to keep commercial-scale column pressure drops below 10.7 kPa (15 psi), as buffer tanks would need to meet American Society of Mechanical Engineers (ASME) codes if the system exceeds this pressure (12).

A quick estimation will illustrate how particle size can dictate column length. Using Equation 10.110, for stationary-phase particles of 200 μm diameter and a linear velocity of approximately 2.5 cm/min (a typical protein ion-exchange linear velocity), we can calculate that the pressure drop will be roughly 28 kPa per meter of column. Thus, under these conditions, the column length should be less than 380 cm if the pressure drop is to be less than 10.7 kPa.

With an increase in particle size, resolution may suffer. From the van Deemter model, we can see that column efficiency decreases as the height of a theoretical tray increases. The following methods show how scale-up can proceed rationally based on the equations developed in this chapter. They assume that a key ratio or ratios should remain constant during scale-up to obtain similar capacities and resolutions in larger-scale columns.

10.11.1 Scale-up method 1: no change in stationary-phase particle size

In protein separations, the jump from bench to commercial scale may be only a small hop, and column scale-up can be accomplished by varying column dimensions and conditions. We must consider four chromatography parameters: column volume, column diameter, superficial velocity or flow rate, and, in the case of gradient usage, gradient volume. These parameters relate to an increase in sample size during scale-up. Table 10.12 summarizes this scale-up approach.

The suggested scale-up conditions can be met by keeping the column length the same and increasing the column diameter and flow rate so as to maintain a constant superficial velocity; the pressure drop and the peak width are not increased. Many commercial columns have good flow distribution at the inlet, meaning that short, squat columns can yield good performance. Note also that the use of short columns translates into relatively short retention times.

By using several columns, a production facility can stagger the columns, thereby creating a continuous cycle. The combination of a small amount of stationary phase and rapid cycling yields high productivity per unit amount of stationary phase (2),

Table 10.12 Scale-Up Methods

Parameter	Method 1: No Particle Size Change	Method 2: Change Particle Size	Method 3: Most Conservative
Column volume	Increase proportional to increase in sample size.	Increase proportional to increase in sample size. The need for large-scale chromatography, however, may exceed the capacity factor used at the analytical scale, probably leading to lower resolution.	Increase proportional to increase in sample size.
Column length	Keeping the length the same will not increase peak width. Based on the DLRM model, peak width at the baseline is approximately proportional to column length to the 3/2 power when Pe << 1. Increasing the column length may thus dilute the product.	Based on the DLRM model, keeping a constant resolution requires increasing the length by $(d_{p2}/d_{p1})^n$ where n is between 1.415 and 2.	Increase in length according to the increase in diameter so that L/D stays the same.
Superficial velocity	Keep about the same or slightly lower (except when ΔP exceeds recommended value for stationary phase, then use at least 10% below this value).	Keep about the same or slightly lower (except when ΔP exceeds recommended value for stationary phase, then use at least 10% below this value).	Keep $DU\pi/\eta$ the same.
Gradient volume	Increase proportional to increase in sample size.	Increase proportional to increase in sample size.	Increase proportional to increase in sample size.

satisfying cost and product stability concerns. (Stationary phases can be expensive, and the bioactivity of some biological molecules steadily degrades during processing.)

It is generally recommended that scale-up take place in an incremental fashion. In protein chromatography, for example, increments of approximately 500 µg are advisable. As a general rule of thumb, vessel size is increased by one order of magnitude during scale-up. This gradual increase, however, can prove costly and be difficult to schedule when the market demands rapid commercialization if the company is to remain competitive. To decide whether increasing size by two orders of magnitude (at most) is feasible, the process designer should perform extensive analysis at a small-scale level.

10.11.2 Scale-up method 2: increasing stationary-phase particle size

The analyses presented in this chapter can be applied to assess many methods of scaling up separation processes by increasing stationary-phase particle size, and more extensive correlations have been presented by various researchers (13). Here we will simply provide one possible strategy and suggest how column performance would vary with scale. Increasing the particle size decreases the exposed surface area per unit volume of stationary phase. In addition, the path length for diffusion in the stationary phase increases with increasing particle diameter. As a result, the column length must increase—or the superficial velocity must decrease—with any increase in stationary phase diameter.

Using the DLRM model when Pe $\ll$ 1 (the typical case when solutes comprise large molecules), we can assume that the peak resolution is essentially proportional to the square root of the Peclet number:

$$R \propto Pe^{1/2} \propto \left(\frac{L}{U}\right)^{0.5} \frac{1}{d_p} \tag{10.111}$$

To maintain the same resolution at the same superficial velocity, the column length must vary according to the following expression:

$$L_2 = L_1 \left(\frac{d_{p_2}}{d_{p_1}}\right)^2 \tag{10.112}$$

When the Peclet number is large, the resolution depends on the rate parameter b. Thus

$$R \propto \left(\frac{L}{U}\right)^{0.5} \frac{1}{b^{0.5}} \tag{10.113}$$

For film resistance controlling, we have

$$L_2 = L_1 \left(\frac{d_{p_2}}{d_{p_1}}\right)^{1.415} \tag{10.114}$$

For pore diffusion controlling, we have

$$L_2 = L_1 \left(\frac{d_{p_2}}{d_{p_1}}\right)^2 \tag{10.115}$$

For film or pore diffusion control cases, every parameter except length and particle size remains constant. This and other empirical analyses (13) suggest that the column length depends on the particle diameter as an exponent ranging from 1.415 to 2. In any event, it is clear that column length must be increased with column volume.

10.11.3 Scale-up method 3: gel permeation and on-off cycling approach

In gel-permeation chromatography, two ratios must be kept constant to ensure that the elution pattern remains the same (14). This conservative approach should ideally keep the dynamic parameters from varying. The same strategy should be used for elution-binding cycle systems, with the ratios L/D and $DU\pi/\eta$ being held constant. Unfortunately, this method does not account for an increase in particle size. If we assume that the Peclet number should also remain constant, the particle size is constrained, because Pe is given by the following equation:

$$Pe = \frac{192 L D_{mobile}}{U d_p^2} \tag{10.116}$$

The particle size must follow the increase in the superficial velocity if the ratio U/d_p is to remain constant.

This method is only one suggestion for gel permeation and must be tested to ensure its validity. To date, it has been used in chromatography when transfer of solute to the surface of a particle—rather than diffusion within the particle stationary phase—is the rate-limiting step for solid-phase binding.

10.12 SUMMARY

Chromatography is a versatile method for purifying biological molecules. It exploits solute properties to separate the components in a solution. Many types of stationary phases are available that contain immobilized ligands or functional groups capable of interacting with the solute of interest.

This chapter introduced six mathematical schemes for analyzing chromatographic methods for the purposes of characterizing the separation achieved and designing a production-scale method. These analyses range from purely phenomenological ones, which simply characterize the data, to mechanistic analyses, which attempt to model column behavior for predictive purposes. Many more detailed analyses of chromatographic methods are described in the technical literature; myriad computer-aided column design methods have been developed as well.

Figure 10.15 provides an example of the decision-making process involved in choosing a liquid chromatographic method. This decision tree summarizes the different chromatographic methods and emphasizes the unique roles that they play in biological molecule separation. The reader is encouraged to critically review this decision tree, and then revise and refine it to suit his or her particular needs. This tree should also be revised as new methods come to light.

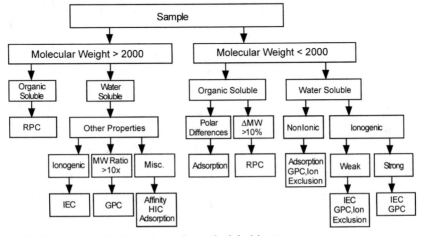

Figure 10.15 An example of a chromatography method decision tree.

10.13 PROBLEMS

10.1 Of the solute mixtures listed below, which can be separated by an ion-exchange packing and under what conditions?

    (a) Glucose, D-malic acid, sodium sulfate
    (b) Lycine, glutamic acid
    (c) Acetic acid, glucose, lactic acid
    (d) Human serum albumin, sodium chloride, human IgG
    (e) Insulin, pro-insulin
    (f) Mannose, arabinose

10.2 Using the DLRM model, provide an equation that gives, in general, the purity of a chromatographic separation between two species when the two peaks overlap as shown in Figure 10.16.

10.3 Assume that by attaching electrodes to a packed bed, ions at a certain potential migrate to the surface and create a net charge, as in ion exchange. By varying the voltage across the column in a sinusoidal fashion, ion-exchange chromatography can be performed in a new way. Assume that local equilibrium theory applies, and the solid-phase concentration varies with time according to the following equation:

$$\frac{\partial q}{\partial t} = K_{o} \sin(\omega t)\frac{\partial C}{\partial t} \tag{10.117}$$

    (a) Based on the local equilibrium theory, how does the residence time vary with column conditions?
    (b) If two species have different K_{o} values, will this technique offer any advantages in separating these components when compared to a constant charge on the solid that does not vary with time?

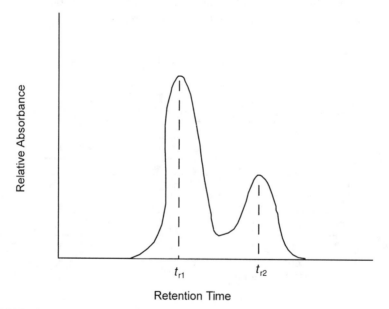

Figure 10.16 Peak overlap.

10.4 In moving-bed chromatography, the solid phase moves in one direction and the mobile phase moves in a countercurrent fashion. Using the staged model as a guide, develop equations as requested in parts (a), (b), and (c) when a perfect pulse is introduced at stage $n = 0$. Assume that the solid and liquid phases are in equilibrium when leaving each stage. Use Figure 10.17 to guide your solution. Assume that the total volume in each stage is V and the void fraction is ε.

(a) Find three equations in Laplace space that enable calculation of liquid-phase concentrations in each of the three stages.

(b) Find two equations in Laplace space that enable calculation of the liquid-phase concentrations in each of the two- and three-stage systems.

(c) Give a generalized formula for calculating stage 2 concentration in Laplace space for an nth-stage system.

10.5 In radial-flow chromatography, we can see that the superficial velocity depends on the radial position. We can use the equation of continuity to find an expression for v_r. The equation of continuity for constant density in radial chromatography is

$$\frac{d}{dr}(rv_r) = 0 \tag{10.118}$$

Since at $r = R$, $Q = \pi R^2 vR$,

$$v_r = \frac{Q}{\pi Rr} \tag{10.119}$$

The equation that describes the phenomenon using local equilibrium theory is

$$\left(\frac{Q}{\pi R}\right)\frac{1}{r}\frac{\partial C_A}{\partial r} + \left(1 + \frac{1-\varepsilon}{\varepsilon}\right)K\frac{\partial C_A}{\partial t} = 0 \tag{10.120}$$

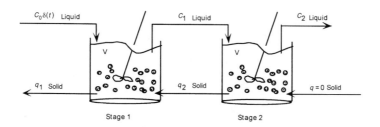

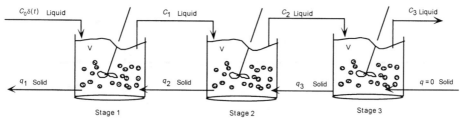

Figure 10.17 Moving-bed chromatography.

(a) Using Laplace transforms and van der Laan's theorem, solve Equation 10.120 for the residence time.

(b) Prove that the residence time is the same only when the radius of the column and the flow rate change proportionately.

10.6 Assume that we can encapsulate a gel within a gel so that large molecules will be totally excluded from the stationary phase, small molecules can penetrate the inner and outer spheres, and molecules that lie in the size range between these extremes can enter the large particles but are excluded from the inner particles. Using the gel-permeation model, develop a model for predicting the retention time of molecules of varying size.

10.7 In a laboratory-scale column (0.5 cm diameter × 5 cm in length), silver ions can be immobilized to resolve histidine and methionine at pH 4. You will be purifying a lysed protein solution rich in methionine (30 mol%) and histidine (40 mol%) with 10 mg of total protein with a molecular weight of 60 kDa. Provide scaled-up designs for the following columns:

(a) A preparative, axial-flow column

(b) A radial-flow column

Figure 10.18 shows the chromatographic response peaks of histidine and methionine on the Ag⁺ form of Bio-Gel P-2 [GA-12 has 0.062 of Ag⁺ mmol Ag(I)/ml resin; GA-24 has 0.141 of Ag⁺ mmol Ag(I)/ml resin] at pH 4.0 with added NaCl.

Column: 0.5 cm × 5.0 cm (1.0 mL)
Mobile phase: 0.05 M sodium acetate buffer with 0.25 M NaCl, pH 4.0

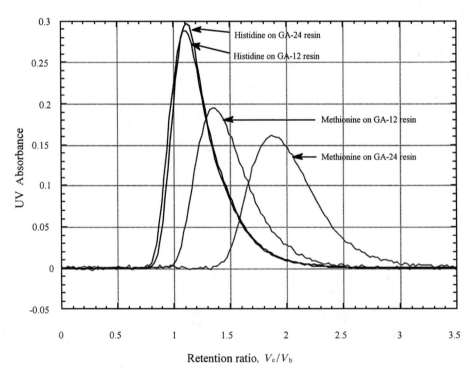

Figure 10.18 Chromatographic response peaks.

Flow rate: 0.2 mL/min

Sample concentration: 0.005 M

Sample loop: 20 μL

Bio-Gel P-2 properties: paricles of 40–80 μm diameter; hydrated volume of 3.5 mL/g dry gel

10.8 For a given column exhibiting N stages, what is the minimum value for the proportionality constant δ for $K_1 = \delta K_2$ so that the peak resolution is 2 when using the staged model? Assume that the mobile and stationary phases are in local equilibrium.

10.9 Are there any advantages in using a column with a binary mixture of stationary phase, such as a cation-exchange material and a gel permeation material, relative to using separate columns? Assume that the total amount of stationary phase used would be the same when combined or when using separate columns. Determine whether a mixed column is advantageous, including in your explanation a consideration of each of the following process parameters.

    (a) Column equipment cost, excluding the cost of the stationary phase

    (b) Mobile-phase chemical cost

    (c) Processing time

    (d) Pump operating costs

    (e) Peak resolution

10.10 In two-dimensional chromatography, a channel can be packed with a gel-permeation packing in the vertical direction while an electric field is imposed in the horizontal direction. In this way, molecules can be separated by charge and size simultaneously. Assume that this process can be modeled with local equilibrium theory using the following equation, where the electric field causes the velocity in the x-direction:

$$U_z \frac{\partial C_A}{\partial z} + U_x \frac{\partial C_A}{\partial x} + \left[1 + \left(\frac{1-\varepsilon}{\varepsilon} \right) K \right] \frac{\partial C_A}{\partial t} = 0 \tag{10.121}$$

U_x can be related to the applied electric field by $U_x = \mu E$.

    (a) Provide a derivation of this expression based on a species balance.

    (b) If the boundary and initial conditions are the same as for local equilibrium theory, but with the added condition that at $x = 0$, $C_A = 0$, find a solution to this equation using van der Laan's theorem.

    (c) When the electric field goes to zero, prove that the solution to part (b) reduces to the answer given above for linear chromatography assuming local equilibrium.

    (d) What are some advantages and disadvantages to performing the separation in this manner, rather than following gel permeation with ion exchange or chromatofocusing?

10.11 Herns, Hadder, and Aguilar (15) state that the capacity factor k' can be related to the ion-exchange distribution coefficient K and the salt concentration in isocratic (i.e., no gradient) IEC using the following semiempirical relationship:

$$\log k' = \log K + Z \log \left(\frac{1}{C} \right) \tag{10.122}$$

Table 10.13 Data for Anion Exchange of dUTP on an Ag^+ Column

Retention Time* (min)	NaCl (M)
66	0
44	10^{-6}
36	10^{-6}
32	10^{-6}
30	10^{-6}
30	10^{-2}
30	10^{-1}

* $t_o = 30$ min.
Source: Adapted from Agarwal S. Separation of biotin-labeled dUTP from its nonlabeled counterpart using immobilized Ag(I) metal affinity chromatography. Arizona State University; M.S. thesis, 1996:76.

Table 10.14 Ion-Exchange Chromatography of IgG

Chromatography Parameter	Value
Diffusivity	$4 \times 10^{-7}\,cm^2/s$
Column length	20 cm
U	$0.5\,cm^3/min$
d_p	$100\,\mu m$
K	0.01
Injection concentration	$10^{-2}\,M$

(a) Based on the data in Table 10.13 for the oligonucleotide dUTP on an Ag^+ column, find the parameters K and Z.

(b) What is the effect of doubling the column length while changing the NaCl concentration from 10^{-4} to 10^{-5}?

10.12 Using the DLRM model for the separation of IgG from BSA with ion exchange when the Peclet number is very large, what column conditions should be chosen so that the peak can be removed at a concentration greater than $10^{-4}\,M$? Use the data in Table 10.14.

10.13 Read a biochemistry or medical textbook to find out about the components that make up animal or human blood serum. Develop a chromatography decision tree specifically for separating different components of serum.

10.14 REFERENCES

1. Giddings JC. Dynamics of chromatography. New York: Marcel Dekker, 1965.
2. Wankat PC. Rate-controlled separations. New York: Elsevier, 1990.
3. Warzecha L, Piwowar Z, Bodzek D. Gel permeation chromatography for the determination of the molecular mass distribution of some industrial proteins. J Chrom 1990;509:227–231.
4. Flapper W, Theeuwes AGM, Kierkels J, Steenbergen J, Hoenders HJ. Separation of serum proteins by high-performance gel-permeation column systems. J Chrom 1990;533:47–61.
5. Fuchs MJ, Keim V. Separation of rat pancreatic secretory proteins by cation-exchange fast protein liquid chromatography. J Chrom 1992;576:287–295.
6. Bio-Rad Laboratories. Life science research products 1997. Bio-Rad Laboratories 1997:46.
7. Kang KA, Ryu DDY. Studies on scale-up parameters of an immunoglobulin separation system using protein A affinity chromatography. Biotechnol Prog 1991;7:205–212.

8. Brooks CA, Cramer SM. Steric mass-action ion exchange: displacement profiles and induced salt gradients. AIChE J 1992;38:1969.
9. Saxena V, Dunn M. Solving scale-up: radial flow chromatography. Bio/Technology 1989:250–255.
10. Thömmes J, Kula M-R. Membrane chromatography—an integrative concept in the downstream processing of proteins. Biotechnol Prog 1995;11:357–367.
11. Mohammad AW, Stevenson DG, Wankat PC. Pressure drop correlations and scale-up of size exclusion chromatography with compressible packings. Ind Eng Chem Res 1992;31:549–561.
12. Prouty WF. How to recover recombinant protein products. Chemtech 1992:608–615.
13. Rudge SR, Ladisch MR. Process considerations for scale-up of liquid chromatography and electrophoresis. In: Asenjo AA, Hong J, eds. Separation, recovery, and purification in biotechnology. Washington: American Chemical Society, 1986.
14. Dechow FJ. Separation and purification techniques in biotechnology. New Jersey: Noyes, 1989.
15. Hearn MTW, Hodder AN, Aguilar MI. High-performance liquid chromatography of amino acids, peptides, and proteins. J Chromatogr 1988;458:27–44.

11

Extraction

The introduction of another phase into an aqueous solution results in an equilibrium-based separation through the movement of the extracted solute into a new solution environment. This new environment does not drastically reduce the solute's conformational and translational energies, as in solid-phase membrane transport or adsorption. This preservation of properties is especially important for proteins, which can lose bioactivity or become completely denatured after interacting with a solid phase. The new fluid environment can also be manipulated to facilitate end-product formulation. In this chapter, we examine the basic principles of extraction and assess various extraction systems.

11.1 CHEMICAL THERMODYNAMICS OF PARTITIONING

At equilibrium, each component in a solution has the same chemical potential:

$$\mu_{solute}^{I} = \mu_{solute}^{II} \tag{11.1}$$

Using the definition of the chemical potential, we obtain

$$\mu_2^{I,0} + RT \ln(x_2^I \gamma_2^I) = \mu_2^{II,0} + RT \ln(x_2^{II} \gamma_2^{II}) \tag{11.2}$$

where γ_2^I and γ_2^{II} are the activity coefficients of the solute, referred to by the subscript 2. Rearranging to solve for the ratio of mole fractions yields

$$\ln\left(\frac{x_2^I}{x_2^{II}}\right) = \frac{\mu_2^{II,0} - \mu_2^{I,0}}{RT} - \ln\left(\frac{\gamma_2^{II}}{\gamma_2^I}\right) \tag{11.3}$$

If the ratio of activity coefficients is a constant, then at a constant temperature the mole fraction ratio is equal to a partition coefficient:

$$K_2 = \frac{C_2^I}{C_2^{II}} \tag{11.4}$$

The units in which K is expressed can vary, so you should take care when using partition coefficients from published sources. The superscript c is sometimes used to denote concentration ratios, while the superscript x denotes mole fraction ratios. The partition coefficient is often used in partitioning model development through semi-empirical methods; thus

$$K = f(T, I, pH, C_j, \ldots) \tag{11.5}$$

where I is the ionic strength.

Another important quantity to measure is the selectivity, which indicates the effectiveness of the extraction in preferentially selecting one solute over another:

$$\alpha_{1,2} = \frac{x_1^{I}/x_1^{II}}{x_2^{I}/x_2^{II}} \tag{11.6}$$

In terms of the partition coefficient, which is applicable when the ratio of activity coefficients remains constant, the selectivity is defined as follows:

$$\alpha_{1,2} = \frac{K_1}{K_2} \tag{11.7}$$

Many models that purport to account for chemical and physical interactions have been developed for solvent extraction processes. In the following sections, we examine how these specific models can be applied to bioseparations.

11.2 ORGANIC-AQUEOUS EXTRACTION

Organic solvents that form a second phase when added to water have been used to extract low-molecular-weight biochemicals, such as carboxylic acids, amino acids, alcohols, and antibiotics. They have also seen limited use in removing biopolymers during enzyme purification. Table 11.1 lists some of the advantages and disadvantages associated with organic-aqueous extractions.

In general, organic solvents denature proteins, polynucleic acids, and cells. As a consequence, they are rarely used for purifying these biomolecules (except in the case of reverse micelles, discussed in Section 11.4). Because polar and unsaturated organic solvents have trace to moderate water solubility, extraction of biochemicals from whole-cell suspensions is not recommended as well. In such cases, cell lysis can occur; besides killing the cells, this rupture may contaminate the organic phase with lipophilic membrane biomolecules.

By itself, transferring a solute to the organic phase is usually insufficient to ensure product recovery. In addition to being separated from contaminants in the aqueous phase, the solute must be crystallized from the organic phase or back-extracted to a clean aqueous phase to undergo further product finishing. Low-molecular-weight organic solvents typically have a lower latent heat of vaporization than water—a desirable property when evaporative crystallization is performed from the organic phase.

Table 11.1 Advantages and Disadvantages of Organic Solvent Extraction

Advantages	Disadvantages
Organic solvents are more volatile than water	Organic solvent environments can denature proteins and biomolecules
Crystallization may produce larger, purer crystals	Flammability can become an issue
Inorganic salts are mostly left behind in the aqueous phase	Toxicity risks and hazardous waste disposal costs must be assessed

Water plays a complex role in the organic phase. Solutes in the organic phase that form complexes with water can improve the distribution coefficient and control solubility to manage crystallization from the predominantly organic phase. The presence of water in the organic phase, however, can lead to emulsions. If the solute has a high affinity for water, then water may complicate crystallization. Also, water can be an undesirable impurity in the final product. If the solute needs to be back-extracted into an aqueous phase for further processing, however, organic-phase water has little effect on the separation.

The old chemistry adage that "like dissolves like" applies to organic solvent extraction, as can be seen in Table 11.2, which lists the partition coefficients and conditions for extraction of various biochemicals. More extensive databases are available in the references cited at the end of this chapter.

One process tool that may be overlooked in organic-aqueous extraction involves the judicious control of temperature. Using the equations derived earlier, the effect of temperature on the partition coefficient can be written as follows:

$$\ln(K_i^x) \propto \frac{1}{T} \tag{11.8}$$

Because many biochemicals are labile, increasing temperature much higher than the general physiologic temperature of 35 °C will not improve the separation. Equation 11.8, however, suggests that lowering the temperature to near the freezing point of water can significantly increase the distribution coefficient. Moreover, after low-temperature extraction, increasing the temperature to 80 to 90 °C during back-extraction can change the distribution coefficient significantly without exposing most biochemicals to conditions that might increase their lability. In the extraction

Table 11.2 Partition Coefficients for Selected Biochemicals Using Pure Solvents

Biochemical	Organic Solvent	Conditions	K
Carboxylic acids			
Citric acid	n-Butanol	25 °C	0.29
	MIBK	25 °C	0.009
Shikimic acid	Hexane	25 °C	0.01
	Propyl acetate	25 °C	0.06
Succinic acid	n-Butanol	25 °C	1.20
	n-Octanol	25 °C	0.26
Alcohols			
Ethanol	n-Octanol	25 °C	0.49
n-Butanol	n-Octanol	25 °C	7.6
Ketones			
Acetone	n-Octanol	25 °C	0.58
Methyl ethyl ketone	n-Octanol	25 °C	1.95
Antibiotics			
Penicillin F	Amyl acetate	pH 4	32
		pH 6	0.06
Erythromycin	Amyl acetate	pH 6	120
		pH 10	0.04

of antibiotics, for example, maintaining lower temperatures and a shorter duration of exposure to organic solvents is considered desirable.

To transfer ionogenic species into the organic phase, it is necessary to neutralize the charge, because the dielectric constants of most organic phases are very low. In one simple approach, we can adjust the pH so that the neutral species will predominate. When extracting a weak acid such as a monocarboxylic acid, the partition coefficient can be rewritten as

$$K_{HA} = \frac{C_{HA}^{org}}{C_{HA}^{aq} + C_{A^-}^{aq}} \tag{11.9}$$

because the aqueous phase includes both the ionized and un-ionized species. Using the definition of the acid dissociation constant, we find the following relationship:

$$K_a = \frac{C_{A^-}^{aq}(10^{-pH})}{C_{HA}^{aq}} = 10^{-pK_a} \tag{11.10}$$

This expression can be solved for the concentration of A^- and combined with Equation 11.9 to obtain the useful forms:

$$K_{HA} = \frac{C_{HA}^{org}}{C_{HA}^{aq}(1 + 10^{pH-pK_a})} \tag{11.11}$$

$$K_{HA} = \frac{K_0}{1 + 10^{pH-pK_a}} \tag{11.12}$$

In Equation 11.12, K_0 is the partition coefficient based on a non-ionizing solute. When the pH is less than the pK_a of the acid, the partition coefficient approaches the value for an un-ionized species. As the pH increases to exceed the pK_a, however, the partition coefficient goes to zero. This effect, which is illustrated in Figure 11.1, has been documented in a number of systems.

In another approach, we can pair the desired cation, anion, or Zwitterion with an ionized molecule, forming an ion pair. The ideal ion-pairing agent is water-soluble when ionized, but forms a very hydrophobic species when paired with the target molecule. Because the ion pair rarely constitutes the desired product, the pair should be easily dissociated to yield the original molecules. In Section 11.2.1, we explore the engineering of the organic phase with the use of ion pairs and other complexing agents in more detail.

11.2.1 Extractant/diluent systems

By introducing a chemical in the organic phase that can be tailored to specifically bind to the desired biomolecule, we can realize a greater degree of specificity and obtain higher distribution coefficients. Such a two-component organic phase is referred to as an extractant/diluent system. In this type of system, a low concentration of organic-soluble extractant provides the complexation chemistry needed to move the biomolecule into the organic phase. The diluent is normally relegated to the role of controlling the density and viscosity of the organic phase and maintain-

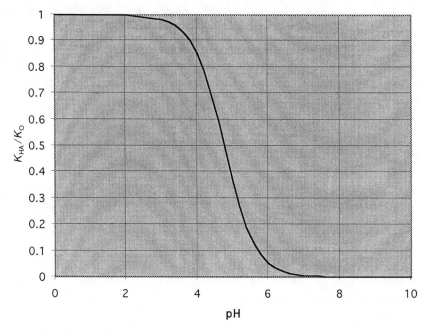

Figure 11.1 The effect of pH on the partition coefficient when only the un-ionized form of a weakly ionic solute with $pK_a = 4.76$ is extracted.

ing easy phase disengagement. Both components should be recovered and reused, given the environmental and cost concerns associated with their use.

Engineering a suitable extractant/diluent system can blur the roles played by the two components, however. For example, diluents may actively participate in the complexation of the biomolecule in the organic phase. Water may also play an active role in solubilizing the solvent/extract complex in the diluent phase.

A variety of complexation chemistries have been employed in bioseparations, and the need for improved selectivity and yield continues to fuel research efforts. Among the classes of interaction chemistries relevant to extracting biochemicals from aqueous solution are hydrogen bonding, ion pairing, and Lewis acid-base complexation.

11.2.1.1 Hydrogen bonding

When biosolutes contain carboxylic acid, alcohol, ketone, ester, or ether functional groups, complexation by hydrogen bonding represents a good strategy for targeting the solute. Hydrogen-bond energies (typically about 2 kJ/mol) are generally weaker than conventional biosolute acid-base interactions, such as those between amines and carboxylic acids. Biological systems usually possess multiple hydrogen bonds, as in double-stranded DNA. In contrast, most organic extractants and biochemical solutes exhibit only one hydrogen bond per molecule.

Hydrogen bonding can also play a significant role in extracting more complex solutes that are coupled with another, stronger functional group. We will discuss examples of this synergy in the next sections.

When using extractant/diluent systems, the concentration of extractant is an important process parameter. Thus the chemical equilibrium (also known as the law of mass action) approach provides a useful model for analysis. In one simple approach, an unknown number of molecules of extractant complex with the bio-chemical (denoted as Bio):

$$\text{Bio}^{\text{aq}} + p\text{S}^{\text{org}} \rightleftharpoons \text{BioS}_{\text{p}}^{\text{org}} \tag{11.13}$$

$$K_{\text{S}} = \frac{C_{\text{BioS}_{\text{p}}}^{\text{org}}}{C_{\text{Bio}}^{\text{aq}} C_{\text{S}}^{\text{p, org}}} \tag{11.14}$$

Although the biosolute and extractant exist in different phases, we assume that they can still come within close proximity to form a complex. We can rationalize this assumption by suggesting that complexation occurs at the interface between the organic and aqueous phases.

If the biochemical contains a weak acidic or basic functional group, then we must introduce these species into the equation for modeling the partition coefficient:

$$K_{\text{HB}} = \frac{C_{\text{BioS}_{\text{p}}}^{\text{org}}}{C_{\text{Bio}}^{\text{aq}} C_{\text{Bio}^-}^{\text{aq}}} = \frac{K_{\text{S}} C_{\text{S}}^{\text{p, org}}}{1 + 10^{\text{pH}-\text{p}K_a}} \tag{11.15}$$

Obviously, when p is greater than 1, the partition coefficient is greatly dependent on the concentration of extractant. To find p, we conduct experiments in which the pH remains well below the $\text{p}K_a$ of the biochemical. An experimental value of p can be found by determining the slope of a plot of $\ln(K)$ versus $\ln(C_{\text{s}})$.

11.2.1.2 Ion pairing

As noted previously, the simple mineral salts of ionized molecules do not easily transfer to organic solvents that have a much lower dielectric constant than water. By pairing the biochemical with an organic ion that is water-soluble when ionized but hydrophobic when uncharged or paired, however, we can produce dramatically higher partition coefficients.

An even more useful environment arises when the extractant is always charged in aqueous solution, as in the case of a quaternary amine. As before, we must analyze the dissociation reaction as well as the complex formation and distribution for both the biosolute and (if necessary) the extractant. Using the previous model of a weakly acidic biosolute and assuming the use of a quaternary amine as the extractant, we have the following reaction:

$$\text{A}^{-,\,\text{aq}} + \text{RN}^{+,\,\text{aq}} \rightleftharpoons \text{RN}^+\text{A}^{-,\,\text{org}} \tag{11.16}$$

$$K_{ip} = \frac{C_{\text{RN}^+\text{A}^-,\text{org}}}{C_{\text{A}^-}^{\text{aq}} C_{\text{RN}^+}^{\text{aq}}} \tag{11.17}$$

The partition coefficient becomes

$$K_{\text{HB}} = \frac{C_{\text{RN}^+\text{A}^-,\text{org}}}{C_{\text{A}^-}^{\text{aq}} + C_{\text{HA}}^{\text{aq}}} = \frac{K_{ip} C_{\text{RN}^+}^{\text{aq}}}{1 + 10^{\text{p}K_a-\text{pH}}} \tag{11.18}$$

Partitioning is favored at high pH, because the ionized solute will complex with the extractant. Equation 11.18 assumes that the unpaired ionized extractant and biosolute will not be present in the organic phase. Note, however, that this assumption is not valid for long-chain ion-pairing agents of sufficient hydrophobicity in polar organic solvents.

11.2.1.3 Acid-base interactions

Lewis acid-base reactions encompass a wide range of chemical interactions, all of which involve electron sharing. Several acid-base scales (as discussed in Chapter 4) are available to guide extractant selection. Some of these scales overlap with hydrogen-bonding systems, where extractants such as tributyl phosphate (TBP) and trioctylphosphine oxide (TOPO) are more generally classified as oxygen donor extractants.

Water is both acid and base. Under Pearson's hard soft acid base (HSAB) theory, it is classified as a hard acid. In a separation, adjustment of pH involves the addition of mineral salts that act as hard acids or bases. According to the HSAB theory, hard acids prefer to complex with hard bases, and soft acids prefer to complex with soft bases. Thus a good overall strategy with acid-base interactions is to pick, if possible, an extractant that can complex with the solute through soft or moderately hard acid-base interactions. Removing mineral salts and water—and thereby the competitive effect—broadens the conditions in which extraction can be conducted and eliminates the need for desalting operations.

We can clearly distinguish between acid-base interactions and other specific chemical effects by requiring 1) the existence of a regular stoichiometry (usually 1:1), and 2) a correlation between the partition coefficient for a particular solute and the inherent acidity or basicity of the extractant. Based on our previous discussion of ion-pair extractants and assuming a 1:1 stoichiometry, the partition coefficient can be described as follows:

$$A^{aq} + B^{org} \rightleftharpoons AB^{org} \tag{11.19}$$

$$K_{a-b} = \frac{C_{HAB}^{org}}{C_A^{aq} C_B^{org}} \tag{11.20}$$

If the biochemical being extracted or the extractant is a weak Brønsted acid or base, then any variation in pH will alter the partition coefficient because of competition from mineral salt, hydronium, or hydroxide ions. It may be quite difficult to study this effect when we use a cell culture medium, as such a solution usually contains sulfate, phosphate, ammonium, chloride, sodium, and trace metal ions. In some cases, methanol or acetone may be used to precipitate these unwanted salts prior to extraction.

Linear free-energy relationships (LFERs) have been extensively used as semi-empirical formulas for predicting catalyst strength, reaction kinetics, and thermodynamic parameters. They can also relate the acid-base properties of extractants to partition coefficients (see Chapter 4). Using this approach, we can tailor extractants for specific separation applications.

11.2.2 Removing biochemicals from the organic phase

Although this chapter has so far focused on the movement of the biosolute into the organic phase, nearly all recovery processes require that the end product be crystalline or in aqueous solution. As organic solvent extraction involves mostly low-molecular-weight biochemicals, the final product usually takes a crystalline form. Consequently, two possibilities for processing exist: 1) crystallization from the organic phase and 2) back-extraction into the aqueous phase followed by further purification and then crystallization. Crystallization from the organic solvent can be desirable if the process uses evaporatively driven supersaturation, as organic solvents have low latent heats of vaporization. When the process relies on extractant/diluent systems, however, the complex must be reversed prior to crystallization. (This step is unnecessary in the rare case in which the complex is the desired end product.)

11.2.2.1 Temperature swing

Increasing the temperature generally lowers the partition coefficient for organic extraction. When extractant/diluent systems are used, a temperature increase generally decreases acidity or basicity. Temperature modifications have little effect on the dissociation constant of carboxylic acids; in contrast, the dissociation constants for amines are quite responsive to temperature changes. Perrin, Dempsey, and Sarjeant (1) provide a useful set of relations for predicting dissociation constants and forecasting the effect of temperature on the dissociation constant.

Example 11.1
Perrin, Dempsey, and Serjeant (1) give the following equation for the prediction of temperature effects on the dissociation constant when the process involves amines:

$$-\frac{d(pK_a)}{dT} = \frac{pK_a - 0.9}{T} \qquad (11.21)$$

If tridodecylamine (pK_a = 10.5 at 25 °C) is used to extract shikimic acid from aqueous solution, predict the change in the pK_a of tridodecylamine if the extraction is performed at 5 °C, followed by a back-extraction procedure at 80 °C.

Solution
We can transform Equation 11.21 by noting that

$$\frac{d(\log T)}{dT} = \frac{1}{T}dT \qquad (11.22)$$

When we plot pK_a versus log T, the slope is equal to 9.6 for tridodecylamine. Thus pK_a = 10.8 at 5 °C, and pK_a = 9.8 at 80 °C. The temperature increase decreases the dissociation constant by a factor of 10. It also dramatically reduces the strength of the acid-base solution in extraction. Consequently, a temperature swing is an effective means for back-extraction of the acid into water.

11.2.2.2 Displacement (push) and aqueous-phase reaction (pull)

We can also borrow from a popular technique in liquid chromatography during extraction. A compound that can effectively compete with the solute complex can be introduced into the organic phase. Displacement could result in direct crystallization from the organic phase in the absence of an aqueous phase. Alternatively, if it comes into contact with fresh aqueous phase, the addition of a displacer compound may result in mass transfer to the aqueous phase. Transferring through displacement allows the solute to be recovered in its original form.

A chemical equilibrium model for this phenomenon is constructed as before:

$$AB^{org} + C^{org} \rightleftharpoons AC^{org} + B^{aq} \tag{11.23}$$

$$K_{dis} = \frac{C_{AC}^{org} C_B^{aq}}{C_C^{org} C_{AB}^{org}} \tag{11.24}$$

Because we can also measure the back-extraction (the total amount of extractant is known) assuming that the amount of free extractant is negligible, we can derive the following relationships:

$$K_{BE} = \frac{C_B^{aq}}{C_{AB}^{org}} \tag{11.25}$$

$$A_{total} = C_{AB_{(org)}} + C_{AC_{(org)}} \tag{11.26}$$

$$C_{A_{total}}^{org} = \frac{C_B^{aq}}{K_{BE}} + K_{BE}K_{dis}C_C^{org} \tag{11.27}$$

Figure 11.2 shows oleic acid, an organic-phase displacer. Oleic acid ($pK_a = 3.33$) has an acidity similar to that of shikimic acid ($pK_a = 4.25$), and it is soluble in the organic phase. Its addition can therefore readily displace shikimic acid from the organic phase:

$$K_{BE} = \frac{C_{acid}^{aq}}{C_{acid}^{org}} \tag{11.28}$$

Alternatively, if the solute is a carboxylic acid, adding a water-soluble base to the aqueous phase will increase the driving force and thereby move the solute out of the organic phase. This procedure, however, produces an aqueous-phase complex that must undergo further processing to yield the final product. It is also possible to simply pull the solute out of the organic-phase complex through an aqueous-phase reaction such as an acid-base reaction.

11.3 TWO-PHASE AQUEOUS EXTRACTION

The addition of one or two water-soluble polymers along with a mineral salt can cause two distinct phases to form. These phases will typically contain 80% to 90% water. This interesting phenomenon, which was first observed by Beijerinck in 1896, has been well documented with a number of ionic and nonionic polymers. Some

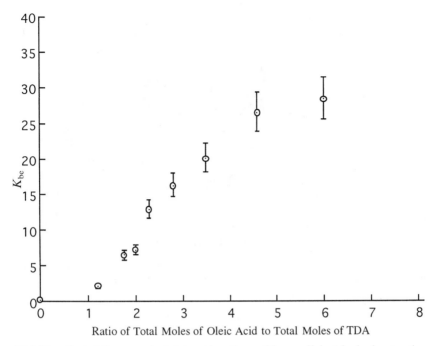

Figure 11.2 The effect of the amount of oleic acid on the partition coefficient for back-extraction of quinic acid using tridodecylamine (TDA) at 22 °C.

reports have indicated that as many as 22 aqueous phases can be generated using a variety of polymers possessing slightly different chemical structures!

Two-phase aqueous extraction's major attraction is its ability to provide a gentle environment for the purification of proteins and cells, thereby preventing denaturation. This environment primarily results from the low interfacial tension between the predominantly aqueous phases.

One interesting feature of these systems is that the partitioning of biomolecules in the solution occurs because of several different phenomena. We will briefly examine two of the more important effects—size and charge—using a simplified framework before delving into more detailed models.

11.3.1 Partitioning due to size

In an idealized model, proteins or cells can be considered to behave as solid colloidal spheres. When a sphere moves from one aqueous phase to another, the change in free energy reflects the difference between the surface energy of the sphere-liquid interface in each phase. The surface energy of this interface is the product of the surface area and the sphere surface tension:

$$G_1 = 4\pi R^2 \gamma_{sl_1} \tag{11.29}$$

A similar equation can be written for the second phase. Thus the change in free energy for moving an idealized protein or cell sphere is proportional to the change in surface energy and surface tension:

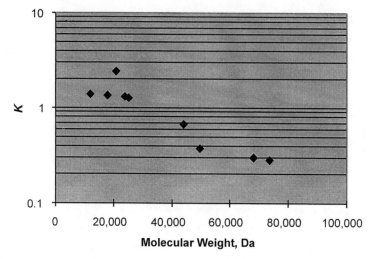

Figure 11.3 The effect of molecular weight on the partition coefficient of proteins in two-phase aqueous extraction, illustrating a trend predicted by the Brønsted equation.

$$\Delta G = 4\pi R^2 (\gamma_{sl_2} - \gamma_{sl_1}) = -RT \ln(K) \tag{11.30}$$

The logarithm of the partition coefficient is clearly proportional to the surface area of the sphere. Given that the surface area of globular proteins is proportional to molecular weight, the Brønsted equation holds for a variety of proteins (Figure 11.3). In Figure 11.3, the experimental data points include insulin, lysozyme, papain, trypsin, α-chymotrypsin, ovalbumin, bacterial α-amylase, BSA, and human transferrin. Adding data points for other proteins or enzymes, such as β-galactosidase (molecular weight = 500,000), can scatter the trend noted in the figure. To obtain the data in Figure 11.3, partitioning was performed with a solution of 7% dextran 500, 4.4% poly(ethylene oxide) 8000, 0.1 M NaCl or 0.05 M NA$_2$SO$_4$, and 10 mM phosphate or glycine buffer at 20 °C.

11.3.2 The effect of protein charge on partitioning

If the system includes a higher concentration of ionic buffer salts, we can modify the protein partition coefficient by changing the pH, ionic strength, or the type of electrolyte in the solution. In particular, salts distribute unequally between the two aqueous polymer phases. This difference in salt concentrations establishes a Donnan-type electrochemical potential difference across the phases that greatly affects the partitioning of charged biomolecules. In addition, because the surface charge of a protein depends on pH, a change in the solution's pH can significantly alter the protein's partitioning behavior. The partitioning behavior of biomolecules also depends on the physical and chemical nature of the phase-forming polymers.

11.3.3 Other effects

Experimental investigations of the partitioning of biomolecules in aqueous two-phase systems (2) have shown that the partitioning behavior of biomolecules is affected by the size and conformation of the partitioned particle, the ionic composition and the salt present, the number of hydrophobic and hydrophilic groups on the surface of the biomolecule, and the concentrations and structures of the phase-forming polymers.

Several theoretical descriptions of the partitioning of biomolecules in aqueous two-phase systems have been published. For example, Alberstone (3) and Brooks (4) treated the protein as a polymer and used the Flory-Huggins theory (5) to describe the phase equilibrium. King (6) and Haynes and Cabezas (7) used the osmotic virial expansion of the McMillan-Mayer theory (8), truncated at the second osmotic virial coefficient, to calculate protein partitioning. Most recently, Haynes (9) produced a very detailed but complicated model that combined statistical mechanics elements to describe aqueous two-phase partitioning. Kang and Sandler (10) used the UNIQUAC solution model to describe nonideal polymer solutions, developing a molecular-weight distribution function for each polymer that incorporated the effect of polydispersity.

Recent data obtained from aqueous two-phase systems, along with useful models, are summarized below.

11.3.3.1 Flory-Huggins

The first attempt at writing a molecular-level theory for the partitioning of macromolecules in a two-polymer aqueous system took advantage of the Flory-Huggins formalism (5), with the partitioning material being treated as a third polymer. The chemical potentials of the partitioned species in the two phases were equated, and only the first-order terms in the polymer concentration differences were retained. The resulting expression gave the partition coefficient, K, in terms of the polymer concentration differences between the phases, the molecular weights of the two polymers, the added macromolecule, and χ_{ij} parameters for the macromolecule interaction with the solvent and each polymer:

$$\ln(K) = P_p\left((\phi_1^t - \phi_1^b)(1 - \chi_{1p}) + (\phi_2^t - \phi_2^b)\left(\frac{1}{P_2} - \chi_{2p}\right) + (\phi_3^t - \phi_3^b)\left(\frac{1}{P_3} - \chi_{3p}\right)\right)$$

(11.31)

In Equation 11.21, P_i is the molecular volume of component i divided by the molecular volume of solvent; i is equal to 1, 2, 3, or p, referring to the solvent, polymer 1, polymer 2, or protein, respectively; ϕ_i is the fraction of solution volume occupied by component I; t or b refer to the top or bottom phase, respectively; and χ_{ij} describes the i-j interaction.

11.3.3.2 Diamond-Hsu

Diamond (11) and Hsu derived a simplified version of the Flory-Huggins expression:

$$\ln(K) = A^*(\omega_2^t - \omega_2^b) \tag{11.32}$$

In this equation, A^* is an empirical parameter that depends on the molecular weights of the polymers and protein, and ω_2 is the weight fraction of polyethylene glycol (PEG) in the indicated phase.

11.3.3.3 King-Haynes

King (6) and Haynes (7) have extended the constant-volume equations to other systems. Although the Flory-Huggins and Diamone-Hsu methods produce equivalent equations for noninteracting solvents, significant differences in the coefficients can arise under some conditions. Thus the constant-pressure approach appears more realistic. The basic equation for isoelectric proteins in the absence of salts can be expressed as

$$\ln(K) = A_{2p}(m_2^b - m_2^t) + A_{3p}(m_3^b - m_3^t) \tag{11.33}$$

In this equation, m_i is the molality of component i in the indicated phase, and A_{2p} and A_{3p} are the virial coefficients to be evaluated.

11.3.3.4 Huddleston

Huddleston (12) examined how the protein partition coefficient varied with the volume ratio for a single, pure protein. The implications of this work for protein partitioning in PEG-salt two-phase aqueous systems and some observations from the literature as well as other experiments focusing on the salting out of proteins and their distribution between a mildly hydrophobic solid phase and a mobile phase containing salt in hydrophobic interaction chromatography are discussed below.

The partition coefficient can be described by Equation 11.34:

$$\ln(K) = \ln(k_0) - \beta' - \Lambda'm + \Omega'\sigma m \tag{11.34}$$

In this expression, β' is derived from the Debye-Hukel theory, which treats proteins in aqueous solution at low ionic strength as simple ions. Λ' is derived from Kirkwood's treatment of proteins as dipolar ions, which is applicable to the prediction of protein solubility in high concentrations of salt, and is related to the dipole moment of the protein by $\Lambda' = D\mu/RT$, where D is a constant derived from Kirkwood's theory. Ω', which is proportional to the hydrophobic surface area of the protein, represents the energy required for the removal of water from hydrophobic regions of the protein.

11.3.3.5 Hartounian-Sandler

The Hartounian-Sandler theory, a thermodynamic model combining the UNIQUAC and extended Debye-Hukel equations, predicts how changes in salt type and concentration will influence protein partitioning (13). A simple model for predicting the protein partition coefficient has been developed based on this detailed thermodynamic analysis:

$$\ln(K) = A(\text{TLL}) + 2\Sigma\beta_{ij}(m_j^{\text{b}} - m_j^{\text{t}}) + C\Delta\Psi \tag{11.35}$$

In Equation 11.35, A is a constant, $C = z_iF/RT$, F is the faraday constant, z_i is the charge on ion i, β_i is the short-range interaction coefficient, m_i is molality, Ψ_i is electrostatic potential in phase I, and TLL is the tie-line length.

This simplification reduces the number of parameters, and allows us to expresss the partition coefficient in terms of a single parameter that is representative of the total polymer composition in the aqueous two-phase system.

11.4 REVERSE MICELLES

Cell membranes deploy amphiphilic molecules to sequester biochemicals, proteins, and inorganic ions within the confines of a cell. In a similar, albeit less sophisticated way, reverse-micelles mimic this approach through the use of an ionic surfactant [such as sodium bis(2-ethylhexyl) sulfosuccinate, didodecyldimethyl ammonium bromide, or trioctylmethyl ammonium chloride] dissolved in an aqueous solution; this solution can generate reverse-micelles when it comes in contact with a nonpolar organic solvent. The interior of the reverse-micelle is a polar environment that can support ions, proteins, nucleic acids, double-stranded DNA, and even bacterial cells, while still maintaining a minimum amount of water, depending on the concentration of the aqueous-phase salt.

Reverse-micelle formation is promoted by several factors. For example surfactant molecules tend to stay at the organic solvent–water interface, as they have both a nonpolar and a hydrophilic section. In addition, water molecules can take advantage of the greater number of opportunities to form hydrogen bonds. Reverse-micelle formation overcomes the free-energy barrier that hinders the formation of a structure with a high liquid-liquid interfacial area.

The size of reverse-micelles is dictated by several considerations:
- The concentration of the surfactant and/or cosurfactant used (higher concentrations lead to smaller reverse-micelles)
- The type of surfactant used, with the chemical group and hydrophobic chain length being important factors (a decrease in the organic solvent–water interfacial tension permits the formation of smaller reverse-micelles)
- The type and concentration of salt used in the aqueous phase
- The type of nonpolar solvent used (which influences the interfacial tension)

The distribution of biomolecules is affected by the same factors that influence reverse-micelle size, plus the environmental pH.

The two most important effects that have been studied are pH and salt type and concentration. Reverse-micelles will solubilize and extract proteins that have a net charge opposite to that of the polar group of the surfactant. This property can be exploited to separate proteins based on their pI values.

The solubilization of proteins is also strongly influenced by the type and concentration of salt in the aqueous phase. Because the reverse-micelles transport proteins through electrostatic interactions, any increase in the ionic strength will satisfy the charge on the proteins and shield them from the electrostatic attraction to the surfactant. This effect can be exploited to move the extracted protein from the reverse-micelles to the aqueous phase for further purification and product formulation. To carry out protein extraction, we need a low concentration of ions in the aqueous phase (roughly 0.1 M for simple salts)—stable macroemulsions are formed at low ionic strengths and effective separation occurs only in submicron-scale, relatively water-free micelles. Macroemulsions contain the aqueous solution and proteins, but at a concentration equal to the concentration in the aqueous feed, which does not provide any advantage in separation procedures.

Like other factors that affect extraction, temperature and surfactant concentration can be explained in an empirical fashion. Without experimentation, it is difficult to predict the effect that the type of salt or surfactant will have on a specific protein separation, because specific chemical interactions greatly influence the extraction behavior. Increasing the temperature, however, increases the solubility of the protein in the reverse micelle. For protein recovery, the useful surfactant concentration is generally the minimum needed for extraction. Any increase in the surfactant concentration—although it might improve protein solubilization—will nevertheless complicate recovery of the purified protein.

Commercial applications of reverse-micelles have not emerged as yet, despite the obvious utility of this technique and the greater selectivity provided by the use of biospecific ligands. Many times, manufacturers may prove reluctant to adopt new processes because of factors unrelated to process effectiveness or efficiency—a situation that has plagued reverse-micelles. A major concern in protein purification is the use of organic solvents, as these hazardous materials require special product deactivation, contamination, and special occupational safety procedures.

Interestingly, one approach that might overcome some of these concerns involves the replacement of the organic solvent with supercritical methane or ethylene. Although these gases are still flammable, their high vapor pressures enable their easy removal from the solution; consequently, these gases do not contaminate the product. Section 11.5 discusses the use of supercritical fluid extraction (SFE) in bioseparation processes.

11.5 SUPERCRITICAL FLUIDS

When a pure liquid in equilibrium with its vapor phase is placed in a sealed environment and heated, our everyday observations lead us to expect that the liquid will vaporize. As pressure builds, more energy should be needed to continue to vaporize the liquid. As early as the 1800s, however, researchers observed that the vapor-liquid meniscus would disappear upon heating when certain liquid volumes were used. This phenomenon resulted when the liquids were placed at a specific temperature and pressure condition—the critical point—at which liquid and vapor do not exist as two distinct phases in equilibrium. At this point, the temperature and pressure are at critical values (T_c and P_c, respectively). Above the critical point, the fluid is considered supercritical. As researchers have found, the fluid density (i.e.,

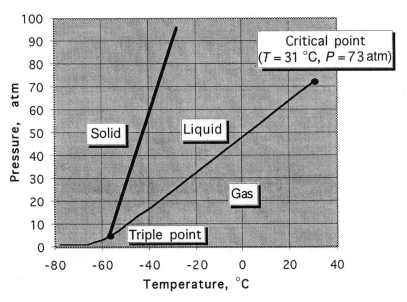

Figure 11.4 Phase diagram for carbon dioxide.

the number of molecules per unit volume) can vary near the critical point. The number of closely spaced molecules affects the interactions between a continuous fluid and a solute dispersed within the fluid. Because the supercritical fluid density is easily manipulated near the critical point, solubility in a supercritical fluid is quite sensitive to temperature and pressure changes. Temperature and pressure thus become controllable parameters that can be manipulated to extract or back-extract solute from or to aqueous solutions.

Other properties also make supercritical fluids valuable for extraction purposes. For example, while the densities of these fluids and hence their extraction properties can resemble those of liquids, the solute diffusion coefficients and supercritical fluid viscosities are one to two orders of magnitude lower than those of liquids. As a result, supercritical fluids are ideal for extracting solutes from solids, as such a fluid can more easily penetrate a solid matrix. These fluids have, for example, been used to remove caffeine from coffee beans and oils from seeds.

Supercritical Fluid Extraction (SFE) with carbon dioxide, whose phase diagram is shown in Figure 11.4, has found application in the recovery of relatively non-polar, non-ionic compounds having therapeutic value, such as alkaloids. Recognizing the challenge of extracting proteins and more hydrophilic biochemicals, researchers have also attempted to use additives (known as modifiers) in the supercritical fluid or aqueous feed. In one approach, a cosolvent such as methanol is added; this type of cosolvent can dramatically increase the solubility of more polar, oxygen-containing compounds of pharmaceutical interest. Predicting solubilities as a function of pressure becomes difficult with a single-component supercritical fluid, however, and even more difficult with a supercritical mixture. A significant amount of experimentation is required to establish process conditions suitable for these extractions.

Another scheme would use supercritical ethane, ethylene, or methane as a dispersed phase for reverse-micelle extraction. Reverse micelles in supercritical fluids require the addition of an ionic surfactant and a cosurfactant, such as octanol, or a cosolvent, such as isooctane. Not surprisingly, pressure affects the size and number of micelles. Extraction and back-extraction with this technique have been demonstrated for some amino acids and proteins.

A more recent and creative departure from the surfactants commonly used in reverse-micelle extraction is the addition of a fluorocarbon surfactant, ammonium carboxylate perfluoropolymer, for protein extraction. This perfluoropolymer allows the process to use supercritical CO_2 and lowers the fluid's critical point. This advance could potentially popularize the use of SFE to separate high-value proteins.

Only a few commercial SFE processes have been developed to date—a reflection of the relatively high expenses associated with high-pressure equipment and the high operating costs of compressors. Recent advances in supercritical fluid technology, such as the use of reverse micelles, should increase the use of this extraction technique when the recovery of high-value products is at stake, or when environmental considerations limit separation process selection.

11.6 LARGE-SCALE VESSELS FOR EXTRACTION

Extraction equipment must perform a number of operations:
- It must promote contact between phases to facilitate mass transfer.
- It must allow for the separation of phases after mass transfer.
- It must transport the process streams to other equipment.

Typically, liquid-liquid extraction equipment takes the form of a mixer-settler, column, or centrifugal separator. Although other types of equipment, such as rotating film contactors, have been proposed, they have not been widely adopted by industry. The choice of equipment depends strongly on the process streams involved, as well as the equipment's inherent characteristics.

Most liquid-liquid extraction operations take place in single- or multi-stage units and are equilibrium-controlled. On the other hand, certain separations are kinetically controlled—for example, separations involving the formation of the complexes between individual components and extractant that permit the isolation of a more quickly forming complex from other, more slowly forming complexes. In any event, the rates of transfer and complex formation are usually rapid and reversible. Equipment selection must therefore account for these aspects of the bioseparation.

11.6.1 Mixer-settlers

If mixing and settling operations are performed in batches, both operations will likely occur in the same vessel. If a process is continuous, it is customary (but not necessary) that mixing and settling take place in different vessels. Mixing ordinarily involves the dispersion of one liquid, in the form of small droplets, in another liquid phase.

Mixer-settlers have many advantages, such as a high stage efficiency that allows the reaction to approach equilibrium. Consequently, they can be designed with ease, and we can predict their performance with a high degree of accuracy. Reasonably reliable scale-up from laboratory to industrial scale is also possible, at least for the simpler designs.

Unlike countercurrent continuous-contact extractors, which usually comprise gravity-operated towers, mixer-settlers can establish the degree of dispersion or intensity of turbulence at any level through mechanical agitation without reducing the flow capacity; emulsions resulting from excessively fine dispersions, however, must be avoided. Any proportion of two liquids can be handled easily, with either phase being dispersed. We can control the ratio of the two liquids in the mixer independently of the ratio of the flow through the cascade as a whole, thereby ensuring that the system achieves the optimal results.

Continuously-operated mixer-settlers may also be shut down for relatively long periods of time without undue problems. When they are restarted, steady-state concentrations are established rapidly, and products may be withdrawn immediately. Tower-type devices cannot match this performance. In addition, mixer-settlers handle suspended solids more successfully than do most other extractors.

Mixer-settlers also have some disadvantages. For example, they require a large capital investment and are costly to operate and maintain. These high expenses reflect the need of mixing, which requires greater power and the inclusion of agitation equipment. Also, the settlers frequently have a large volume, which can be costly not only in terms of equipment, but also in solvent inventory.

11.6.2 Extraction columns

Literally hundreds of different types of columns have been built, studied, or proposed for extraction over the years. We will discuss only columns that have had fairly extensive application. Continuous-contact extractors may be classified into categories based on the complexity of their internal construction (Table 11.3). In the following subsections, we will discuss these types in order of increasing complexity.

11.6.2.1 Gravity-operated extractors

In wetted-wall towers, a film of one liquid (usually the more dense fluid) flows along the inside of a narrow vertical circular tube. The other liquid passes in counter-

Table 11.3 Classification of Gravity-Operated Extractors According to Increasing Complexity of Internal Construction

No Mechanically Operated Parts	Mechanically Agitated Extractors
Wetted-wall towers and similar types	Towers agitated with rotating stirrers
Spray towers	Pulsed towers
Baffle towers	
Packed towers	
Perforated (sieve)-plate towers	

current flow in a central core. The mass transfer rate of each liquid is influenced only by its respective flow rate (14).

Spray towers are based on a simple design involving the dispersion of one of the liquids. As such, they are merely empty shells with provisions for introducing and removing the liquids [Figure 11.5(a)]. Because there is no internal structure over the active portion of the tower, the continuous phase remains relatively free to circulate from top to bottom. This flexibility reduces extraction rates because it destroys the true countercurrent concentration differences between the liquids. On the other hand, the absence of internal structure allows these towers to handle liquids containing suspended solids.

Baffle towers take the form of cylindrical shells containing horizontal baffles that direct the liquid flow [see Figure 11.5(b)]. Another possible arrangement relies on "disc and doughnut" baffles, a design in which annular rings are attached to the shell of the tower, and centrally located circular discs supported by vertical rods or arms extend to the shell. Baffles are ordinarily spaced four to six inches apart. Their primary purpose is to reduce vertical circulation and thereby provide somewhat longer residence times for the dispersed phase. The frequent coalescence and redispersion of one of the liquids also accelerates mass transfer (15). Baffle towers have long been used to extract products such as acetic acid from pyroligneous liquors.

The extraction tower's shell may be filled with packing material. This material

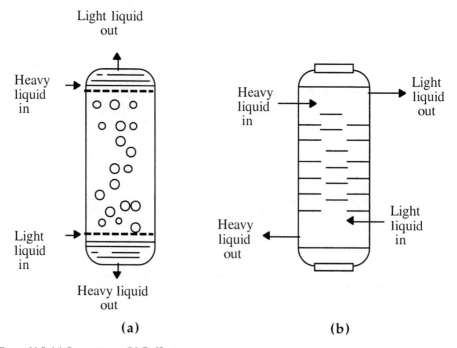

(a)   (b)

Figure 11.5 (a) Spray tower. (b) Baffle tower.

simultaneously reduces axial mixing and jostles and distorts the droplets of dispersed liquid, increasing mass transfer rates. By reducing the space available for liquid flow, packing diminishes baffle towers' flow capacities per unit of cross-sectional area relative to the flow capacities of spray towers. Furthermore, baffle towers are not suitable for the separation of liquids containing suspended solids. Nevertheless, because they can hold the equivalent of many stages, these towers have found wide application in almost every type of commercial extraction processing.

When using standard manufactured packings, it is important to choose a material that is preferentially wetted by the continuous liquid. This selection will ensure that the drops of dispersed liquid do not extensively coalesce within the packed volume. Stoneware and porcelain packings are almost always preferentially wet by aqueous—rather than organic—liquids; carbon and plastic packings are usually preferentially wet by organic liquids. The relative wetting of metal packings, however, can prove difficult to predict (16).

In perforated-plate towers, one of the liquids is repeatedly dispersed and coalesced as it flows through a series of trays in which small holes have been drilled or punched. The rate of extraction is enhanced by the repeated dispersion, which essentially multiplies the end effects. The most important advantage offered by this type of tower, however, is its elimination of vertical back-mixing.

In a simply constructed perforated-plate tower, the plates resemble those in the side-to-side baffle tower, except that they are perforated. The continuous phase flows across the horizontal perforated plates and from plate to plate by means of downspouts. The dispersed phase collects on the trays in coalesced layers and then bubbles through the perforations of the plate (Figure 11.6).

An alternative arrangement relies on center-to-side and side-to-center plates. A special bypass allows interfacial accumulation to pass through the tray, although this setup diminishes tray efficiency. It is also common practice to disperse both liquids at each tray (16). Alternatively, ordinary column packing may be placed in the downspouts to act as coalescers for the entrained dispersed phase. A simple arrangement in which a wire screen substitutes for the perforated tray has been used with success as well (17, 18).

11.6.2.2 Mechanically agitated extractors

A rotating disc extractor includes several horizontal stator rings—that is, ring-shaped baffles fixed to the tower shell that divide the extractor into a number of small compartments (Figure 11.7). The rotation of a series of flat, circular discs, arranged on a central shaft and centered in each compartment, provides the mechanical agitation. The light liquid is dispersed and the principal liquid-liquid interface occurs at the top of the tower; the heavy liquid could just as easily be dispersed, however.

Rotating disc extractors have been widely used in petroleum extraction processes, as well as in the by-product coke, food, metal separation, and organic

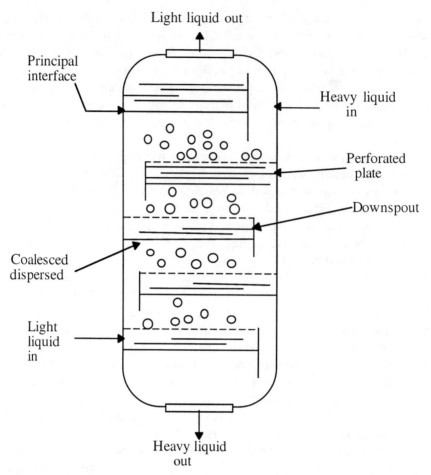

Light liquid out

Principal interface

Heavy liquid in

Perforated plate

Downspout

Coalesced dispersed

Light liquid in

Heavy liquid out

Figure 11.6 A perforated-plate extractor.

chemical industries. Laboratory-sized towers are usually four or six inches in diameter; the diameter of the largest industrial tower is 10 feet (15).

In pulsed extractors, a rapid reciprocating motion of relatively short amplitude is applied to the liquid contents. Because the extractor does not use moving parts to create the mechanical agitation, the frequency with which worn and corroded parts must be replaced is decreased. The practical application of these systems appears to be limited to atomic energy materials processing, largely because of their relatively high power requirements (16).

11.6.3 Centrifugal contactors

Substitution of the force of gravity by a large centrifugal force, several thousand times as large as the gravitational force, creates high liquid velocities and correspondingly small extractor volumes. The small extractor volumes, which can achieve high flow capacities, offer advantages other than space savings. These systems need

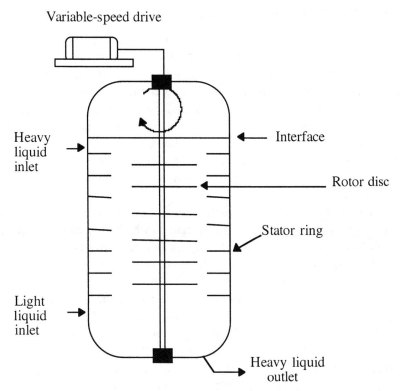

Variable-speed drive

Heavy liquid inlet →

Interface ←

Rotor disc ←

Stator ring

Light liquid inlet →

Heavy liquid outlet →

Figure 11.7 A rotating-disc contactor.

a smaller amount of expensive solvent, and continuous extraction processes reach the steady state quickly, with relatively little reprocessing required to bring early product to final specifications (15). Disadvantages of centifugal contractors include the relatively high initial, operating, and maintenance costs associated with high-grade machinery (19).

11.6.4 Comparison

Mixer-settlers require a large amount of solvent. Columns, on the other hand, require an intermediate amount. As noted earlier, centrifugal extractors have minimal solvent requirements.

Mixer-settlers can be shut down and restarted without the loss of the concentration gradient or profile across a multiple-stage battery. In contrast, a column shutdown results in profile loss, meaning that the column's contents must be reprocessed after the system is restarted. Centrifugal contactors do not have such problems and can be cycled as desired because the holdup is small.

Any change in the feed to a centrifugal contactor will be reflected immediately in the streams leaving the contactor. By comparison, it will take some time before a real change in the streams leaving a mixer-settler becomes apparent.

11.7 CONFIGURATIONS FOR STAGE-WISE CONTACTING

In a popular mode of extraction, two phases are mixed and then allowed to reach equilibrium. Each process vessel or section of a vessel that allows the two phases to come to equilibrium is aptly called an equilibrium stage. By definition, the streams leaving an equilibrium stage are assumed to be at equilibrium.

Three basic configurations exist for bringing process streams in contact through equilibrium stages: cocurrent, countercurrent, and cross flow. By far the most popular contacting mode is countercurrent staging. In this section, we will discuss the characteristics of the various contacting modes, provide a general approach for analyzing variations on these three modes, and illustrate why countercurrent staging is the preferred mode.

From a process design standpoint, it is important to note the parameters involved in equilibrium staging and the extent to which they can be controlled by design specifications. This step should be taken before analyzing contacting modes, so as to enable the process to achieve the goals of high yield, active product, and high purity at minimum cost. The designer dictates the number of stages for a particular extraction—but note that the inclusion of more stages inevitably leads to higher costs. The designer also specifies the flow rate of each phase, though practical limitations in process vessel design, operability, or partial miscibility of the phases can constrain the ranges and ratios of flows. The designer can manipulate equilibrium conditions, usually through his or her choice of extractant phase composition; again, however, practical considerations such as cost or product quality may significantly limit design flexibility. Although the initial concentration of solute in the aqueous phase is often considered the starting point of an extraction process, the designer may include a preconcentration or other pretreatment step to improve the extraction process's efficiency.

Clearly, equilibrium-stage modeling is important because the designer must specify many parameters. One of the first design considerations should take into account the number of stages needed to obtain a target purity or yield. In the approach described here, we will first study the function of the three contacting modes by modeling a simplified, ideal case. From this starting point, we can then move on to more realistic systems, using this ideal case as a reference point.

The ideal case includes three assumptions:
- Each phase is completely insoluble in the other phase.
- There is linear equilibrium ($y = Kx$).
- Each stage is at the same temperature and pressure.

Later, we will introduce another simplification—equal flow rates for each phase—to make some conclusions regarding the performance of each mode.

11.7.1 Cocurrent contacting

Figure 11.8 depicts a cocurrent process involving one, two, or three stages. From this process, we can generalize how the number of stages affects the concentration of solute produced in the extractant phase (y_1, y_2, or y_3). In going from one to three stages, a pattern becomes readily apparent, providing a generalized solution.

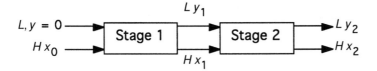

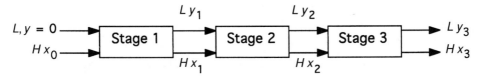

Figure 11.8 Cocurrent contacting processes.

Performing a species balance around stage one yields the following equation, which can be solved for y_1:

$$Hx_0 = Ly_1 + Hx_1 = \left(L + \frac{H}{K}\right)y_1 \tag{11.36}$$

$$y_1 = \frac{Hx_0}{L + \dfrac{H}{K}} \tag{11.37}$$

To simplify this result, we introduce the extraction factor, $E = KL/H$, giving the following result:

$$y_1 = \frac{Kx_0}{E+1} \tag{11.38}$$

The same approach can be taken for two and three stages, yielding

$$y_2 = y_3 = \frac{Kx_0}{E+1} \tag{11.39}$$

Equation 11.39 demonstrates that, in this case, yield and purity are not affected by the number of stages.

11.7.2 Crosscurrent contacting

In a crosscurrent (or cross flow) contacting system, we can track the concentration of solute remaining in the aqueous phase to obtain the extraction yield. Using the same definitions as before, a species balance for a one-, two-, and three-stage crosscurrent contactor gives the following equations for the solute concentration in the aqueous phase:

For one stage:

$$x_1 = \frac{x_0}{E+1} \tag{11.40}$$

For two stages:

$$x_2 = \frac{x_0}{(E+1)^2} \tag{11.41}$$

For three stages:

$$x_3 = \frac{x_0}{(E+1)^3} \tag{11.42}$$

These results can be easily generalized to obtain a formula for n stages:

$$x_n = \frac{x_0}{(E+1)^n} \tag{11.43}$$

As $E > 0$, increasing the number of stages will decrease the concentration of solute left in the aqueous phase, thereby increasing the yield. If the flow rates remain constant for each stage, however, the concentration of the solute in the extracting phase diminishes in each subsequent stage. This product dilution effect can increase the cost of extraction.

11.7.3 Countercurrent contacting

Having the extracting and aqueous phases flow in opposite directions through the series of stages, creates a more complex analysis. The following equations give the concentration of solute in the extracting phase for one-, two-, three-, and four-stage systems:

For one stage:

$$\frac{Kx_0}{y_1} = E+1 \tag{11.44}$$

For two stages:

$$\frac{Kx_0}{y_2} = \left[E+1 - \frac{E}{E+1} \right] \tag{11.45}$$

For three stages:

$$\frac{Kx_0}{y_3} = \left[E+1 - \frac{E}{E+1-\dfrac{E}{E+1}} \right] \tag{11.46}$$

For four stages:

$$\frac{Kx_0}{y_4} = \left[E + 1 - \cfrac{E}{E + 1 - \cfrac{E}{E + 1 - \cfrac{E}{E + 1}}} \right] \tag{11.47}$$

Rearranging these equations gives the general equation for countercurrent contacting:

$$y_n = \frac{Kx_0 \sum\limits_{j=1}^{n} E^{j-1}}{\sum\limits_{j=1}^{n} E^j + 1} \tag{11.48}$$

This pattern illustrates an important property of the extraction factor—that is, as the number of stages goes to infinity, the result becomes

$$\text{as } n \to \infty, \ y_\infty = \frac{Hx_0}{L} \tag{11.49}$$

Thus the countercurrent system can overcome equilibrium constraints and enrich the solute in the extractant phase by increasing the ratio of aqueous to extractant flow rates.

11.7.4 A comparison of contacting modes

For crosscurrent, cocurrent, and countercurrent stagewise contacting methods, letting the number of stages go to infinity in each case allows the comparison of the intrinsic properties of these contacting modes.

Cocurrent:

$$y_\infty = \frac{Kx_0}{E + 1}, \ x_\infty = \frac{x_0}{E + 1} \tag{11.50}$$

Crosscurrent:

$$x_\infty = 0, \ y_n \text{ varies with } y_\infty = 0 \tag{11.51}$$

Countercurrent:

$$y_\infty = \frac{Kx_0}{E} = \frac{Hx_0}{L}, \ x_1 = 0 \tag{11.52}$$

Countercurrent contacting has an advantage in that it provides the highest concentration, but does not dilute the product in the extractant phase. In fact, countercurrent contacting can actually increase the product concentration in the extractant phase. In contrast, the concentration in the extractant phase varies with the stage number in crosscurrent contacting, creating a more diluted product.

More complex contacting systems offer more flexibility in dealing with multi-component aqueous feeds from which several products must be isolated or for increasing product purity. The curious nature of aqueous-phase systems also opens up opportunities for contacting more than two phases at once. Multiphase contacting schemes are currently used only for research purposes, however. We mention them here to illustrate how extraction stage design can be influenced by the introduction of more than two phases.

Example 11.2

Compare crossflow and countercurrent contactors in a two-solute extraction process when $K_1 > K_2$ in terms of the following: 1) fraction extracted, 2) concentration of solute in the extract phase, and 3) solvent-free product purity. Which scheme would you choose when the cost of crystallizing a component from the extract phase is proportional to its concentration in that phase?

Solution

The fraction extracted can be determined from a mass balance for each of the systems.

Cross flow:

$$f_1 = 1 - \frac{H\left(\dfrac{x_0}{(1+E_1)^n}\right)}{Hx_0} = 1 - \frac{1}{(1+E_1)^n} \tag{11.53}$$

Countercurrent flow:

$$f_1 = \frac{\dfrac{LK_1 x_0 \displaystyle\sum_{j=1}^{n} E_1^{j-1}}{\displaystyle\sum_{j=1}^{n} E_1^j + 1}}{Hx_0} = \frac{\displaystyle\sum_{j=1}^{n} E_1^j}{\displaystyle\sum_{j=1}^{n} E_1^j + 1} \tag{11.54}$$

Plotting the fraction extracted versus the extraction factor at different numbers of stages helps to illustrate the behavior of these two systems. From this analysis, we conclude that the fraction extracted is always higher for cross-flow contacting; the difference is minimal for higher values of the extraction factor.

For cross flow contacting, the concentration of solute 1 in the extract phase varies with the stage number. To compare these contacting systems, we can pool these extracts together. The resulting single stream has a flow rate of nL and an extract concentration of

$$y_{\text{pooled}} = \frac{L \displaystyle\sum_{j=1}^{n} j \frac{Kx_0}{(1+E_1)^n}}{nL} = \frac{1}{n} \displaystyle\sum_{j=1}^{n} j \frac{Kx_0}{(1+E_1)^n} \tag{11.55}$$

The concentration for countercurrent contacting was given previously. It is readily apparent that the concentration in the extract phase is higher in the countercurrent contactor. Hence countercurrent contacting is preferred when the cost of crystallization depends on the concentration in the extract phase.

The solvent-free purity can be easily determined for each stage in cross-flow extraction by the following equation:

$$\text{Purity}'_{\text{component 1}} = \frac{y_1}{y_1 + y_2} = \frac{\dfrac{K_1 x_0}{(1 + E_1)^n}}{\dfrac{K_1 x_0}{(1 + E_1)^n} + \dfrac{K_2 x_0}{(1 + E_2)^n}} = \frac{1}{1 + \dfrac{K_2}{K_1} \dfrac{(1 + E_1)^n}{(1 + E_2)^n}} \tag{11.56}$$

Equation 11.56 shows that the purity depends on the ratio of the flow rates as well as the ratio of the equilibrium constants. For countercurrent contacting, after applying some algebra we find the solvent-free purity of component 1 to be

$$\text{Purity}^{\text{countercurrent}}_{\text{component 1}} = \frac{1}{1 + \dfrac{K_2}{K_1} \dfrac{\displaystyle\sum_{j=1}^{n} E_2^{j-1} \left(\sum_{j=1}^{n} E_1^{j} + 1 \right)}{\displaystyle\sum_{j=1}^{n} E_1^{j-1} \left(\sum_{j=1}^{n} E_2^{j} + 1 \right)}} \tag{11.57}$$

Equation 11.57 also illustrates that both the ratio of flow rates and the ratio of equilibrium constants affect product purity in countercurrent extraction.

In both countercurrent and crosscurrent contacting, the product purity varies in a complex fashion with the stage number. We can make one general statement: If the equilibrium constants are similar, cross-flow contacting can yield higher purities for a given number of stages. Another observation is that, when the equilibrium constants differ substantially, countercurrent contacting can provide higher levels of purity. These observations are important, for example, in two-phase aqueous extraction where the equilibrium constants are very similar for proteins with nearly the same isoelectric points and similar molecular weights.

Example 11.3
Consider a two-dimensional contacting system, like the one illustrated in Figure 11.9. Compare the yield, solute concentration, and purity of this two-dimensional system with one-dimensional, three-stage countercurrent and crosscurrent contactors.

Solution
Splitting the aqueous feed and extract streams into three parts allows us to compare one-dimensional countercurrent and cross-flow contactors. In the two-dimensional system, the stages are one-third the size of the stages in the one-dimensional systems because the flow rates to each two-dimensional stage are one-third of the one-dimensional flows. Examining the two-dimensional system, we note that the extract concentrations from stages 3 and 6 should be higher than the extract concentrations from stages 2 and 3 in the one-dimensional cross-flow contacting, yet lower than similar stages in the one-dimensional countercurrent system. The concentration of the extract leaving stage 6 is given by

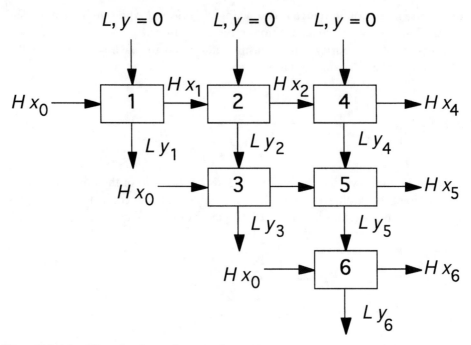

Figure 11.9 A two-dimensional cross-flow cascade.

$$y_6 = Kx_0 \left(\frac{E^2}{(1+E)^5} + \frac{E}{(1+E)^4} + \frac{1}{(1+E)} \right) \tag{11.58}$$

Plotting this result, along with the fraction extracted with the one-dimensional systems, shows that the extract concentration of the two-dimensional system approaches that of the one-dimensional system, countercurrent system as the extraction factor increases (Figure 11.10). For the purpose of drawing these curves, the ordinate is taken to be the extract concentration normalized by dividing by the equilibrium constant times the feed concentration.

We must now track the yields for each contacting configuration. Using the procedures described earlier for writing mass balances around each stage, we develop an equation for the fraction extracted:

$$f = 1 - \frac{1}{3} \left(\frac{E^2}{(1+E)^5} + \frac{E}{(1+E)^4} + \frac{1}{(1+E)^3} + \frac{1}{(1+E)^2} + \frac{2}{(1+E)} \right) \tag{11.59}$$

Figure 11.11 shows that the yield is lower in the two-dimensional cross-flow configuration. In this system, some of the extract phase contacts fresh aqueous feed after having extracted some solute from previous contact with an aqueous-phase solution.

The third, and perhaps most important, comparison involves the purity achieved with these systems. This calculation is complex when multicomponent systems are under consideration. A simple comparison is to limit the analysis to a two-component aqueous feed at a specific L/H ratio. Matching the third-stage extract

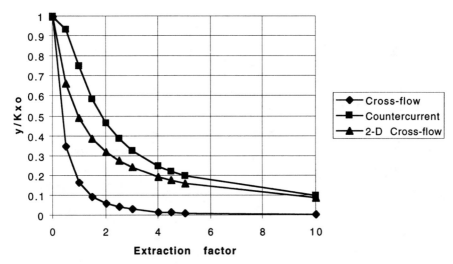

Figure 11.10 Effect of the extraction factor on the concentration in the extract phase for a three-stage two-dimensional cross-flow cascade.

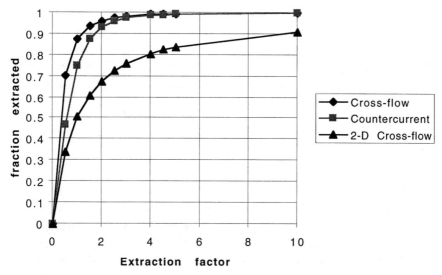

Figure 11.11 Effect of the extraction factor on the fraction extracted for a three-stage two-dimensional cross-flow cascade.

stream in the cross-flow system and the sixth stage of the two-dimensional system against the countercurrent extract stream also helps to compare one-dimensional to two-dimensional cross flow.

As Figure 11.12 illustrates, even this simplified case involves complex behavior. In this case, $K_1 = 10$. It can be easily seen, however, that the general trends described here apply to any value of K_1. When $K_2 < K_1$, the two-dimensional cross-flow and one-dimensional countercurrent systems provide higher purity than one-dimensional cross-flow system. The opposite is true when $K_2 > K_1$. In addition, the

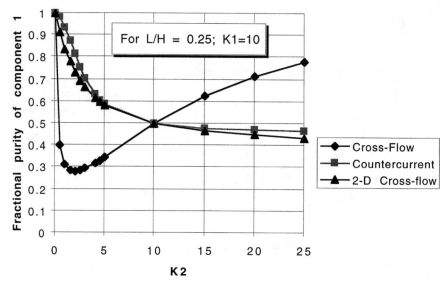

Figure 11.12 Effect of the partition coefficient of solute 2 on the fractional purity of solute 1 for a three-stage two-dimensional cross-flow cascade.

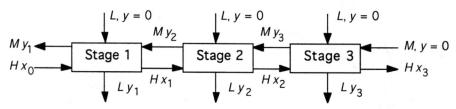

Figure 11.13 A three-phase, three-stage countercurrent contactor.

countercurrent system approaches a limiting purity as K_2 increases, whereas the cross-flow systems do not exhibit this behavior. These observations can be explained by realizing that, in the first stage of the six-stage cross-flow system when $K_1 > K_2$, most of component 1 is extracted in the first stage. Thus the purity of component 1 is higher at the end of the process train only when $K_2 > K_1$. It is also apparent that the purity achieved with a two-dimensional cross-flow system is similar to that found with a one-dimensional countercurrent system.

Example 11.4

Consider an extraction system that consists of three phases created by using two water-soluble polymers in aqueous two-phase extraction. In Figure 11.13, the three-stage system includes two phases that are introduced countercurrently and a third phase that is in cross flow with each of the countercurrent phases. Even when all three phases are in equilibrium, we can define an equilibrium constant for two phases at a time so that $K^I = y/x$ and $K^{II} = z/x$. How would this three-stage, three-phase system compare with three-stage, two-phase countercurrent and cross-flow systems with respect to yield, extract concentration, and purity?

Solution

Before tackling the three-phase system, it is useful to derive equations for the one- and two-stage systems so as to compare them with two-phase cross-flow and countercurrent staging. We can use the previous definition of equilibrium in three-phase systems, and define R as the flow rate of the third phase in a one-stage system so that $E^{\mathrm{I}} = K^{\mathrm{I}}L/H$ and $E^{\mathrm{II}} = K^{\mathrm{II}}R/H$. Using mass balances, we derive the following equations for the aqueous phase (i.e., the feed stream) leaving the contactor system:

For a one-stage system:

$$x = \frac{x_0}{1 + E^{\mathrm{I}} + E^{\mathrm{II}}} \tag{11.60}$$

For a two-stage system:

$$x = \frac{x_0}{\left(1 + E^{\mathrm{I}} + E^{\mathrm{II}}\right)^2 - E^{\mathrm{I}}} \tag{11.61}$$

It should be readily apparent that when $E^{\mathrm{II}} = 0$, these equations give a result equal to that for two-phase countercurrent contacting.

Returning to the question posed, the fraction extracted for the three-stage, three-phase system is given by

$$f = 1 - \left(\frac{1}{\left(1 + E^{\mathrm{I}} + E^{\mathrm{II}}\right)^3 - 2\left(E^{\mathrm{I}} + [E^{\mathrm{I}}]^2 + E^{\mathrm{I}}E^{\mathrm{II}}\right)} \right) \tag{11.62}$$

This equation can be graphed along with the equations for two-phase countercurrent and cross-flow systems (Figure 11.14). Given that the total extract flow rate to each stage of the three-phase system is double that for the two-phase systems, it is not surprising that the three-phase system exhibits a greater fraction extracted. We can normalize this effect by comparing the fraction extracted for two-phase systems with the fraction for a three-phase system and taking two times the extraction value of the three-phase system. The fraction extracted lies between the value for the cross-flow and countercurrent extraction even when $E^{\mathrm{I}} = E^{\mathrm{II}}$, because the three-phase system combines both of these contacting modes.

We can graph the concentration in the extract phase in a similar manner (Figure 11.15). Once again, the three-phase system falls between the two-phase cross-flow and countercurrent systems.

Purity in binary systems depends on two sets of values for K, E^{II}, and E^{I}. To compare countercurrent and cross-flow contacting, we can manipulate both component values for E^{I} based on the prior graphs. Figure 11.16 shows that, when $E^{\mathrm{I}} = E^{\mathrm{II}}$, two-phase countercurrent contacting at higher values of K_2 is preferable. When E^{I} does not equal E^{II}, purity decreases dramatically at higher values of K_2. This divergence implies that higher levels of purity can be obtained with three-phase systems under certain equilibrium conditions by manipulating the flow rates of the extract phases.

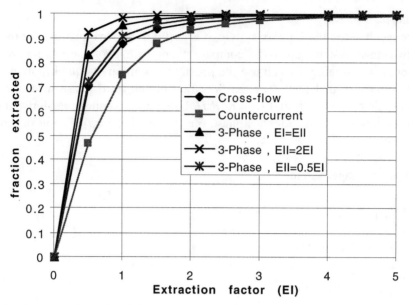

Figure 11.14 A comparison of the influence of the extraction factor on the fraction extracted in cross-flow, countercurrent-flow, and three-phase cross-counter contactors.

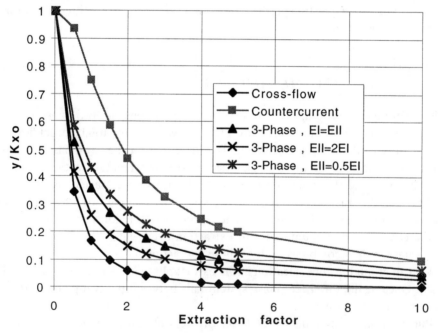

Figure 11.15 A comparison of the influence of the extraction factor on the extract concentration for cross-flow, countercurrent-flow, and three-phase cross-counter contactors.

Examples 11.2 through 11.4 have illustrated in detail the ramifications of the choices that can be made in stagewise contacting. In later discussions, we will extend the results obtained with these analyses to other configurations, such as fractional extractional systems. Fractional extraction adds another aqueous-phase stream to

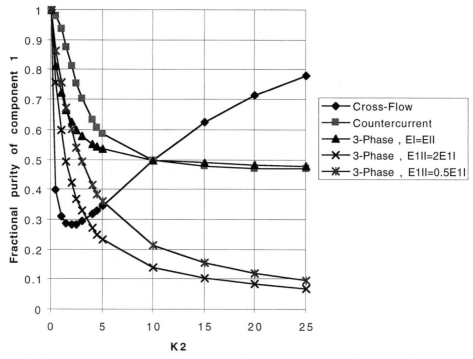

Figure 11.16 A comparison of the influence of the partition coefficient of solute 2 on the fraction purity of solute 1 for cross-flow, countercurrent-flow, and three-phase cross-counter contactors.

better isolate components based on their relative tendencies to partition into the extract phase. Necessarily, this type of system cannot process as much feed as do other contacting systems for a given process vessel size. The goal of higher purity, however, vastly offsets the lower volumetric productivity provided with fractional extraction.

We will also briefly review a graphical approach that allows us to easily calculate the number of stages needed in countercurrent extraction when the equilibrium is nonlinear. To this point, the mathematical methods described have relied on linear equilibrium. Although we can reanalyze the contacting systems for nonlinear equilibrium by using polynomial expressions for $y = f(x)$ at a specific temperature, pressure, and composition, the equations generated are best solved with computers. The graphical method, on the other hand, offers a welcomed simplicity for visualizing complex properties.

11.7.5 Graphical solution

In one convenient method for discerning the number of stages required when the equilibrium relation is nonlinear, we can graph the equilibrium curve and operating line together. In the McCabe-Thiele method (20), the operating line is a visual representation of the mass balances around each stage, and the equilibrium line repre-

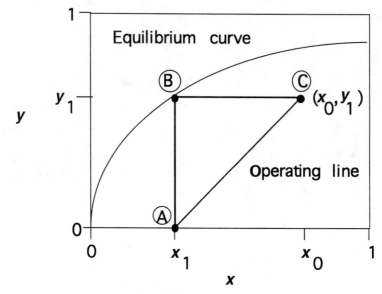

Figure 11.17 One-stage graphical solution method.

sents the phase concentrations leaving a stage. Overall and individual stage mass balances provide general equations for the operating line of countercurrent contactors:

$$y_{\text{cascade exit}} = \frac{H}{L}(x_0 - x_1) \tag{11.63}$$

For other extract streams,

$$y_n = \frac{H}{L}(x_{n+1} - x_1) \tag{11.64}$$

Starting with a one-stage extraction process, Figure 11.17 shows an equilibrium line and an operating line with a single step; this step represents the stage. Point A indicates the entering extract stream, which has no solute in this case, and the exiting aqueous stream. As the system includes only one stage, C represents the extract stream leaving the stage (y_1) and the aqueous feed concentration (x_0). The only point that lies on the equilibrium curve, B, shows that there is only one stage and gives the equilibrium concentrations leaving the stage.

For a two-stage process, we can redraw Figure 11.17 for the specific case where the feed and extract concentrations are identical to those in the one-stage system. This new system is depicted in Figure 11.18. The extra stage produces a higher yield if the *H/L* ratio decreases.

In most design situations, we know the feed concentration and have designated a target yield. Thus we know the point on the operating line (x_0, y_1). If a target aqueous stream flow rate has also been identified, then the point (x_0, 0) can be determined from the overall balance and the operating line can be constructed. With the operating and equilibrium lines in place, we draw the steps starting from (x_0, y_1) until the horizontal lines drawn meet or exceed the feed concentration x_0.

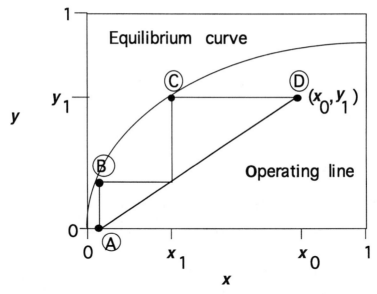

Figure 11.18 Graphical solution method for a two-stage countercurrent contactor.

Economic analyses, a requirement that existing equipment be used for a new extraction application, or fractional extraction designs may dictate different graphical methodologies than those described in the simple cases presented here.

11.7.6 Fractional extraction

Two-dimensional extraction methods bear some similarities to one-dimensional fractional extraction systems. The latter approaches are best illustrated by two contacting methods: Craig and double-countercurrent extraction. These modes represent variations on the one-dimensional countercurrent mode. The remainder of this section analyzes these systems. These fractional extraction processes are often applied in the pharmaceutical industry because they can successfully isolate individual compounds in a multicomponent feed solution.

11.7.6.1 Craig extractor

Rather than move both phases, one type of fractional extractor keeps the starting, heavy phase stationary and moves only the light phase (Figure 11.19). In the beginning of this contacting pattern, equilibrium vessel 1 has an aqueous feed, and fresh extract phase is introduced. Upon completing the first equilibration, the extract phase is separated from the aqueous solution and moved to a new vessel. Fresh extract is again introduced in vessel 1, while fresh aqueous solution containing no solutes is added to vessel 2. Both vessels are then allowed to equilibrate, and the switching and introduction of fresh phases continue.

Repetition of this protocol distributes the solute concentration in a manner analogous to that achieved by a chromatographic separation. After a specific number of

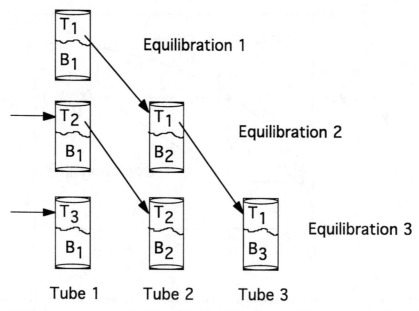

Figure 11.19 Craig extraction method. B and T refer to the bottom and top phases, respectively. The subscripts refer to the order in which they appear in the extraction.

transfers, the vessel number may be plotted against solute concentration, yielding a peak that resembles a distribution. A Poisson distribution can be used to determine the fraction of solute extracted:

$$f = (r)^{n+1}(1-r)^{t-n}\frac{t!}{n!(t-n)!} \tag{11.65}$$

In this equation, $r = E/(E+1)$, t is the number of transfers, and n is the vessel number.

Example 11.5
Shikimic acid and its derivatives can be extracted using an organic phase consisting of a long-chain tertiary amine dissolved in heptanol. The partition coefficient is 2.2 for shimikic acid and 1.4 for dehydroshikimic acid. If a bioreactor solution contains an equimolar solution of shikimic and dehydroshikimic acids, plot the fraction extracted after 50 and 150 transfers. Discuss how many transfers you would choose if the goal is 99% solvent-free purity.

Solution
By plotting the tube number versus the fraction extracted in Figure 11.20 and Figure 11.21, we can see that increasing the number of transfers improves peak separation, but that the fraction extracted and hence the peak concentration decrease. After 150 transfers, a reasonable number of tubes contain 95% or better purity of shikimic acid; few tubes achieve the same purity after only 50 transfers. To increase the yield, 200 to 250 transfers are recommended. Note that once baseline separation has been achieved and the peaks no longer overlap, additional transfers will not improve purity.

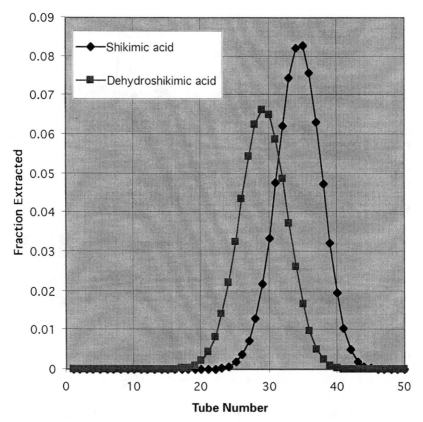

Figure 11.20 Craig extraction of shikimic and dehydroshikimic acids using 50 transfers.

11.7.6.2 Double-countercurrent extractor

When aqueous and extract streams flow countercurrently, the resulting extraction system moves solutes that exhibit a preference for one of the phases to one end of the cascade. To analyze such a double-countercurrent extraction, we separate the system into two parts. The end of the cascade that lies upstream from the feed is simply a countercurrent contactor; the stage at which the feed is introduced and stages downstream from the feed stage demonstrate increased heavy-phase flow rates. These two sections are regarded as the enriching and stripping sections, respectively. In the enriching section, solutes that have a high affinity for the extract become concentrated at that end of the contactor. The stripping section concentrates the solutes that have an affinity for the heavy, or aqueous, phase.

When the equilibrium is nonlinear, double-countercurrent contacting dictates that we replace the single operating line used in countercurrent contacting with two operating lines. (The problems at the end of this chapter give practice in using this concept by following the graphical technique described in Section 11.7.5.)

For linear equilibrium, it is useful to define another extraction factor, $E_{DC} = KL/(H + A)$. This extraction factor, along with the previously defined E, can be used to analyze fractional extraction when we would like to account for the placement

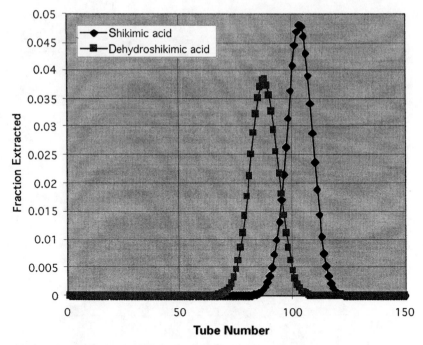

Figure 11.21 Craig extraction of shikimic and dehydroshikimic acids using 150 transfers.

of the feed as well as to describe how flow rates and the equilibrium constant affect the separation. The feed stage is called stage f and the total number of stages is n. Breaking the analysis into two parts, we find that from stage 1 to f, the following equation holds:

$$x_{\mathrm{f}} = x_1 \sum_{j=0}^{f-1} E_{\mathrm{DC}}^j \tag{11.66}$$

Now taking the right-hand side and determining how y_{f} varies with y_n, we have

$$y_{\mathrm{f}} = \frac{y_n}{E^{n-f}} \sum_{j=0}^{n-f} E^j \tag{11.67}$$

Using the overall balance and a guiding equation,

$$Ax_0 = Ly_n + (H + A)x_1 \tag{11.68}$$

we can combine the equations for the stripping and enriching sections to obtain the desired result:

$$y_n = Kx_1 E^{n-f} \frac{\displaystyle\sum_{j=0}^{f-1} E_{\mathrm{DC}}^j}{\displaystyle\sum_{j=0}^{n-f} E^j} \tag{11.69}$$

Example 11.6

Using the partition coefficients for shikimic and dehydroshikimic acids provided in Example 11.5, find the ratio of mole fractions of the extract- and aqueous-phase concentrations using 21 stages, where the feed entered at stage 11. Assume that the ratio of the phase flow rates is 1 and that $H = 0.1 \times A$.

Solution

Inserting the data into Equation 11.69, we have

$$\frac{y_{21}}{x_1} = KE^{11} \frac{\sum_{j=0}^{10} E_{DC}^j}{\sum_{j=0}^{10} E^j} = \frac{K^{11}}{1.1} \frac{\sum_{j=0}^{10} K^j}{\sum_{j=0}^{10} K^j} = \frac{K^{11}}{1.1} \tag{11.70}$$

Solving this equation gives $y_{21}/x_1 = 5312$ for shikimic acid and $y_{21}/x_1 = 36.8$ for dehydroshikimic acid.

11.7.7 Continuous countercurrent extraction

Phases can be brought into contact with one another in a single vessel and processed without waiting for equilibrium. This type of continuous contacting allows for rapid extraction, usually at lower cost. The driving force for solute transfer from one phase to another is related to the equilibrium state. An adequate representation of the transfer rate is given by the following linear difference equation:

$$\text{rate} = \frac{\text{moles}}{\text{time}} = ka(x - x^*) \tag{11.71}$$

This equation represents a simplification of a complex series of events that involve interfacial solute transfer. The rate of transfer (given in units of moles per second) is proportional to the experimentally determined rate constant k, the interfacial area a, and the difference between the solute mole fraction and the equilibrium solute mole fraction $x - x^*$. In many cases, ka is determined experimentally, because separating the effect of transfer rate from the surface area available for transfer can be difficult. If the contacting vessel or column can be treated as having uniform properties across the diameter, a solute mass balance can be performed on a single differential column slice:

$$\begin{bmatrix} \text{Rate of accumulation} \\ \text{of A in one phase} \end{bmatrix} = \begin{bmatrix} \text{rate of} \\ \text{A in} \end{bmatrix} - \begin{bmatrix} \text{rate of} \\ \text{A out} \end{bmatrix} - \begin{bmatrix} \text{rate of transfer} \\ \text{to other phase} \end{bmatrix} \tag{11.72}$$

$$0 = H(x|_h - x|_{h+\Delta h}) - Aka\Delta h(x - x^*) \tag{11.73}$$

As $\Delta h \to 0$,

$$\frac{dx}{dh} = \frac{kaA}{H}(x - x^*) \tag{11.74}$$

The resulting equation is usually integrated over the entire column height Z, using an overall column mass balance:

$$Hx = Hx_{out} + Ly \tag{11.75}$$

We can then obtain two parameters that are convenient in design: the height of a transfer unit (HTU) and the number of transfer units (NTU). If the equilibrium is nonlinear, the design equation becomes

$$Z = \frac{H}{Aka} \int_{x_{out}}^{x_{inlet}} \frac{dx}{\left(x - \frac{(x - x_{out})}{E} \right)} \tag{11.76}$$

This equation can be solved for linear equilibrium:

$$Z = \left(\frac{H}{Aka} \right) \left(\frac{E}{E-1} \right) \ln \left(\frac{1 + \frac{x_{out} - x_{inlet}}{E}}{x_{out}} \right) \tag{11.77}$$

In both equations,

$$HTU = \frac{H}{Aka} \tag{11.78}$$

and the second term is the NTU. This dichotomy allows the designer to isolate the effects created by changing the column's cross-sectional area and increasing the interfacial rate of solute transfer from the flow rate and equilibrium parameters that are incorporated in the extraction factor. Note that, according to L'Hôpital's rule, when the extraction factor goes to infinity, the height of the column depends only on the HTU.

11.8 SUMMARY

Extraction offers a versatile array of methods for recovering and purifying a wide range of biological components—from small biochemicals to proteins and cells. The first consideration in selecting an extraction process is to find a solvent that does not degrade the biological molecule. Another important design issue is that, although simplicity is favored, mixed solvent systems could offer improved selectivity for the biosolute of interest. Because extraction adds another fluid phase, better process economics will arise when the solvent or solvent mixture can be easily recovered and reused. A successful design will be more likely when the designer understands how the partition behavior reflects both aqueous solution composition and process conditions.

Besides taking advantage of favorable equilibrium, successful bioseparation processes can be realized by manipulating the manner in which the phases come in contact with one another. Countercurrent contacting provides a powerful, multi-faceted method for processing large amounts of feed; it can also be adapted to provide resolution rivaling that achieved with liquid chromatography. Although the systems created are more complex, two-dimensional and multiphase contacting offer some unusual opportunities for increasing purity. Nonlinear equilibrium can limit extraction efficiency, as can be demonstrated by the graphical solution method.

11.9 PROBLEMS

11.1 You have been instructed to design an antibiotic recovery process that will achieve a 98% yield. The product should also be obtained at the highest possible concentration in aqueous solution. You begin with 100 L of aqueous solution at an antibiotic concentration of 0.01 M. Assume that the distribution coefficient K is 1.2. The maximum flow rate of the aqueous solution is 10 L/h and the organic solvent-to-aqueous solution flow rate ratio is 1 for countercurrent contactors. For a cocurrent contactor, the maximum aqueous to organic flow rate ratio is 4.

(a) Would you select a cocurrent or countercurrent contactor?

(b) How many equilibrium stages would you need to achieve the process goals?

(c) What is the final yield and product concentration in the organic phase?

11.2 Protein X can be recovered by two-phase aqueous extraction from an aqueous mixture containing another protein, Y. Given the effect of pH on selectivity and K_X for protein X shown in Figure 11.22, answer the following questions. The molecular weight of protein X is 68,000 Da, while the molecular weight of protein Y is 123,000 Da.

(a) For a single-stage extractor, what is the maximum purity obtainable if protein X has a starting concentration of 5 mg/L, while protein Y's concentration is 1 mg/L? Remember that purity can be expressed in many ways. Specify a useful definition of purity in your answer. Assume that the two-phase system is created by adding dry PEG and salt to the bioreactor solution, yielding two phases of equal volumes.

(b) Assuming that you can use a four-stage extractor with equal heavy- and light-phase flow rates, what is the maximum purity obtained for a countercurrent system?

(c) What would you suggest as separation processes for achieving higher purity than in part (a) or (b)?

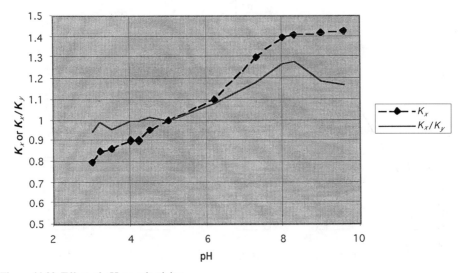

Figure 11.22 Effect of pH on selectivity.

11.3 Consider the use of a three-phase system in a Craig extractor based on Figure 11.23. For want of a better name, we will call this configuration a double Craig extractor (DCE). At the top of the pyramid, after the first equilibration the top phase moves to the right (tube 2) and the middle phase moves to the left (tube 3). Just as in the two-phase Craig extractor, the bottom phase is stationary. The index refers to the tube number and equilibration where the phase is introduced in the cascade.

(a) Given that the equilibrium distributions can be provided in terms of mole fraction ratios,

$$K_a^I = \frac{y_a}{x_a} = 2.2; \quad K_a^{II} = \frac{z_a}{x_a} = 1.2 \tag{11.79}$$

$$K_b^I = \frac{y_b}{x_b} = 1.6; \quad K_b^{II} = \frac{z_b}{x_b} = 3.4 \tag{11.80}$$

plot the results of DCE for five equilibrations if the volume of each phase is 10 mL.

(b) Based on the result from part (a) and the equation for Craig extraction given in the chapter, derive an expression for DCE that relates the fraction extracted to the extraction factors, tube number, number of equilibrations, and any other pertinent parameters.

11.4 In complexation extraction, partitioning between the phases can be primarily dictated by the concentration of the unreacted extract in the organic phase and the equilibrium constant for the complexation reaction:

$$K = K_{complex} y_e \tag{11.81}$$

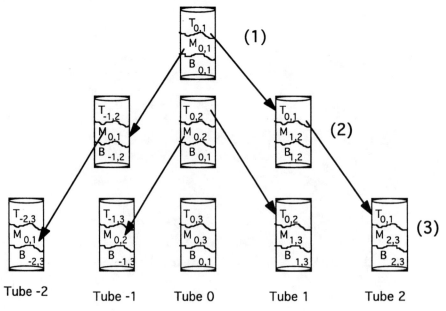

Figure 11.23 Double Craig extractor.

For example, in the case of tridodecylamine complexation with an organic acid,

$$(HA)_{aqueous} + TDA \rightleftharpoons (HA-TDA)_{organic} \tag{11.82}$$

and

$$K_{complex} = \frac{[HA-TDA]}{[HA][TDA]} \tag{11.83}$$

K is defined as the ratio of the acid-base complex concentration over the concentration of the free acid, as the acid can exist only in the organic phase when complexed with TDA, and only free acid exists in the aqueous phase.

(a) Derive an equation for a continuous contactor that accounts for complexation extraction.

(b) When the aqueous feed includes two acids, how can the equation for a continuous contactor be modified to determine the height needed?

11.5 Given the data in Table 11.2 for the extraction of organic acids, determine the number of stages needed to resolve acetic acid (component 1) at 95% purity from a feed mixture containing $x_1 = x_2 = x_3 = 0.1$ using a fractional extractor. In developing your design, consider the effects of varying the flow and feed rates, as well as the feed location. Present your design as a series of choices depending on these variables.

11.6 Two-phase aqueous systems can separate DNA from its denatured form. Using a 5 mM NaH_2PO_4 and 5 mM Na_2HPO_4 buffer in a 5% w/w dextran 4% w/w PEG 6000 extraction solution, $\log K$ of denatured DNA is equal to 1, while for DNA, $\log K = 1.6$.

(a) For a 10 mg/L solution of DNA where 20% is denatured, what percentage of DNA is recovered and what is the purity when a five-stage cross-flow contactor is used? Assume that the flow rate ratio is $L/H = 2$.

(b) Using a countercurrent contactor, what is the purity and yield for a five-stage system?

(c) Using a fractional extractor with five stages, what is the purity and yield when the feed is introduced at stage 3? Use a feed rate of $1/4H$.

11.7 A Podbielniak extractor is a centrifugal extractor that is used extensively for antibiotic recovery. It is also valuable in aqueous two-phase extraction, as low interfacial tensions can make it difficult to separate the phases. The largest-capacity Podbielniak extractors have capacities of 260 gal/min. If a 50,000 gal yeast fermentation is used to produce 3 g/L of both β-galactosidase ($K = 62$) and alcohol dehydrogenase ($K = 8.2$), how many stages are needed if the enzymes must be purified to 95%? Assume that a PEG/salt system is used.

11.8 Penicillin K is being extracted using a Podbielniak extractor in which amyl acetate is the extract phase. If the distribution coefficient K is 12 at pH 4 and 0.1 at pH 6, design a process where a 100 m³/day facility produces penicillin K at a concentration of 22 g/L with a product yield of 90%. Use the maximum Podbielniak capacity of 260 gal/min in your design. Specify how many separate Podbielniak staged series will be needed to keep up with the production and yield requirements.

Table 11.4 Equilibrium Data

x	y
0.01	0.03
0.05	0.09
0.1	0.13
0.2	0.24
0.3	0.4
0.4	0.55
0.6	0.72
0.8	0.88
0.9	0.98

11.9 Given the equilibrium data in Table 11.4, use the graphical McCabe Thiele technique to determine the minimum number of stages needed to recover 95% of the solute in a fractional extraction when the feed has $x = 0.4$, the feed rate is 1 L/h, and the aqueous and extract phase flow rates are 10 L/h.

11.10 Many organic solvents used in extraction can be soluble in the aqueous phase. Likewise, water can have a limited solubility in an extract phase.

(a) Develop a scheme for modeling the effect of water solubility in the extract phase in a staged, cross-flow extractor. Use a diagram and label the individual parameters needed for a mathematical analysis.

(b) Derive an expression for the fraction extracted when water has a limited solubility in the extract phase in a cross-flow contactor.

(c) How would you deal with the case in part (b) when the extract phase has a limited solubility in water?

(d) What would be the effect of limited, mutual phase solubility in a countercurrent contactor?

11.11 The liquid extraction of proteins using reverse micelles shows a maximum extraction as a function of pH and ionic strength at low salt concentrations. This extraction can be modeled as

$$K = K_{max}\exp[pH - pH_{max}]^2 \tag{11.84}$$

$$pH_{max} = pH_0 + \alpha I \tag{11.85}$$

where K_{max} and pH_{max} are constants for a particular protein, I is the ionic strength, pH_0 is the pH maximum with no salt added, and α is a constant. The pH maximum increases proportionally with increasing ionic strength as long as the salt concentration is less than approximately 0.2 M.

(a) Assuming that the feed is a protein mixture, how would you exploit this behavior in a Craig extractor to obtain purified proteins in the aqueous, rather than in the extract, phase?

(b) Would manipulation of pH and/or ionic strength reduce the number of tubes required for a given purity?

(c) Assuming that an enzyme at a feed concentration of 10 mg/L has $K_{max} = 2.4$, $pH_0 = 9.6$, $I = 40$ mM, and $pH_{max} = 10.7$, what is the maximum percent recovery in the aqueous phase for a four-stage Craig extractor when you also wish to maximize the concentration of the enzyme in one of the tubes? Assume that the extract and aqueous solution volumes are both 50 mL.

11.12 Supercritical CO_2 has been applied to extract pharmaceuticals used in chemotherapy. Data are usually provided in terms of solubility (mole fraction of weight percent) as a function of supercritical-phase density.

(a) If a chemotherapeutic obtained by crushing plant stems or seeds is 1.2 wt% of the biomass, how many stages are needed in a countercurrent contactor to process 100 g/min of biomass using 5 L/h of supercritical CO_2 to recover 90% of the chemotherapeutic? Assume that its solubility is 0.07 wt%.

(b) As moving solids in a staged system is too cumbersome, how might you simulate countercurrent motion without actually moving the seeds?

11.13 Partition coefficients can be measured in the following way: Starting with an aqueous solution of 10 wt% oxalic acid, 10 mL of this solution is added to a scintillation vial containing 4 mL of heptanol. After letting the phases come to equilibrium at 22 °C, the aqueous phase was found to contain 8.4 wt% oxalic acid using ion-exclusion HPLC.

(a) What is the value of the partition coefficient under these conditions?

(b) Design an experiment that would determine how the partition coefficient changes with aqueous-phase concentration, given that the aqueous solution oxalic acid feed concentration is 10 wt%.

(c) What steps would you take to ensure that the phases are in equilibrium before recording the data needed to calculate the partition coefficient?

11.14 Partition coefficients can be calculated based on the ratio of the mole fractions, molar concentration, molal concentration, and weight percents. In many scientific articles and reference books, partition coefficients are provided in tables without a clear definition of the units used in calculating them. Can this lack of explantion be a problem? If so, when is it problematic?

11.15 For a staged cross-flow contactor, how would we use the graphical solution method to determine the number of stages needed? Assume that the equilibrium curve has an arbitrary shape. Graph the operating line and show how the overall solute balance is used in the graphical solution.

11.16 A major cost in solvent extraction is the disposal or processing of the extract after most of the solute and other components are removed by back-extraction or crystallization. Extract can also be lost because it has limited solubility in the aqueous phase.

(a) What is the cost per milligram of protein recovered for disposing of PEG in aqueous two-phase extraction when a 5% solution of PEG 400 is used to recover 10 mg/L of a protein? State the source used for the price of PEG 400 along with your answer. What processes would you suggest for recovering PEG 400 if the protein has a molecular weight of 180,000?

(b) Amyl acetate is a popular solvent used in antibiotic extraction. What is the solubility of amyl acetate in water? How much amyl acetate would be lost per day in a 100 m³/day antibiotic production facility? What would be a useful method for recovering amyl acetate from an aqueous solution?

11.10 REFERENCES

1. Perrin DD, Dempsey B, Sergeant EP. PK_a prediction for organic acids and bases. London: Chapmann and Hall, 1981.
2. Walter H, Johansson G, Brooks DE. Partitioning in aqueous two-phase systems: percent results. Anal Biochem 1991;1:1–18.
3. Alberston PA. Partition of cell particles and macromolecules: separation and purification of biomolecules, cell organelles, membranes, and cells in aqueous polymer two-phase systems and their use in biochemical analysis and biotechnology, 3rd ed. New York: Wiley-Interscience, 1986.
4. Walter H, Brooks DE, Fisher D. Partitioning in aqueous two-phase systems: theory, methods, uses, and applications to biotechnology. Orlando: Academic Press, 1985.
5. Flory PJ. Thermodynamics of high polymer solutions. J Chem Phys 1942;10:51–61.
6. King RS, Blanch HW, Prausnitz JM. Molecular thermodynamics of aqueous two-phase systems for bioseparations. AIChE J 1988;34:1585–1594.
7. Haynes CA, Blanch HW, Prausnitz JM. Separation of protein mixtures by extraction: thermodynamic properties of aqueous two-phase polymer systems containing salts and proteins. Fluid Phase Equilibr 1989;53:463.
8. McMillan WG, Mayer JE. The statistical thermodynamics of multicomponent systems. J Chem Phys 1945;13:276–305.
9. Haynes CA, Benitez FJ, Blanch HW, Prausnitz JM. Application of integral-equation theory to aqueous two-phase partitioning systems. AIChE J 1993;39:1539–1557.
10. Kang CH, Sandler SI. A thermodynamic model for two-phase aqueous polymer systems. Biotechnol Bioeng 1988;32:1158–1164.
11. Diamond AD, Hsu JR. Fundamental studies of biomolecule partitioning in aqueous two-phase systems. Biotechnol Bioeng 1989;34:1000–1014.
12. Huddleston JG, Wang R, Flanagan JA. Variation of protein partition coefficients with volume ratio in poly(ethylene glycol)-salt aqueous two-phase systems. J Chromatography A 1994;668:3.
13. Hartounian H, Kaler EW, Sandler, SI. Aqueous two-phase systems. 2. Protein partitioning. Ind Eng Chem Res 1994;33:2294.
14. Warner RK. Extraction of uranyl nitrate in a disc column. Chem Eng Sci 1954;3:161–174.
15. Treybal RE. Mass transfer operations, second edition. McGraw-Hill, 1968.
16. Marcus Y. Solvent extraction reviews. NY: Marcel Dekker, 1971.
17. Humphrey IW. Solvent refining of wood rosin. Ind Eng Chem Res 1943;35:1062.
18. Lister DA. U.S. Patent No. 2,401,432. 1936.
19. Blumberg R. Liquid-liquid extraction. San Diego: Academic Press, 1988.
20. McCabe WL, Smith JC, Harriott P. Unit operations of chemical engineering, 5th ed. New York: McGraw-Hill, 1993.

12

Electrophoresis

The application of an external electric field is often used in the analytical separation of charged biological molecules. Scale-up and commercial applications of this technology, however, are hampered somewhat by equipment design limitations for processing large solution volumes. Nonetheless, electrophoresis remains a predictable technique that can be run at mild conditions and can provide very high resolution. During the scale-up to preparative or production scale, however, it may be difficult to match the resolution found at the analytical scale. In addition, the electrical power requirements and the cost of buffers for production-scale operations can place electrophoretic methods at an economic disadvantage relative to other separation technologies. Despite these potential problems, new devices and applications using electrophoretic methods for bioseparations continue to be developed.

Electrophoretic methods are similar to chromatographic methods in that fixed barrier phases (ranging from paper to polymer gels) are employed to facilitate separation. Unlike in other bioseparation processes, however, the electrophoretic driving force for creating motion—the electric field—can create unwanted heating due to the Joule effect. Many electrophoresis devices are designed to minimize this type of heating.

Myriad methods and devices have been developed to exploit the simple idea of using an applied electric potential to produce directed movement of charged molecules. Toward the end of this chapter, the four main modalities in electrophoresis—zone, moving boundary, isoelectric focusing, and displacement (isotachophoresis)—are analyzed in order to predict separation time, concentration, and purity in these methods.

12.1 A BRIEF INTRODUCTION TO SOME POPULAR ELECTROPHORETIC METHODS

Before describing the fundamental principles of electrophoresis, we will briefly survey the designs used in many electrophoretic separations. Some of these methods have been employed at production scale; others are relegated to analytical determination of concentration and purity. Electrophoresis can be performed using a barrier such as gel or paper, or it can take place in solution (i.e., "free-flow" electrophoresis). Some popular electrophoretic methods are described in the immediately following subsections. Mathematical approaches to help design equipment and procedures are discussed later in an attempt to weave together the recurring strategies used in electrophoretic methods.

12.1.1 Gel electrophoresis

Gel electrophoresis is known as a zonal electrophoresis method because the apparatus separates the sample into well-defined zones. The applied sample migrates through a polymer matrix—usually composed of polyacrylamide, starch, agar, or agarose (Figure 12.1 shows the chemical structures of polyacrylamide and agarose). Although paper and membranes can replace this gel, only cellulose acetate membranes are regularly used for this purpose because paper usually carries a fixed charge that leads to unwanted electro-osmotic effects (see Section 12.2.1). The gel is sandwiched between two slabs and can be mounted vertically or horizontally. A horizontal orientation is preferred when working with fragile gels. The resulting systems can be automated to perform quantitative analyses and to handle large-volume samples.

In gel electrophoresis, samples are applied in distinct locations known as lanes. With the slab configuration, the components travel in well-defined lanes. Specific biopolymers are recovered in gel electrophoresis in one of two ways:

- In laboratory procedures, the biomolecule is located based on chemical staining or radioisotope label detection. The spot is then cut out and the sample leached from the gel.
- In preparative work, the biomolecule must travel the length of the apparatus to elute from the gel.

12.1.1.1 Pulsed-field gel electrophoresis

The pulsed-field variation of gel electrophoresis is useful in separations involving high-molecular-weight (greater than 20 kilobases) DNA. In this application, the choice of gel is agarose. For low-molecular-weight DNA, the mobility in a constant field is simply related to size because of the rodlike structure of these molecules. As the DNA size increases, a limit is reached where all DNA molecules have the same mobility—that is, the molecules are coiled and project nearly the same surface area.

To overcome this limitation, several researchers developed the idea of changing the field's orientation with respect to the gel and applying the new potential for a limited time (1). Pulsed-field gel electrophoresis (PFGE) was the first electrophoretic method shown to break the DNA size barrier. In the initial experiments,

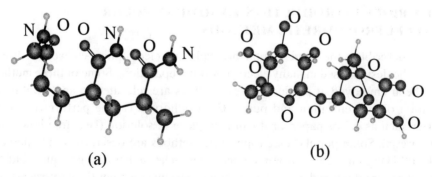

(a)   (b)

Figure 12.1 Molecular structures of commonly used electrophoresis gels: (a) polyacrylamide and (b) agarose (nitrogen and oxygen atoms are labeled N and O respectively).

researchers pulsed the field and turned it at a 90° angle. Later, it was shown that turning the field at a 180° angle (hence the nomenclature "field-inversion gel electrophoresis" or FIGE) could also work as long as the forward and reverse pulses had unequal duration or field strength (1).

The original explanation of this phenomenon suggested that separation in a pulsed field was due to the DNA's ability to orient itself and move toward the positive electrode when the potential is applied. According to this hypothesis, DNA could move only when it coiled itself back into a compact shape during the "off" portion of the cycle.

Recently, this model has been shown to be incorrect (1). Instead, the process is now considered to effect high-molecular-weight DNA separation. The basic idea is similar to the phenomena observed when we place marbles of two different sizes in a jar. Even if the larger marbles are denser than the smaller ones, shaking the jar will cause the smaller marbles to fall to the bottom because they can better maneuver through the crevices between the larger marbles. The analogy is not a perfect one, as the DNA strands must wend their way through the spaces between the agarose bundles. Nevertheless, this idea of "shaking" the DNA strands is acceptable because pulsing the field at different orientations provides the energy and the opportunity for smaller strands to move faster through the agarose maze. In PFGE, two different DNA strands become elongated, which enables them to maneuver around the thick agarose bundles. At the end of a pulse, the field is switched at an angle greater than 90° and the back ends of the strands move around some agarose bundles. At the end of the second pulse, the smaller strand has moved farther down the gel axis.

A simple mathematical analysis can provide additional insight into how PFGE works. The distance that a strand will move depends on the duration of a pulse and the velocity that the strand attains:

$$d = vt \tag{12.1}$$

With respect to the gel axis, the distance traveled (x) is given by

$$x = d \cos(\theta) \tag{12.2}$$

as the field is pulsed at an angle θ from the gel axis. For a strand of length l, the distance along the axis of the gel between the front and back end of a strand after one pulse is calculated as follows:

$$J = (d - l)\cos(\theta) \tag{12.3}$$

The separation between two strands (the ith and jth strands) after one pulse is given by the following equation:

$$\Delta J = -(l_i - l_j)\cos(\theta) \tag{12.4}$$

A simple way of looking at the cumulative effect of n pulses is to multiply J and ΔJ by n. This manipulation yields the final equation for predicting the separation as a function of pulse orientation and the number of pulses:

$$\Delta S = -n(l_i - l_j)\cos(\theta) \tag{12.5}$$

While this analysis may seem very simple, its result is important from a design standpoint. Experimental work has verified that the separation is proportional to the number of pulses and that, for a given separation time, $n = (\text{total time})/t_{\text{pulse}}$. Thus a shorter pulse duration improves separation.

The size limit for a separation is given by d because when l is greater than d, $(d - l) < 0$. Consequently, the back end will not move past the location held by the front end on the previous pulse. Although shortening the pulse length may improve the separation, the pulse should not be too short because it dictates d. To circumvent this limitation, we can increase the field strength.

12.1.1.2 Polyacrylamide gel electrophoresis

By far the most popular laboratory-scale gel electrophoresis method is polyacrylamide gel electrophoresis (PAGE). Acrylamide can be polymerized in place so as to vary the gel pore size. Because polyacrylamide does not carry a charge and produces very small pores (less than 80 nm), this technique minimizes unwanted effects such as electro-osmosis and Joule heating-related convection. Moreover, polyacrylamide pore size can be controlled by varying the monomer concentration or the degree of cross-linking (2). This control makes possible several variations in gel electrophoresis techniques, such as pore limit electrophoresis and stacking or disc gel electrophoresis (Figure 12.2).

Disc gel electrophoresis combines isotachophoresis and zone electrophoresis. The upper buffer reservoir contains tris-glycine buffer at a pH of 8.3. The sample gel and dilute stacking gel (approximately 2.5% acrylamide), which make up the

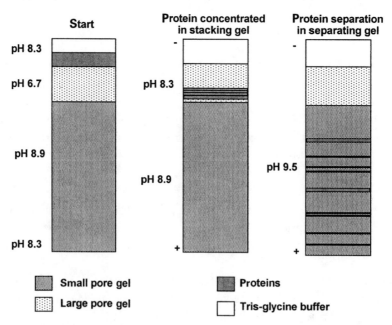

Figure 12.2 Stacking gel electrophoresis. (Adapted from Ornstein L, Davis BJ. Disc electrophoresis: parts I and II. Rochester: Distillation Industries, 1962.)

middle region, exist in a tris-chloride buffer at a pH of 6.7. The separating gel lies at the bottom (approximately 15% acrylamide). As it has a high mobility compared with glycine at the stacking gel pH, the chloride ion becomes the leading front; the glycine ion brings up the rear. Reflecting the fact that conductivity is inversely proportional to voltage, the movement of the glycine ion accelerates to match the speed of the leading chloride ion front. A steady state is formed where these two types of ions move at the same speed but remain separated by their differences in mobility. Proteins are swept along between the two fronts in the stacking gel. When the fronts meet the separating gel, the higher pH causes the glycine front to move behind the chloride ion front. Proteins can then be separated in this gel based on their size, as the gel pores are very small. The proteins do, however, experience an increasing voltage gradient as each fraction moves past the boundary.

To ensure that proteins of particular sizes are detected with high accuracy, sodium dodecyl sulfate (SDS) may be added to the protein gel electrophoresis process. The addition of SDS, along with the injection of mercaptoethanol (which disrupts disulfide bonds), to proteins yields denatured, rodlike molecules with practically equal charge-to-mass ratios. Thus protein separation relies solely on protein size. SDS-PAGE is a widely utilized technique for analytical procedures. Approximately 1.4 g of SDS will bind to 1 g of protein (2).

12.1.2 Capillary electrophoresis

Capillary electrophoresis (CE) does not require the use of a supporting medium (gel or paper) and can take place in free aqueous solution. The most commonly used capillary material is fused silica, although glass and Teflon have been used for this purpose in the past. Capillary tubes must be small—they range in size from 25 to 75 μm—as larger capillaries show poor performance due to convection from Joule heating. The silanol groups on the silica capillary walls acquire a negative charge, with positive counter-ions residing in the aqueous electrolyte buffer. These hydrated counter-ions become attached to the cathode, giving rise to an electro-osmotic flow toward the cathode. Hence, if an electric field is applied across the fused-silica capillary, the mobile ions of the solution will migrate with their water of hydration toward the cathode, initiating a flow within the entire solution. The presence of the electro-osmotic flow enables the separation of both negatively and positively charged species in the same run (3).

Corporate and academic laboratories face economic and social pressures to minimize solvent wastes, which unfortunately many analytical chromatography techniques generate. CE, which uses aqueous salts as the buffers, is therefore a very popular analytical technique. It represents a very powerful tool for separating charged biomolecules.

Consider, for example, the separation and quantification of low-molecular-weight carboxylic acids by capillary zone electrophoresis (CZE) with an on-column conductivity detector. The addition of 0.2 mM TTAB (tetradecyltrimethylammonium bromide) controls the electro-osmotic flow, enabling all of the carboxylate anions to pass through the detector. Unlike other CZE detection methods, conductivity detection shows a direct relationship between retention time and peak area. Conductivity detection in CZE therefore offers a unique advantage: Use of an internal standard permits the accurate determination of absolute concentrations in a mixture without requiring a separate response calibration for each component.

Some carboxylic acids have very similar structures but mobilities that, under suitable conditions, differ sufficiently to allow full resolution. CZE has proved useful for separating optically active isomers of fumaric and related acids (3).

12.1.3 Isoelectric focusing

Isoelectric focusing capitalizes on the maintenance of a pH gradient between two electrodes (Figure 12.3). Proteins applied at the anode will migrate toward the cathode, as they have a net positive charge at low pH. Proteins travel through the electrophoresis media until they reach a pH equal to their isoelectric point (pI). When pH = pI, the protein will not migrate—it does not have a net charge and hence no net force acts on the protein because of the imposed electric field. Isoelectric focusing offers high resolution (it can separate proteins with a pI difference of as little as 0.02) and can concentrate the sample by restricting the protein to a small volume.

The pH gradient is created when soluble ampholytes are used as carriers in free-flow electrophoresis methods; immobilized pH gradients can be constructed by modification of acrylamide. The focusing effect of this technique allows for useful free-flow methods, whereas zone electrophoresis leads to spreading bands. One such free-flow apparatus has been used to separate cells (4).

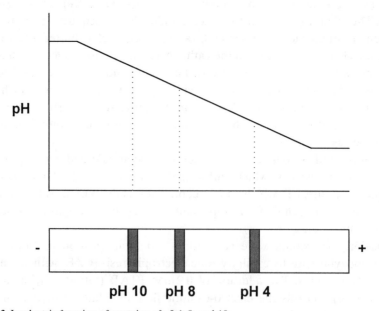

Figure 12.3 Isoelectric focusing of proteins of pI 4, 8, and 10.

12.1.4 Isotachophoresis

Dual-buffer systems can produce a situation in which a protein sample becomes sandwiched between a leading and trailing electrolyte (see Figure 12.2). We previously discussed this situation as it applied to stacking or disc gel electrophoresis, but the same concept can be used in gel, paper, or free-flow electrophoresis, and in one- or two-dimensional electrophoresis. These techniques are collectively known as isotachophoresis.

In isotachophoresis, the leading band appears at the lowest field strength, and the trailing band lies at the highest field strength. The field strength is constant within each band. The drop in field strength is approximately proportional to the ratio of the electrophoretic mobility of the component.

Isotachophoresis offers an advantage in that the field gradient created with a judicious use of leading and trailing electrolytes (5) produces self-sharpening zones. The development of sharp zones makes this method very desirable for large-scale, continuous electrophoretic separation of proteins. The only drawback to large-scale separation is that we must use spacer ions when a particular component is present in a small amount; the inclusion of such ions, however, can dramatically narrow the component's band. Spacer ions are needed because, while each component moves in a separate band, the bands themselves are not separated by any space. The spacer component has a mobility between the small band and the next band, thereby providing band separation. The appropriate spacer for a given separation is determined through experimentation.

12.1.5 Moving boundary

In 1937, Arne Tiselius developed the moving boundary method of electrophoresis, which provided the first illustration of how electric fields could be used in separations. Today, this free-solution technique is popular for measuring electrophoretic mobilities with a high degree of accuracy.

In the moving boundary approach, a U-tube is filled with a protein solution through one of the open arms (Figure 12.4). A buffer solution is added at the other arm. Only the fastest-moving proteins, however, are separated at the tube arms. Minimal convection occurs because the protein solution is denser than the buffer solution. Next, the electrodes are immersed in the solution at each arm and the voltage is applied. As the proteins migrate to the cathode or anode at various velocities depending on their electrophoretic mobility, interference or schlieren optics is employed to track the migration of species. Although this detection system is expensive, it provides very accurate measurements of electrophoretic mobilities.

12.2 BASIC CONCEPTS OF ELECTROPHORESIS

Introductory chemistry courses use the migration of ions to an oppositely charged electrode as a definition in describing the ionic conductivity phenomenon. This point charge model, however, is too simplistic a model for cells, proteins, DNA, and other

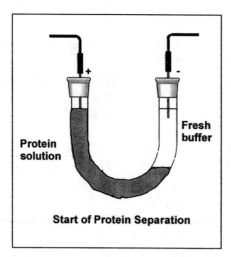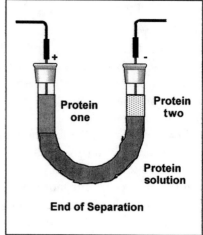

Figure 12.4 A schematic representation of moving boundary electrophoresis separation of proteins in free solution.

charged biopolymers. An interpretation of biopolymer and cell mobility must consider that proteins and cells have both positive and negative charged species on their exposed outer surface. Thus net charge is the important parameter in describing biopolymer and cell mobility in an electric field. Also, when dealing with colloidal particles (whose sizes range from roughly 0.1 to 5 μm), a successful model of colloidal behavior uses a tripartite structure: 1) a surface charge distribution, 2) a bound layer of counter-ions, and 3) a diffuse layer of ions.

The model given in Figure 12.5 allows us to examine the effects of solution conditions, biopolymer or cell type, and electric field strength on the movement of colloidal structures. Note that this model is still rather simplistic, because solution conditions can affect protein and cell charge distributions, and it is impossible to describe charge heterogeneity exactly. Nevertheless, we can use this model to drive home several key points about the design of electrophoresis devices.

The first point is that electrophoresis can separate biopolymers based on net charge and/or size and shape differences. Larger particles will experience resistance to flow caused by viscous drag and by the ions associated with these particles that reside between the bound layer and the shear plane boundary.

Using a force balance for the case in which a colloidal particle is moving at a constant velocity in an electric field,

$$\sum F = 0 \tag{12.6}$$

$$F_{\text{electric field}} - F_{\text{drag}} = 0 \tag{12.7}$$

$$EQ - C_{\text{D}}U = 0 \tag{12.8}$$

we can see that the velocity of the particle is directly proportional to the electric field (E) and the net charge of the particle (Q), and inversely proportional to the coefficient of drag (C_{D}). Before deriving a more general mathematical description of electrophoresis with biological molecules, we will take advantage of this simple

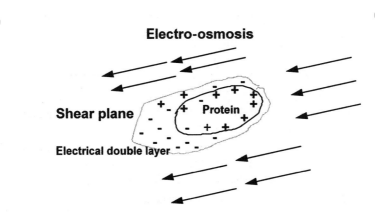

Figure 12.5 A model of protein charge structure and accompanying ion layers in a potential field. The arrows show the retarding force attributable to the flow of water because of the presence of the electric field (electro-osmosis).

analysis to define a key parameter and to provide a basis for rough calculations. The electrophoretic mobility is an important parameter in that it normalizes the effect of the electric field:

$$\mu = \text{Mobility} = \frac{U}{E} \tag{12.9}$$

Assuming that the biopolymer is spherical, the mobility can be related to biopolymer properties in a simple way:

$$\mu = \frac{Q}{6\pi\eta R} \tag{12.10}$$

We must regard this equation with caution, however. Because solution conditions and biopolymer structure can affect the electrophoretic mobility in so many complex ways, mobility is best measured experimentally. Nevertheless, Equation 12.10 illustrates that we must consider both size and charge when choosing the conditions for separating biopolymers. It also shows that the media viscosity (η) plays an important role in dictating mobility. To minimize Brownian motion, which will lower the resolution of the electrophoretic separation, highly viscous phases such as polymer gels are used instead of aqueous solutions.

Researchers in capillary electrophoresis have developed an expression to predict peptide electrophoretic mobility based on the charge and size (3). In free-solution electrophoresis, mobility is proportional to this charge and inversely proportional to the particle size:

$$\mu = D\frac{\ln(q+1)}{n^{0.43}} \tag{12.11}$$

This semi-empirical relationship has been shown to hold for peptides in various buffers with $n = 3$ to 39 amino acid residues and charge values (q) ranging from 0.33 to 14. The slope D is determined experimentally.

12.2.1 Electro-osmosis and the relaxation effect as retardation forces

We must also consider two additional forces before completing our description of the model system. The first important phenomenon, known as electro-osmosis, is the movement of water caused by the imposed electric potential field. Ions transfer a force to the water molecules, which hydrate the ions (6). The second important force, known as the relaxation effort, follows from the distortion of the electrical double layer from the ion atmosphere around the particle.

Reviewing Appendix B, we can see that the electro-osmosis effect can be taken into account by considering the Huckel and Helmholtz-Smoluchowski equations. Describing the relaxation effect mathematically, however, is a more complicated task. In any event, both retardation effects suggest that

$$\mu \, \alpha \, \frac{\varepsilon_0 \varepsilon \zeta}{\eta} \tag{12.12}$$

where ε_0 and ε are the permittivity of a vacuum and water, respectively, and ζ is the zeta potential.

The zeta potential is related to solution conditions and the surface potential Ψ_0 (see Appendix B for more details). A useful shorthand method for describing this effect is to employ a capacitor analogy. One plate is the surface, and the other is the ion atmosphere closely associated with that surface. The gap between the capacitor "plates" is the double-layer thickness, $1/\kappa$ (see Appendix B). Increasing the ionic strength brings the double layer closer to the surface, effectively shielding the surface charge. Hence, as the ionic strength increases, the double-layer thickness decreases and the zeta potential decreases.

Using our refined model for mobility, we can now better assess how changes in solution conditions affect mobility (Table 12.1). In our analysis, we must note that the zeta potential—not the surface potential—dictates the charge interactions between the particle and the imposed electrical field. Consequently, important electrophoretic methods such as SDS-PAGE, which enables ionic species to bind to the particle, can dramatically alter the zeta potential. Also, increasing the ionic strength and thereby decreasing the double-layer thickness produces a decrease in the zeta potential, because its position is dictated by hydrodynamic effects involving the shear plane.

12.2.2 Situations that can hamper electrophoretic separation

Although Table 12.1 lists an increase in temperature as a beneficial effect (it increases mobility), Joule resistance heating created by the imposition of an electric field is considered undesirable because it causes convection. Convection currents sweep along all biopolymers in a solution at the same velocity, potentially obliterating a separation based on size or charge.

Heating produces convection by inducing density gradients in the fluid. A simple, but accurate description of temperature's effect on density is to say that density normally decreases linearly with increasing temperature, and the coefficient of volumetric expansion, β, dictates the slope of the graph of ρ versus T. Water represents an important exception to this linear relationship. At 4 °C, it has a

Table 12.1 Effect of Solution Conditions on $1/\kappa$, ζ, and μ

Condition	Direction of Change	Effect on $1/\kappa$	Effect on ζ	Effect on μ
pH	$\uparrow\downarrow$	Depends on ionic strength	Ion- and chemistry-specific	Ion- and chemistry-specific
Ionic strength	$\uparrow$	$\downarrow$	$\downarrow$	$\downarrow$
Presence of protein-binding ions	$\uparrow$	No effect at constant ionic strength	Ion-specific	Ion-specific
Temperature	$\uparrow$	$\uparrow$	$\uparrow$	$\uparrow$

density maximum and $\beta = 0$. As it offers the added benefit of minimizing biomolecule degradation at lower temperatures, electrophoresis at 4 °C is a popular mode of operation.

A closer look at convection by Joule heating provides more insight into electrophoretic device operation. The Rayleigh (Ra) and Grashof (Gr) numbers are dimensionless groups of parameters that serve as useful design guides. Adjusting a system to maintain a low Rayleigh number prevents convection; systems with larger values of this number tend to experience convection. The intensity of the convection cells is dictated by the Grashof number. As can be seen in Equation 12.14, the Grashof number measures the ratio of the buoyancy effect to viscous forces:

$$Ra = \frac{\beta \rho C_p g h^3 \Delta T}{k_T \eta} \tag{12.13}$$

$$Gr = \frac{\beta \rho^2 g h^3 \Delta T}{\eta^2} \tag{12.14}$$

Besides operating at 4 °C, conducting electrophoresis under microgravity conditions can greatly minimize convection. Although performing this type of operation in space is not practically feasible at this time, we can nevertheless increase the fluid viscosity and keep the gap distance h small by employing capillary tubes or porous gels in the separation. Geometry also affects the Rayleigh number, and horizontal equipment produces larger critical Rayleigh numbers for the onset of convection.

Another important consideration involves the initiation of electro-osmosis by fixed charges on the walls of the electrophoretic chamber. Although capillary electrophoresis takes advantage of electro-osmosis to carry solutes through the tube with a flat velocity profile, the separation itself should not be based on electrophoretic mobility. To minimize electro-osmosis, we can coat the chamber walls with a nonionic material.

12.3 ZONE ELECTROPHORESIS

Zonal methods characterize both barrier and free-flow electrophoresis. Conceptually, this technology is the simplest type of electrophoresis. The mobility equation gives the time needed for biopolymer A to move a distance x:

$$t = \frac{x}{\mu E} \tag{12.15}$$

To this point, our analysis has assumed an idealized case in which conditions remain perfectly uniform throughout the system. In reality, charged biomolecules do not move as a perfect front; rather, they form a band that widens with distance traveled. This effect appears in chromatography (see Chapter 10) as well. We can therefore define resolution in zone electrophoresis much as we did in our discussion of chromatographic resolution:

$$R = \frac{l_B - l_A}{2(\sigma_B - \sigma_A)} \tag{12.16}$$

In this equation, l_B and l_A are the distances traveled by the middle of the band in time t, and σ_B and σ_A are the band widths. The latter two parameters indicate the point by which the detector signal falls off to 5% of the signal from the middle of the band (i.e., one standard deviation). Using the mobility equation and the fact that complete band separation occurs at $R = 1.5$, we find that the relationship

$$t = \frac{3(\sigma_B + \sigma_A)}{E\left(\dfrac{1}{\mu_B} - \dfrac{1}{\mu_A}\right)} \tag{12.17}$$

can help determine how long to run a zonal electrophoretic separation if we can estimate the band widths. As $\sigma_B = \sigma_A$ in most cases, this relationship can be further simplified:

$$t = \frac{3\sigma_A}{2E\left(\dfrac{1}{\mu_B} - \dfrac{1}{\mu_A}\right)} \tag{12.18}$$

As the electric field increases, the time needed to carry out the separation decreases.

Summarizing our analyses thus far, we have identified several important parameters that are used to predict separation by electrophoretic methods: the overall net charge of the molecule, the electric field strength, the solution viscosity, the ionic strength, and temperature. Some of these variables are interrelated.

12.3.1 Band dispersion

Realizing that solutes move with a concentration profile that can be modeled as a Gaussian peak, we can derive a band dispersion that enables us to predict the effect of operating conditions. A Gaussian distribution is usually written as follows:

$$C(x) = \frac{1}{\sigma\sqrt{2\pi}} \exp\left(\frac{-(x - x_{peak})^2}{2\sigma^2}\right) \tag{12.19}$$

In this equation, x_{peak} is the position at which we find the highest concentration of the solute. Because x_{peak} also corresponds to the position of the peak midpoint, we can use Equation 12.15 to develop an expression that describes mobility in zone electrophoresis:

$$x_{\text{peak}} = \mu E t \tag{12.20}$$

It is evident that zone electrophoresis mirrors pulse chromatography with a trace injection. In Chapter 10, we saw that the Gaussian distribution can be compared to the result of modeling convection and diffusion. Fahien (7) has already provided this result:

$$C(x) = \frac{M}{S\sqrt{4\pi D_{\text{eff},i} t}} \exp\left(\frac{-(x - x_{\text{peak}})^2}{4 D_{\text{eff},i} t}\right) \tag{12.21}$$

A comparison of the equations for chromatography and electrophoresis generates a value for band dispersion based on the system's particular properties:

$$\sigma_D^2 = 2 D_{\text{eff},i} t \tag{12.22}$$

This variance accounts only for diffusion-related band spreading; it is proportional to the effective diffusivity, D_{eff}. To introduce the effect of the applied electric field, we substitute for t and acknowledge that the solutes traverse the entire length of the device (L) before being collected:

$$\sigma_D^2 = \frac{2 D_{\text{eff},i} L}{\mu E} \tag{12.23}$$

Note that the band-broadening effect created by increasing the distance between electrodes can be offset by increasing the field strength. We must, however, take care to stay within the power limits of the equipment and avoid convection currents.

For cases where more complex phenomena increase band width, we can define a continuous chromatographic or electrophoretic separation in terms of a discrete system of theoretical plates. By definition, the number of theoretical plates is given by

$$N = \frac{L^2}{\sigma_T^2} \tag{12.24}$$

and the height of a theoretical plate is easily calculated:

$$H = \frac{L}{N} \tag{12.25}$$

In Equation 12.24, σ_T is the total variance, which includes the variance due to diffusion:

$$\sigma_T = \sigma_D + \sum_j \sigma_J \tag{12.26}$$

This approach is both practical and phenomenological. After running the zone electrophoresis system, we can fit a Gaussian distribution to the peak to obtain σ_T. More detailed knowledge of the zone electrophoresis system is needed before we can determine the other variance values in terms of system parameters. The references at the end of this chapter provide additional information on equations that would apply in capillary, gel, and other zonal configurations.

12.4 ISOELECTRIC FOCUSING

Once a medium containing a pH gradient reaches steady state, a balance is achieved between the movement of solutes caused by the focusing effect of the pH gradient and the spreading by diffusion due to the concentration gradient. Using the law of conservation of mass, we can then write an equation for the mass rate of a solute moving through an interface (i.e., flux, which has units of mass/time-area). A convenient coordinate system specifies that $x = 0$ at the peak midpoint, which is the focal point for solutes that undergo isoelectric focusing (IEF). With this system, we can equate the flux generated by the electric field with the diffusion-related flux:

$$D\frac{dC}{dx} = \mu CE \tag{12.27}$$

A useful simplification is to assume that μ is proportional to the distance from the focal point:

$$\mu = -\frac{d\mu}{dx}x \tag{12.28}$$

The proportionality constant is the mobility gradient, which we can separate into two parts:

$$\frac{d\mu}{dx} = \frac{d\mu}{d(\text{pH})}\frac{d(\text{pH})}{dx} \tag{12.29}$$

The first term is considered to be constant because of the narrowness of the bands. The second term is the imposed pH gradient, which is approximately linear. Combining Equations 12.29 and 12.27, we have

$$\int_{C_{\max}}^{C}\frac{dC}{C} = -\frac{E}{D}\frac{d\mu}{d(\text{pH})}\frac{d(\text{pH})}{dx}\int_{0}^{x}x\,dx \tag{12.30}$$

Equation 12.30 can be solved to give a Poisson distribution:

$$C = C_{\max}\exp\left(\frac{-x^2}{2\sigma_{\text{IEF}}^2}\right) \tag{12.31}$$

We define σ_{IEF} by realizing that this equation defines the band width generated by dispersion in isoelectric focusing:

$$\sigma_{\text{IEF}}^2 = \frac{D}{E\dfrac{d\mu}{d(\text{pH})}}\frac{d(\text{pH})}{dx} \tag{12.32}$$

Proteins have a low diffusivity and experience large changes in mobility with a change in pH. Consequently, isoelectric focusing will produce sharp bands that can resolve proteins having small differences in pIs. Sharp pH gradients decrease band widths as well.

12.5 ISOTACHOPHORESIS

Unlike zone electrophoresis and isoelectric focusing, isotachophoresis involves high concentrations of ions. Also, the mobile phase consists of leading and trailing electrolytes introduced before and after the sample—that is, a heterogeneous solution phase. Because of the high electrolyte concentration and buffer changes, the apparent electric field felt by solutes changes with position. These conditions enable the system to achieve both concentration and focusing. As a result, isotachophoresis provides a valuable method for recovering trace solutes.

In isotachophoresis, the pH is adjusted until all solutes have the same charge. The sample is then placed between the leading and trailing electrolytes. A constant current is applied throughout the separation, and the solutes aggregate in bands based on their mobility.

To illustrate the governing principles of isotachophoresis, let us consider a simplified model based on a dilute solution with fully ionized salts. At steady state, all bands move at a constant velocity. The product of the mobility and the local electric potential are therefore constant for each band:

$$\mu_L = \mu_1 = \mu_2 = \mu_n = \mu_T \tag{12.33}$$

$$\mu_L E_L = \mu_n E_n = \mu_T E_T \tag{12.34}$$

Electrochemically speaking, the total current incorporates the flow of positive and negative ions. Electroneutrality is maintained in isotachophoresis, giving the following expression for the current per unit of the device's cross-sectional area:

$$i_x = C_n z_n F(\mu_n^+ + \mu^-) \tag{12.35}$$

In Equation 12.35, z is the valence, F is Faraday's constant, and each solute has a common negative counter-ion. Using Equations 12.34 and 12.35, we can derive a useful relationship to predict the concentration profile of a solute based on the concentration and the choice of leading or trailing electrolyte. If we use the leading electrolyte, we derive the following expression for solute n:

$$C_n = C_L \frac{z_L(\mu_L^+ + \mu^-)}{z_n(\mu_n^+ + \mu^-)} \tag{12.36}$$

An equation expressed in terms of mobility, rather than velocity, has more utility, however. To derive this expression, we substitute the velocity with the electric field strength and mobility, and then equate the result with the current:

$$C_n = C_L \frac{z_L E_L(\mu_L^+ + \mu^-)}{z_n E_n(\mu_n^+ + \mu^-)} = C_L \frac{z_L \mu_n^+(\mu_L^+ + \mu^-)}{z_n \mu_L^+(\mu_n^+ + \mu^-)} \tag{12.37}$$

The term on the right shows the final result after we substitute the field strengths for the ratio of mobilities.

Equation 12.37 indicates that the solute concentration within its own band is independent of the feed concentration. An additional benefit appears when we calculate the band width with this simplified model. To obtain the band width, we write a simple mass balance:

$$C_{feed}L_{feed} = C_{n,exit}L_{band} \qquad (12.38)$$

The band width is then easily calculated from the feed width and the ratio of feed-to-exit concentrations. Here we can see that the focusing effect is directly related to the concentrating power of the leading electrolyte. Note, however, that band concentration and sharpening will be constrained by the presence of solutes whose solubilities fall below the concentration achievable by the leading electrolyte.

12.6 TWO-DIMENSIONAL ELECTROPHORESIS

Adding another dimension to electrophoresis enables us to separate very complex mixtures into many components. This technique "fingerprints" a complex mixture and offers a way to rapidly determine the minimum number of components present in the solution.

Two-dimensional electrophoresis combines differing separation modes, such as affinity and size. To date, scale-up efforts in electrophoresis have focused on two-dimensional, continuous-flow systems. In this type of separation, the flow moves at an angle perpendicular to the applied electric field. Because the two-dimensional continuous-flow system has drawn so much interest for its process-scale potential, this section will describe only this type of electrophoresis. The end-of-chapter references discuss other high-resolution separations that combine two electrophoretic steps, such as isoelectric focusing and SDS-PAGE, in laboratory-scale systems.

In two-dimensional, continuous-flow electrophoresis, solutes can have different trajectories depending upon whether one of the dimensions comprises zone electrophoresis, isoelectric focusing, or isotachophoresis. As shown in Figure 12.6, the solution is introduced at the top of the device, and the electric field is applied at a right angle to the flow. Note how the trajectories differ for the three types of electrophoresis. (In the figure, the letters denote three different pairs of proteins being separated in each device.) Zone electrophoresis provides a linear path for solutes with positively and negatively charged components, creating a ray pattern. Isotachophoresis produces a trajectory in which the leading electrolyte exhibits the most deflection toward the cathode, assuming that all solutes are positively charged. Isoelectric focusing is associated with exponential-like trajectories on either side of the midpoint—the solutes slow their horizontal speed as they approach a pH zone near their respective pIs.

In the case of zone electrophoresis, the net velocity is easily calculated as the vector addition of the vertical flow component and the horizontal motion attributable to the applied electric field. We can easily determine the displacement from the center point. In the device depicted in Figure 12.7, the displacement angle from the centerline is given by

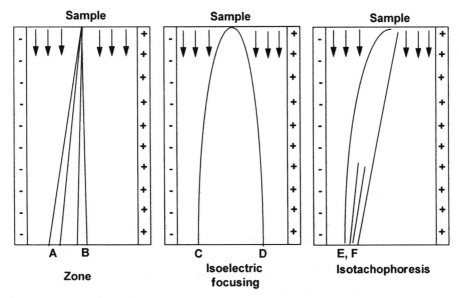

Figure 12.6 A schematic diagram of solute trajectories in two-dimensional continuous-flow electrophoresis for zone, isoelectric focusing, and isotachophoresis.

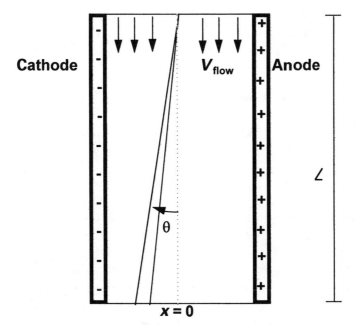

Figure 12.7 Determining the solute position when exiting the device in two-dimensional continuous-flow zone electrophoresis.

$$\theta_i = \tan^{-1}\left(\frac{\mu_i E}{V_{\text{flow}}}\right) \tag{12.39}$$

The displacement is then calculated:

$$x = L \tan(\theta) \tag{12.40}$$

The trajectories for isotachophoresis and isoelectric focusing are not linear, making these calculations more complex. Especially in the case of isotacho-phoresis, these trajectories may be approximated.

12.7 SUMMARY

This chapter introduced the four principal electrophoretic methods: zone, moving boundary, isoelectric focusing, and isotachophoresis. Examples of the use of each mode in specific applications, such as SDS-PAGE and capillary electrophoresis, are discussed. To explain the protein and biopolymer movement produced by an applied electric field, we used a solid-particle model taken from colloid chemistry. This model gave rise to the definition of the electrical double layer, the zeta potential, and the concept of a surface-bound ionic layer. In biopolymer electrophoresis, electro-osmosis and viscous drag forces can hinder the separation. Mathematical models for dilute solutions can be applied to determine separation time and band width in zone, isoelectric focusing, and isotachophoresis.

12.8 PROBLEMS

12.1 Consider a mixture of three proteins (A, B, and C). After titrating each protein individually, the curves are plotted in Figure 12.8.

(a) What would be an appropriate pH range to separate these proteins using zone electrophoresis? What type of zone electrophoresis would you recommend, and why?

(b) If you were to separate these proteins using isotachophoresis, proteins A and C dictate that you use a high field strength. Why?

12.2 Write an expression for the separation time and band width for two-dimensional, continuous zone electrophoresis in a slab geometry.

12.3 In preparing the separation media for gel electrophoresis, gel concentration is plotted versus $\log \mu$ (known as a Ferguson plot) to determine the appropriate gel composition. Higher gel concentrations translate into a more tortuous path for proteins to traverse during their migration. The Ferguson plots are usually straight lines. Figure 12.9 gives these plots for ferritin and ovaltransferritin.

(a) Why do the lines in Figure 12.9 cross? Do you expect this pattern with all proteins? When would the lines not cross?

(b) What gel concentration is not desirable for separation? What gel concentration would be best for separating these proteins?

12.4 A hybrid of gel and capillary electrophoresis, appropriately called capillary gel electrophoresis, was introduced as a commercial analytical product in 1990. This

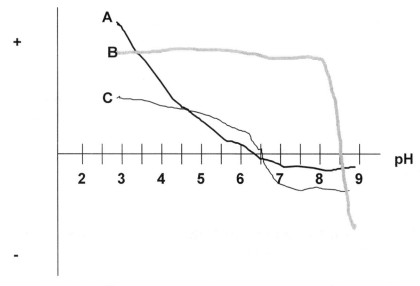

Figure 12.8 Protein titration curves.

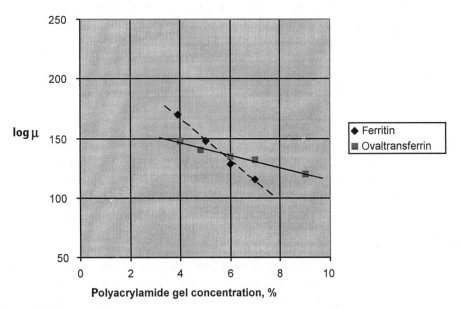

Figure 12.9 Ferguson plots.

technique reportedly offers an improvement over traditional gel slab electrophoresis (Figure 12.10), largely because the resolution is maintained with increasing field strength.

(a) Why does this situation arise?

(b) What other advantages in detection and analysis time can be realized with
capillary gel electrophoresis?

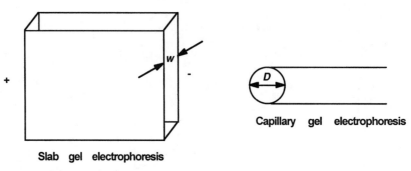

Slab gel electrophoresis

Capillary gel electrophoresis

Figure 12.10 Comparison of slab and capillary gel electrophoresis.

12.5 In zone electrophoresis using PAGE, the protein BSA is separated from the antibody IgG. Their diffusivities are approximately the same ($D = 1 \times 10^{-7} \, cm^2/s$) and their mobilities are $\mu_{BSA} = 1.04 \times 10^{-5} \, cm^2/Vs$ and $\mu_{IgG} = 1.02 \times 10^{-5} \, cm^2/Vs$, respectively. To facilitate the removal of the bands from the gel and then redissolve them, the bands should be separated by at least 5 mm. What voltage would you use if a slab of 20 cm length is available?

12.6 Immobilines are weak acids and bases that can be immobilized onto acrylamides. Using immobilines, a fixed pH gradient can be created for IEF gel electrophoresis. If a linear gradient of 0.25/cm is used, what field strength is needed to separate two proteins whose pIs are 4.5 and 5.0, respectively? Assume that the effective diffusivity $D_{eff} = 1 \times 10^{-5} \, cm^2/s$. Base your definition of adequate separation on the normally used criteria in IEF—that is, that the peak-to-peak distance should be three times the average value of σ.

12.7 Isotachophoresis can be used to separate two types of penicillin that differ in terms of their side-chain chemistry. If the penicillins' electrophoretic mobilities are $\mu_1 = 3.0 \times 10^{-4} \, cm^2/Vs$ and $\mu_2 = 3.5 \times 10^{-4} \, cm^2/Vs$, respectively, what would be good choices for leading and trailing electrolytes for isotachophoresis based on an examination of mobilities in references such as (8)? Using 20 kV and a current of 3 μA, how would you design two-dimensional, continuous isotachophoresis equipment for the separation?

12.8 Scaling up electrophoresis is quite difficult, mainly due to unwanted convection effects. In an effort to process large amounts of proteins, developers have focused their attention on continuous two-dimensional systems. Several concepts have achieved rates of protein separation of several liters per hour. One such geometry, which is commercially available as the Biostream separator, employs two concentric cylinders (Figure 12.11). Fluid flows between the cylinders, and the outer cylinder rotates. Feed is introduced along the circumference of the inner cylinder. The inner cylinder is the cathode, and the outer one is the anode.

(a) Why does the outer cylinder, rather than the inner one, rotate?

(b) Trace a path for a protein migrating from the cathode to the anode in this system.

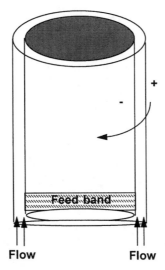

Flow   **Flow**

Figure 12.11 Biostream separator.

(c) Why can this device process at a rate measured in liters per hour, but a two-dimensional continuous device using a slab geometry can process at a rate measured in only milliliters per hour?

(d) Could you perform isotachophoresis with this device? If so, how?

12.9 You are using polystyrene spheres of 200 nanometers in diameter to calibrate an instrument that will measure electrophoretic mobilities. The zeta potential for these particles is 80 mV, and the solvent consists of highly purified, deionized water with a viscosity of 1 cP. Determine the particle mobility by choosing one of the following relationships (see Appendix B):

$$\text{For } \kappa R < 0.1: \quad \mu = \frac{2\varepsilon_0 \varepsilon \zeta}{3\eta} \tag{12.41}$$

$$\text{For } \kappa R > 100: \quad \mu = \frac{\varepsilon_0 \varepsilon \zeta}{\eta} \tag{12.42}$$

12.10 Assume that an inventor presents you with a new, three-dimensional, continuous electrophoresis device using concentric spheres. The inner sphere is the cathode, and the outer sphere is the anode (see Figure 12.12). A separation buffer divides the spheres. The sample is applied to the surface of the inner sphere, and the fluid flows radially outward from the inner sphere to the ports on the outer sphere. The space between the spheres contains particles that help minimize convection. The inventor claims that this device offers a higher throughput than a system using slab geometry because the sample is applied around a spherical surface.

(a) Is the inventor correct? Why or why not?

(b) Are there any advantages to performing gel electrophoresis with spheres instead of slabs? Any disadvantages?

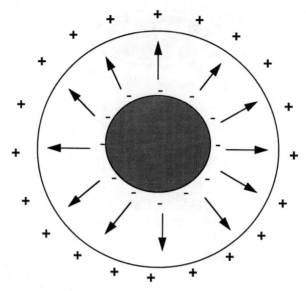

Figure 12.12 Spherical electrophoresis system.

(c) Develop equations for the position and band width when using this system in zone electrophoresis. Do not neglect the curvature of the system.

12.9 REFERENCES

1.   Southern EM, Elder JK. Theories of gel electrophoresis of high molecular weight DNA. In: Monaco AP, ed. Pulsed field gel electrophoresis: a practical approach. Oxford, UK: Oxford University Press, 1995.
2.   Allen RC, Budowle B. Gel electrophoresis of proteins and nucleic acids: Selected techniques. New York: Walter de Gruyter, 1994.
3.   Grossman PD. Free-solution capillary electrophoresis. In: Grossman PD, Colburn JC, eds. Capillary electrophoresis: theory and practice. San Diego: Academic Press, 1992.
4.   Hjertén S. Isoelectric focusing in capillaries. In: Grossman PD, Colburn JC, eds. Capillary electrophoresis: theory and practice. San Diego: Academic Press, 1992.
5.   Demarest CW, Monnot-Chase EA, Jiu J, Weinberger R. Separation of small molecules by high-performance capillary electrophoresis. In: Grossman PD, Colburn JC, eds. Capillary electrophoresis: theory and practice. San Diego: Academic Press, 1992.
6.   Grossman PD. Factors affecting the performance of capillary electrophoresis separations: Joule heating, electroosmosis, and zone dispersion. In: Grossman PD, Colburn JC, eds. Capillary electrophoresis: theory and practice. San Diego: Academic Press, 1992.
7.   Fahien RW. Fundamentals of transport phenomena. New York: McGraw-Hill, 1983.
8.   Everaerts FM, Beckers JL, Verheggen TM. Isotachophoresis: Theory, instrumentation and applications. New York: Elsevier, 1976.

13

Magnetic Bioseparations

Magnetic separations have been used as an alternative to gravitational or centrifugal separations in biological systems. This separation technique's attractiveness arises from its scalability, efficiency, simplicity, and mild conditions. Although magnetic supports or labels have been employed for the selective separation of immobilized enzymes and affinity ligands for more than 20 years, this procedure did not immediately gain widespread acceptance because of its labor-intensive preparation. Only in the 1980s did magnetic bioseparation systems that were feasible for general use become commercially available.

Today magnetite (Fe_3O_4) is most commonly employed as the magnetically susceptible material, although nickel powders coated with NiO and TiO_2 have been used as well. This magnetically susceptible material is either dispersed within or encapsulated by polymeric matrices. To immobilize enzymes or ligands, the material can be treated with coupling agents similar to those used in affinity chromatography.

In this chapter, we will review the science behind the magnetic bioseparation processes. A review of possible magnetic materials will provide the necessary background for understanding of the physical aspects of magnetic bioseparations. Important theoretical aspects, as well as specific applications, will be discussed as well.

13.1 MAGNETIC PROPERTIES OF MATERIALS

Before examining the materials involved in magnetic bioseparations, we will first review some basic concepts of magnetism and magnetic fields (1). The only elemental metals that can be magnetized—and thus produce a strong magnetic field—are Fe, Ni, and Co. These materials are known as ferromagnetic metals (and are discussed later in this chapter in depth).

Magnetism is dipolar in nature, which means that any magnet has two magnetic poles through which the magnetic field lines enter, cross, and exit the magnet. Figure 13.1 shows the lines associated with a magnetized metal bar.

Magnetic fields can be generated in two ways: by magnetic dipoles that are formed by the orbital motions of electrons in atoms, and by the intrinsic spin motion of electrons, protons, and many other such particles. The orbit of an electron about an atom generates magnetic dipole moments, which in turn create the magnetic field. In an unmagnetized solid, the magnetic moments of the electrons cancel one another's external effects. Only when atoms contain unpaired electrons with aligned dipole moments can a magnetic field form (2).

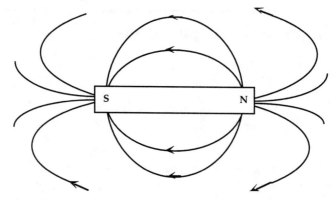

Figure 13.1 Field lines of a magnetized metal bar.

When a demagnetized bar made of a ferromagnetic material is placed inside a solenoid carrying a current, it increases the solenoid's external magnetic field. This enhanced magnetic field is known as the magnetic induction, B (with SI units of Webers per square meter, Wb/m^2, or Teslas, T). It is calculated by summing the applied field H, generated by the solenoid, and the magnetization M, generated by the magnetization of the bar inside the solenoid (both H and M have SI units of amperes per meter [A/m]). Magnetic induction can be defined by the following equation:

$$B = \mu_0(H + M) \tag{13.1}$$

In this equation, μ_0 is the permeability of the vacuum, equal to $4\pi \times 10^{-7}$ Tesla-meters per ampere (Tm/A).

The increase in the intensity of a magnetic field is quantified in terms of the magnetic permeability μ_m:

$$\mu_m = \frac{\vec{B}}{\vec{H}} \tag{13.2}$$

Equation 13.2 implies that easily magnetized materials have a high magnetic permeability. We can also define magnetic permeability in terms of the relative permeability, μ_r:

$$\mu_r = \frac{\mu_m}{\mu_o} \tag{13.3}$$

Substituting Equation 13.3 into 13.2 yields:

$$\vec{B} = \mu_0 \mu_r \vec{H} \tag{13.4}$$

The relative permeability μ_r is dimensionless and varies according to the material.

Molecules and materials may be classified into two groups—namely, those with unpaired electrons and those with spin-paired electrons. Members of the first group are attracted to regions with strong magnetic fields. This attraction is quantified in terms of the volumetric magnetic susceptibility χ_v, which is positive in this case. Members of the second group demonstrate repulsion under the same conditions

Table 13.1 Properties of Various Magnetic Materials

	Diamagnetic	Paramagnetic	Ferrimagnetic	Ferromagnetic	Superparamagnetic
μ_r	Constant	Constant	Variable	Variable	Constant
χ_v	<0	>0	>>0	>>0	>>0

and have a negative χ_v. An electron possesses a magnetic moment because its spin causes it to behave as a magnetic dipole. The magnetization $\vec{M}$ of a material in an applied magnetic field $\vec{H}$ is given by the following equation:

$$\vec{M} = \chi_v \vec{H} \tag{13.5}$$

Magnetization indicates the alignment of the material's individual unpaired electrons when the material is placed in an applied field, and χ_v measures the susceptibility of the electrons to such alignment. Although χ_v is dimensionless, it is called the volumetric magnetic susceptibility because M is measured as the magnetic moment per volume. Other types of susceptibilities can also be defined.

Mass susceptibility:

$$\chi_m = \frac{\chi_v}{\rho} \tag{13.6}$$

Atomic susceptibility:

$$\chi_A = \chi_m A \tag{13.7}$$

Molecular susceptibility:

$$\chi_M = \chi_m M' \tag{13.8}$$

In these equations, ρ is the density, A is the atomic mass, and M' is the molecular weight. Writing Equation 13.4 in terms of volumetric magnetic susceptibility χ_v yields

$$\vec{B} = \mu_0 (1 + \chi_v) \vec{H} \tag{13.9}$$

which implies that

$$\mu_r = 1 + \chi_v \tag{13.10}$$

Materials can be classified into various groups according to the strength of their interaction with applied magnetic fields. Here we will discuss the three most common magnetically susceptible materials (diamagnets, paramagnets, and ferromagnets) and two less common materials (ferrimagnets and superparamagnets). Table 13.1 lists the properties of these various magnetic materials.

In the diamagnetic materials, which include SiO_2, NaCl, graphite, alumina, and most organic compounds, small magnetic dipoles form within the atoms when the orbiting electrons become unbalanced; these dipoles oppose the external magnetic field. Diamagnetism is thus a negative magnetic field that forms in opposition to an applied field. It produces a very small magnetic susceptibility, with small negative values on the order of 10^{-6}, or relative permeabilities close to but less than 1. At any time, this tendency is canceled by positive magnetic effects. For that reason, dia-magnets have no importance in magnetic bioseparations.

The paramagnetic materials include $NiSO_4$, some transition metal complexes, NO, O_2, and organic free radicals. In these materials, the electronic spins of each molecule interact weakly or not at all with the spins of other molecules. Each material behaves as a collection of independent spins and thus possesses a low positive magnetic susceptibility (10^{-6} to 10^{-2}), relative permeabilities slightly larger than 1, and a nonzero magnetic moment. The paramagnetic effect increases with temperature, as thermal agitation randomizes the directions of the magnetic dipoles. Paramagnetic particles are attracted to a magnetic field much more weakly than other materials, such as ferromagnetic or ferrimagnetic materials. Nevertheless, this attraction is strong enough to permit their use in magnetic bioseparations.

In the ferromagnetic materials (Fe, Ni, and Co), each atom has several unpaired electronic spins. These uncompensated spins generate a large dipole moment. Inter-atomic forces prompt these moments to align themselves in a parallel fashion in a magnetic field. The resulting magnetization yields very high magnetic susceptibilities and relative permeabilities, both of which are functions of $\vec{H}$. The parallel alignment of the dipole moments occurs within microscopic regions called magnetic domains.

Above a certain temperature (known as the Curie temperature), ferromagnetic materials lose their ferromagnetic properties and become linear paramagnetic materials. This transformation occurs when the thermal energy crosses a threshold, causing the parallel alignment of the magnetic dipoles to become random. The effect of temperature on the mass magnetic susceptibility is described by the Curie-Weiss law:

$$\chi_m = \frac{C}{T - \theta} \tag{13.11}$$

In Equation 13.11, C is the Curie constant, T is the absolute temperature, and θ is a constant with temperature units. In the special cases of diamagnetic or ideal paramagnetic materials, θ is zero. When T is equal to or larger than the Curie temperature for a specific material, the mass magnetic susceptibility calculated by Equation 13.10 will fall within the range expected for a paramagnetic material, $(10^{-6} - 10^{-2})/\rho$.

Ferrimagnetic materials are ceramics that are created by mixing iron oxide (Fe_2O_3) with carbonates and oxide powders. In these materials, forces between adjacent atoms cause the atomic moments to align in an antiparallel fashion. These moments are unequal, which partially nullifies the net magnetization. This structure produces a weaker response to a magnetic field than that shown by the ferromagnets—but a nevertheless strong response. Like ferromagnets, ferrimagnets show high magnetic susceptibilities and relative permeabilities. The most widely used ferrimagnet in bioseparations is magnetite, $FeO \cdot Fe_2O_3$ (or Fe_3O_4); an oxidized form of magnetite, maghemite (γ-Fe_2O_3), is used as well.

Ferrimagnetic materials can be characterized by magnetization curves (Figure 13.2), which plot the magnetization M versus the applied magnetic field H. In curves (a) and (b) in Figure 13.2, the magnetization of diamagnetic and paramagnetic materials clearly varies in a linear fashion with the applied field, implying constant values

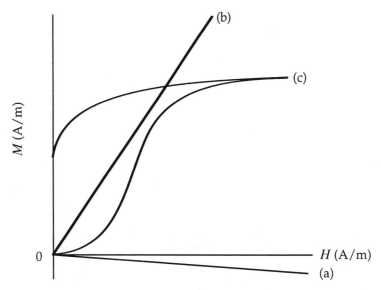

Figure 13.2 Generalized magnetization curves for (a) diamagnetic, (b) paramagnetic, and (c) ferromagnetic or ferrimagnetic material.

of χ_v and μ_r. The material becomes demagnetized ($\vec{M} = 0$) when we remove the applied field ($\vec{H} = 0$). In contrast, the magnetization curves for ferromagnetic and ferrimagnetic materials are nonlinear.

Even with such nonlinearities, these magnetization curves share some common characteristics, such as saturation and hysteresis. Saturation is the value at which the material cannot be further magnetized through increases in the applied field. Hysteresis, or irreversibility, becomes apparent after the material has reached its saturation; in this phenomenon, decreasing the applied field to zero does not remove all of the magnetization from the material. This property can be exploited to create permanent magnets.

Ferromagnetic and ferrimagnetic materials may also be characterized through the construction of *B-H* curves or hysteresis loops. These plots show the effect of an applied field *H* on the magnetic induction *B* of the material during magnetization and demagnetization. Figure 13.3 illustrates a typical *B-H* curve for a ferromagnetic or ferrimagnetic material. When the applied field increases from zero, the magnetic induction increases along the dashed curve 0*X* until it reaches a maximum, or saturation, induction B_{max}. This curve is known as the initial magnetization curve. If the applied field then decreases to zero, the initial magnetization curve is not retraced; instead, a magnetic flux density, known as remnant induction B_r, will persist. Such remnant induction must be present during the formation of a permanent magnet. To decrease the magnetic induction to zero, we apply a negative applied field H_c, or coercive field intensity. Magnetic materials with small values of H_c are known as hard magnetic materials; those with large values of H_c are labeled as soft magnetic materials.

A further increase in the negative applied field H_c results in another saturation induction at point *Y*. When we remove the negative field, the magnetic induction returns to the remnant induction $-B_r$. Increasing the applied field again will cause

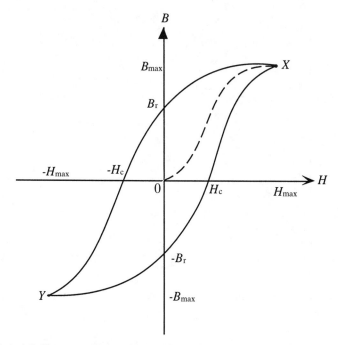

Figure 13.3 Typical B-H curve and hysteresis loop for a ferromagnetic or ferrimagnetic material.

the magnetic induction to reach its saturation value at point X. This process completes the hysteresis loop. The area occupied by this loop offers a measure of the energy lost in a cycle of magnetization and demagnetization.

For ferromagnetic and ferrimagnetic materials, magnetization and B-H curves are used interchangeably. In many cases, the magnetization M is much larger than the applied field H, making the magnetic induction directly proportional to the magnetization (as shown by Equation 13.1).

All of these materials have some beneficial and detrimental properties with respect to successful magnetic bioseparations. For example, paramagnetic materials remain magnetized only while the magnetic field is applied. This property is beneficial for two reasons:

- The magnetic material will be retained while the field is applied, and released upon the field's removal.
- The material will not aggregate after the field is removed.

In contrast, ferromagnetic and ferrimagnetic materials remain magnetized after field removal and can aggregate.

A drawback of the paramagnetic materials is that their magnetic susceptibilities are much lower than those for ferromagnetic or ferrimagnetic materials. As a result, paramagnetic materials give a much weaker response to a magnetic field than do either ferromagnetic or ferrimagnetic materials.

Materials that possess all the requirements for a proper magnetic bioseparation are known as superparamagnetic. These ferromagnetic materials have a crystal diameter smaller than a critical value, usually 30 nm. In single-domain spherical particles, this critical diameter represents the volume at which a magnetization can be

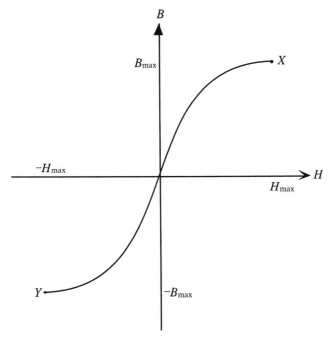

Figure 13.4 Typical *B-H* curve and hysteresis loop for a superparamagnetic material.

reversed by an applied field. The term "single-domain particle" refers to a particle that undergoes uniform magnetization (3). This behavior is equivalent to that displayed by paramagnetic materials, with one important difference—in superparamagnetic materials, the magnetic moment is more than 10^5 times larger than that seen in paramagnetic materials. Consequently, a superparamagnetic material can be highly magnetized in a reversible fashion without hysteresis.

This lack of hysteresis—a requirement for superparamagnetic materials—can be seen in the typical *B-H* curve for such a material (Figure 13.4). This behavior is ideal for magnetic bioseparations, as a high magnetic moment requires a lower magnetic field. The reversibility of the magnetization also prevents aggregation of the particles due to magnetic attraction.

Another requirement for superparamagnetism is that the magnetization and *B-H* curves for an isotropic sample be temperature-dependent. This temperature dependency follows the Curie-Weiss law. Thus, if magnetization data (either *B* or *M* as a function of *H*) taken at different temperatures were plotted as *B* or *M* versus *H/T*, then the curves would superimpose.

13.2 MAGNETIC PARTICLE CLASSIFICATION

The magnetic particles used in bioseparations are typically immobilized on microscopic capsules or particles that range in size from 25 nm to 20 μm. The latter particles are generally constructed from cross-linked polymer strands, such as polystyrene, cellulose, polyvinyl, or alginate. The superparamagnetic material is either evenly dispersed throughout or encapsulated within this matrix. Not surprisingly, the characteristics of the particles determine their potential range of applicability.

Affinity ligands may also be employed in the application of these magnetic particles. In general, to immobilize such ligands, the magnetic material must be encapsulated by, or dispersed in, a rigid, inert, uniform polymeric matrix. Particles are classified initially as porous or nonporous, because this property directly affects their performance. Further distinctions reflect the immobilized ligand's location. If the ligands appear in the interior of the particle, the matrix must be porous if it is to be accessible to the target molecule. Most commercially available particles are nonporous, including either with dispersed or encapsulated superparamagnetic material. Cellulose, agarose, and alginate—all of which are porous—are exceptions (4).

The particle size determines the particle's physical behavior and its susceptibility to manipulation. Relatively large, dense particles tend to settle and aggregate, whereas small, light particles form homogeneous suspensions or colloids. The amount of magnetic material in a particle, known as the magnetic loading, is related to the susceptibility of the particles in an external magnetic field and therefore affects the speed of recovery. Magnetic loadings range from 20 to 50 wt% of the particles. Trade-offs are inherent to high magnetic loadings—while they increase particle susceptibility, they also bolster density. A stronger magnetic field is then required to counteract the greater particle weight.

No matter how ideal the physical properties of a magnetic particle, the particle will be useless in bioseparations unless it possesses the appropriate surface chemistry. Factors such as hydrophobicity, surface functional groups, ion-exchange residues, affinity ligands, and the effective surface density all determine the quantity and quality of material that will adhere to the particle. Built-in detection mechanisms offer advantages when particles are deployed to concentrate a material, rather than to remove the component from a mixture. Common detection methods include fluorescence, luminescence, and colored dyes. Concentrated particles may also be decanted, resuspended, and subjected to enzymatic reactions, resulting in luminescence or color changes.

13.3 THEORETICAL CONSIDERATIONS

In this section, we will integrate some of the basic concepts and definitions related to magnetic materials and their effect on magnetic fields so as to develop a magnetic mobility expression.

In spherical coordinates, the applied field $\vec{H}$ has the components H_R, H_θ, and H_ϕ. If a spherical magnetic particle enters this field, a force will be exerted on the particle in the R direction (Figure 13.5). This R component of the force vector is given by the following equation:

$$F_R = \frac{\chi_v V}{2} \left(\frac{\partial H_R^2}{\partial R} + \frac{\partial H_\theta^2}{\partial R} + \frac{\partial H_\phi^2}{\partial R} \right) \tag{13.12}$$

Substituting $\dfrac{\partial H}{\partial R} = 2H \dfrac{\partial H}{\partial R}$, we have

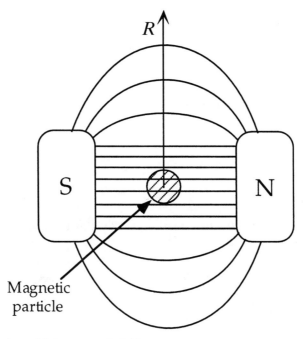

Figure 13.5 Magnetic particle in a magnetic field.

$$F_R = \chi_v V \left(H_R \frac{\partial H_R}{\partial R} + H_\theta \frac{\partial H_\theta}{\partial R} + H_\phi \frac{\partial H_\phi}{\partial R} \right) \tag{13.13}$$

where V is the particle volume.

Because the motion of the particle in the $+R$ direction displaces the same volume of medium in the $-R$ direction, we must include a correction for this effect in our equation. The force on the particle then becomes

$$F_R = (\chi_v - \chi_0)V \left(H_R \frac{\partial H_R}{\partial R} + H_\theta \frac{\partial H_\theta}{\partial R} + H_\phi \frac{\partial H_\phi}{\partial R} \right) \tag{13.14}$$

Alternatively, using Equation 13.12,

$$F_R = \frac{(\chi_v - \chi_0)V}{2} \left(\frac{\partial H_R^2}{\partial R} + \frac{\partial H_\theta^2}{\partial R} + \frac{\partial H_\phi^2}{\partial R} \right) \tag{13.15}$$

where χ_0 is the medium's volumetric magnetic susceptibility.

In Figure 13.5, the field is predominantly in the y-direction in the region occupied by the particle. Thus H_R and H_ϕ, as well as their gradients, are small in comparison. These terms can therefore be canceled, yielding

$$F_R = (\chi_v - \chi_0)VH_\theta \frac{dH_\theta}{dx} \tag{13.16}$$

where dH_θ/dR is the field gradient, and

$$F_R = \frac{(\chi_v - \chi_0)V}{2} \frac{dH_\theta^2}{dR} \tag{13.17}$$

The force on a small segment of the particle dR of volume dV is calculated as follows:

$$dF_R = \frac{(\chi_v - \chi_o)dV}{2} \frac{dH_\theta^2}{dR} \tag{13.18}$$

Because the volume for a sphere is

$$V = \frac{4}{3}\pi R^3 \tag{13.19}$$

then

$$dV = 4\pi R^2 \, dR \tag{13.20}$$

Substituting Equation 13.20 into Equation 13.18 yields Equation 13.21:

$$dF_R = 2(\chi_v - \chi_o)\pi R^2 \, dH_\theta^2 \tag{13.21}$$

Thus the force exerted on the particle is expressed as follows:

$$F_R = 2(\chi_v - \chi_o)\pi R^2 \int_{H_{\theta 0}}^{H_\theta} dH_y^2 = 2(\chi_v - \chi_o)\pi R^2 (H_\theta - H_{\theta 0}^2) \tag{13.22}$$

Generally, $H_\theta \ll H_{\theta 0}$. Thus $H_{\theta 0}^2$ is negligible compared with H_θ, and

$$F_R = 2(\chi_v - \chi_o)\pi R^2 H_\theta^2 \tag{13.23}$$

Because the net force on a particle at constant velocity is zero, the magnetic force must be equal to the drag force. The drag force for a spherical particle of radius R is defined as follows:

$$F_d = 6\pi\eta \, vR \tag{13.24}$$

Here v is the speed at which the spherical particle moves in a medium with viscosity η. Substituting the volume for a sphere into Equation 13.21, equating the result with Equation 13.24, and solving for the ratio v/H_θ^2 yields the following expression:

$$\frac{v}{H_\theta^2} = \frac{(\chi_v - \chi_0)R}{3\eta} \tag{13.25}$$

This ratio is defined as the mobility u_m:

$$u_m = \frac{(\chi_v - \chi_0)R}{3\eta} \tag{13.26}$$

13.4 MAGNETIC PARTICLE SEPARATIONS

To this point, we have described the magnetic particles used in bioseparations and the general theory explaining the effect of a magnetic field on them. We now turn to the specific unit operations used to isolate these supports. These procedures include high-gradient magnetic separations, chromatography, and aqueous two-phase separations. All of these operations are continuous or semi-continuous. It is important to mention, however, that batch separation by the direct application of a powerful rare earth magnet (such as a neodymium-iron-boron magnet) is a widely used technique, as most magnetic bioseparations occur at the bench scale.

13.4.1 High-gradient magnetic separations

As the magnetic susceptibility of the magnetic particles decreases, the strength of the magnetic field gradient $\vec{\nabla H}$ becomes more important. A low magnetic susceptibility results from either a low content of magnetic material or a small particle volume. In these cases, high magnetic gradients are required to generate the necessary force to obtain a rapid separation. Such high-gradient magnetic separations (HGMS) (5) were initially developed to aid in the wet processing of kaolin clay to remove colored magnetic impurities, such as anatase, rutile, and iron pyrite. Purified kaolin clay is a white alumino-silicate mineral that finds application in the paper industry.

The most common application of HGMS is magnetic filtration. In this type of separation, the equipment consists of an electromagnet in the form of an iron box that encloses the energizing coils. Electromagnets typically generate magnetic fields of approximately 2 T; superconducting magnets can generate magnetic fields as high as 4.5 T. The magnetic filtration equipment is filled with a matrix formed of compressed mats of magnetic fibers. It functions in a semi-continuous fashion. First, the electromagnet is turned on, and the system retains the magnetic particles from a magnetic/diamagnetic aqueous mixture. Next, the power to the electromagnet is turned off, releasing the retained magnetic particles and allowing fresh solvent to pass through the equipment. This procedure can be used for positive selection (if the target material is bound to the magnetic particles) or negative selection (if the material to be discarded is bound).

The creation of the high magnetic gradient depends on the usually ferromagnetic fibers or filter. To understand why, consider the effect of the applied field generated by the electromagnet on a single strand of the ferromagnetic fiber (Figure 13.6). In such a system, the total magnetic field represents the sum of two components: the field applied by the electromagnet and the field generated by the fiber. The applied field interacts very strongly with the electrons in the ferromagnetic fiber, aligning the magnetic moments. The fiber becomes magnetized, generating its own magnetic field. In addition, the region close to the wire becomes highly magnetized and creates a nonuniform magnetic field. In Figure 13.6, for example, the close proximity between the magnetic field lines denotes a higher field.

HGMS represents an attractive alternative in magnetic bioseparations. It offers high filtration rates, although the retention capacity of the easily saturated matrix is low. This low retention capacity is rarely a problem because most bioseparations involve dilute solutions. Larger particles, however, may become trapped in the matrix. Given the vast array of magnetic particles and surface ligands that are commercially available, the particle is generally not the problem—rather, the biomaterial to be isolated may pose the challenge. The main drawback is the price of the magnet, and the fact that the magnet size limits the matrix size.

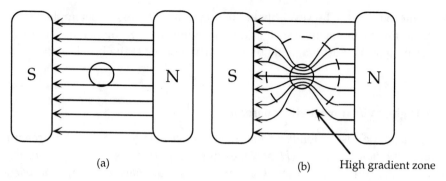

(a)   (b)   High gradient zone

Figure 13.6 Magnetic field around a wire (a) at $t = 0$ and (b) after magnetization at $t = t$.

13.4.2 Affinity chromatography

Affinity chromatography is a highly selective purification technique that employs affinity ligands specific for one or more biomolecules. These ligands are coupled to a solid support so as to form a stable, but reversible complex with its complementary biomolecule. Many types of organic and inorganic supports exist, varying in terms of bulk and surface chemical structure, as well as pore structure. In affinity chromatography, the choice of support depends on the type of ligand employed. Conventional chromatography uses several types of supports, such as dextrans, in conjunction with varying degrees of cross-linking, molecular sieves, and ion exchangers.

Affinity chromatography isolates biomolecules through a sequence involving adsorption, washing, elution, and regeneration. In the first step, the specific biomolecule to the ligand is adsorbed from a multicomponent feed mixture. The column is then washed to remove any components not specifically bound to the ligand. To elute the biomolecule, the binding strengths of the ligand and its corresponding biomolecule are decreased in one of two ways: modification of the pH or the ionic strength, or use of denaturing agents or competing ligands. The final step involves the regeneration of the ligand, which can be reused in another purification cycle.

Although the binding step is critical to the success of affinity chromatography, the elution and regeneration steps also play an important role in determining this process's commercial viability. The choice of elution method will determine not only the amount of product recovered, but also its integrity. Regeneration is a critical factor when the separation uses expensive ligands such as antibodies.

One of the first applications of magnetic particles in this type of bioseparation deployed continuous affinity chromatography in a magnetically stabilized fluidized bed (MSFB). In this case, the magnetic material is magnetite, evenly distributed in an alginate matrix. Cibacron Blue F3GA was immobilized on the surface of these particles and used as an affinity ligand to separate human serum albumin.

In an MSFB, the forces acting on the magnetic particles—that is, magnetic, pressure, gravitational, and drag forces—are balanced so as to obtain a net movement opposite to the solvent flow. The pressure force is related to the feed flow rate, which depends on the amount of feed to be processed. Adjusting the magnetic force allows us to create a stable countercurrent flow of magnetic particles through the control of the magnetic field intensity. Such control can be exerted by varying the current across the coils in electromagnets.

Advantages of an MSFB include improved heat transfer and intraparticle mass transfer. The increase in internal mass transfer offers benefits when the separation relies on porous alginate particles. Most commercial magnetic particles are non-porous, however, so this characteristic has little importance in practical applications. The increase in heat transfer is more significant when the process involves heats of adsorption or desorption. On the other hand, the binding is highly specific in most magnetic bioseparations, so this factor is less important.

Disadvantages of an MSFB include the need for larger columns because of the large void volume and a larger axial dispersion rate than is seen in a packed column.

The application of an MSFB to affinity chromatography is an intriguing idea. It offers excellent contact between the solid (magnetic particles) and liquid (solvent) phase. The recycling of the solid phase acts as a self-cleaning filter, possibly elimi-nating some of the primary purification required in chromatography. The proposed system consists of two columns in series operated countercurrently. Binding occurs in the first column, where recycled magnetic particles come in contact with the feed containing the target biomolecule in this column. The magnetic particles with the bound biomolecule then enter the second column, where elution takes place as the particles come in contact with the elution buffer. The magnetic particles are then recycled to the first column to begin a new cycle.

13.4.3 Aqueous two-phase separations

Aqueous two-phase systems are formed by mixing two aqueous solutions of hydrophilic polymers, or a dilute polymer and a salt. This method is widely used to isolate biomolecules, as the separation takes place under very mild conditions. The most-studied system employs polyethylene glycol (PEG) and dextran. Systems using PEG and salts have high ionic strength, which might affect the interactions between the target biomolecule and ligand.

The distribution of biomolecules between phases is measured in terms of the partition coefficient—that is, the ratio of the concentration of the biomolecule in the upper phase to that in the lower phase. Surface properties of the biomolecule as well as the composition of the system influence this distribution. The biomole-cule will show an affinity for the phase that offers the most electrical, hydrophobic, hydrophilic, or conformational interactions. The degree of these interactions is, in turn, affected by polymer concentration and molecular weight, salt type and con-centration, temperature, and pH.

In one of the first applications of magnetic particles in two-phase separations, PEG was coupled to magnetite and various enzymes (the enzyme choice depended on the application). The isolation of a 0.8 wt% magnetite-PEG-enzyme conjugate took five minutes with the use of this relatively weak magnet.

Another proposed system consists of magnetite immobilized on dextran. The resulting ferrofluid can separate enzymes in a multistage magnetic separator. During the stirring period, the phases become thoroughly mixed and equilibrated. The stirrer is then stopped, and the dextran phase containing the ferrofluid is drawn to the magnet. The continuous PEG phase is pumped out, but the dextran phase remains close to the magnets. After the removal of the continuous phase, the pump stops and a new cycle begins.

A preparative-scale continuous separator has also been applied to separate enzymes from yeast. This system uses Cibacron Blue F3GA as the continuous PEG phase; dextran is still part of the ferrofluid. Although this technique results in enzyme recovery of only 87% to 93% (lower than that achieved with other methods), the settling time is greatly reduced when compared with conventional aqueous two-phase separations.

13.5 APPLICATIONS

In this section, we will examine specific applications in the isolation of biological materials, emphasizing the separation methods as well as the types of particles used.

13.5.1 Cell separation

A popular cell separation method employs nonporous magnetic particles with dispersed or encapsulated superparamagnetic material (6). The magnetic particles are coated with monoclonal antibodies specific for surface groups on the target cells. Particles with dispersed superparamagnetic material form 1 to 5 µm spheres, while those with encapsulated material form 25 to 180 nm spheres. The large superparamagnetic particles are isolated batchwise with a rare earth magnet; colloidal particles require the application of an HGMS filtration system.

In a typical cell separation protocol, the designer must first define a characteristic of the cell to be either positively or negatively selected. Usually, a surface protein or receptor is chosen. A monoclonal antibody specific for that receptor can then be immobilized on the surface of the particle, or commercially available magnetic particles appropriate for the separation can be identified. The cell mixture is incubated with the particle-antibody conjugate, allowing the antibody to bind the corresponding antigen. Typically, one or two cells will attach to a large particle, while hundreds of colloidal particles will bind to a single cell. Magnetic separation then isolates the resulting cell-antibody-magnetic particle complex.

After isolation of the cells is complete, several methods are available to dissociate the complex. For instance, we can add a polyclonal antibody mixture that will be attracted to the monoclonal antibody; such a mixture will compete with the cell receptor binding site. Another possibility involves the addition of an enzymatic reagent to degrade the monoclonal antibody, which releases the cell. Both types of protocols are commercially available in the form of prepared isolation kits, and both offer cell recoveries yield in the range of 85% to 99%.

Immunicon Corporation has patented another type of magnetic cell separator using ferrofluids. In its protocol, the ferrofluids are thick, dense colloidal dispersions of nanoscale magnetic particles. Unlike suspensions formed from 10 to 20 µm particles, the nanometer-scale particles do not settle under the influence of gravity.

Instead, the separation takes place only with the application of a strong magnetic field. A ferrofluid behaves as if the fluid itself were magnetized. The small size of the particles ensures that their binding kinetics are diffusion-controlled. Consequently, affinity binding does not depend on mixing, and simple diffusion and Brownian motion keep the magnetic solute uniformly distributed throughout the solution without the need for agitation. Importantly, this system can be automated.

The cell separation techniques described here have been applied to separate lymphocytes or tumor cells from bone marrow, lymphocytes from peripheral blood, and lymphokine-activated killer cells or tumor-infiltrating lymphocyte cells from mononuclear cell concentrates.

13.5.2 Immunoassays

The use of magnetic particles in conjunction with monoclonal antibodies offers a unique tool for rapidly isolating antigens found on the surfaces of cells, bacteria, viruses, and membranes. Although magnetic particles are ideal for the isolation of particulate antigens such as cells and bacteria, they must undergo modifications before they can be used with immunoassays involving soluble antigens. To obtain a useful particle for this type of immunoassay, the particle size, density, and surface characteristics must be altered.

As yet, no convenient, automated device for the washing of magnetic particles has reached the market. Hence bioseparations must rely on manual magnetic racks devised for certain commercial kits. This option works well only when relatively small-scale assays are performed.

13.6 SUMMARY

Magnetic bioseparations represent a viable alternative to gravitational and centrifugal separations. They typically take advantage of polymeric matrices with either evenly dispersed or encapsulated superparamagnetic material. Magnetic bioseparations can be performed under high gradient conditions, with the most common application involving magnetic filtration. Magnetically stabilized fluidized beds can be used in affinity chromatography, and aqueous two-phase separations can be accomplished with the use of ferrofluids. Magnetic particles have also found application in cell separations and immunoassays.

13.7 PROBLEMS

13.1 Describe the advantages and disadvantages of the different magnetic materials discussed in this chapter in relation to their application in magnetic bioseparations. Of these materials, which is the best choice? Why?

13.2 You want to separate two subpopulations of T cells—those with the surface molecule CD4 (CD4+ cells) from those with the surface molecule CD8 (CD8+ cells). Which separation method and particle type would you use for this task? Explain.

13.3 Determine the time required to separate a dilute suspension of $2\,\mu$m diameter magnetic particles ($\chi_v = 0.077$) in a buffer solution ($\chi_0 \cong 0.000$) using a neodymium-iron-boron magnet ($H = 35,000\,$Oe) if the particles can be as much as $10\,$cm away from the magnetic source at the bottom of the vessel. Assume that the particle and buffer densities are $1300\,$kg/m^3 and $1000\,$kg/m^3, respectively. ($1\,$A/m $= 4\pi \times 10^{-3}\,$Oe.)

13.4 A buffer solution ($\chi_0 \cong 0.000$) flows through a $200\,\mu$m capillary tube at a flow rate Q. The capillary has a neodymium-iron-boron magnet ($H = 35,000\,$Oe) on one end. If a $2\,\mu$m magnetic particle ($\chi_v = 0.077$) flows with the buffer solution, determine the maximum flow rate that will permit recovery of the particle by the magnet. Assume that gravity has no effect, and that the buffer density and viscosity are $1000\,$kg/m^3 and $10^{-3}\,$Ns/m^2, respectively.

13.8 REFERENCES

1. Cullity BD. Introduction to magnetic materials. Reading: Addison-Wesley, 1972.
2. Gerber R, Birss RR. High gradient magnetic separation. Chichester: Research Studies Press, 1983.
3. Halliday D, Resnick R. Fundamentals of physics, 3rd ed. New York: Wiley, 1988.
4. Hayt WH. Engineering electromagnetics, 5th ed. New York: McGraw-Hill, 1989.
5. Smith WF. Principles of material science and engineering, 2nd ed. New York: McGraw-Hill, 1990.
6. Uhlén M, Hornes E, Olsvik O, eds. Advances in biomagnetic separation. Natick: Eaton Publishing, 1994.

14

Solvent Removal and Drying

When cell disruption ends, the desired products are generally in solution. The majority of these solutes, such as cell debris and other solids, are removed through sedimentation, centrifugation, and/or various other separation processes. The next step in product recovery usually involves the removal of the solvent.

After this solvent removal step, the final product can exist as either a liquid or a solid. The fermentation industry typically requires liquid products, and the pharmaceutical industry prefers solid products. If the desired products take the form of solids (such as proteins), they must be dried after their removal from solution. Processes such as extraction, precipitation, crystallization, evaporation, and drying can be employed to achieve solvent removal and final product separation. This chapter examines the process of solvent removal through evaporation and drying as it relates to the bioprocess industry (1).

14.1 METHODS OF SOLVENT REMOVAL

Evaporation and drying are the most common processes employed to enrich liquid solutions or dilute slurries. This enrichment occurs through the vaporization of the solvent from the solution, usually via changes in temperature and pressure. Generally, the product solute shows negligible volatility, especially in the case of most solid solutes. Even if some solute vaporizes, no attempt is generally made to separate the various components in the vapor stream. If heat is added to the solution during solvent vaporization, the process is classified as evaporation. If heat is added directly to the solids, it is called drying. After evaporation, the final product usually takes the form of a concentrated solution or slurry. In drying, however, the end products are solids that lack any residual solvent.

Both evaporation and drying are essentially similar in that they involve the vaporization of the solvent by the addition of heat. A variety of methods—including conduction, convection, and radiation—can supply the heat energy required for these processes. Evaporation and drying can also be accomplished by flash evaporation. In this technique, the system pressure is reduced below the solvent's vapor pressure. The solvent will then vaporize, and the system pressure remains at this level until the solution is enriched to the desired extent (2).

Other commonly used enrichment techniques include distillation and crystallization. In distillation, the disparate volatilities of the solute and solvent are exploited to provide selective enrichment of either the vapor or liquid stream during changes in phase. In crystallization, so much solvent is removed from the solution that the solution becomes supersaturated, resulting in a crystalline precipitate of the solute.

Because the bioproducts to be enriched are usually thermolabile, the temperatures should remain low. If that is not possible, then the duration of exposure to high temperatures should be minimized. Freeze drying also has utility in solvent removal. In this technique, which is also known as sublimation, both the solvent and the solutes are frozen. Evaporation occurs as the solvent goes directly from the solid state to the gaseous state, bypassing the liquid stage. The conditions for sublimation exist only below the triple point, where all three states of aggregation (solid, liquid, and gaseous) can exist.

Another frequently used process involves the extraction of the solute into a more favorable solvent; this second solvent may be either easier to handle in future downstream steps or necessary for end-product formulation. Known as solvent or liquid extraction, this method separates the constituents of a liquid solution by bringing the solution into contact with another immiscible liquid. Extraction, which requires the presence of both feed and solvent phases, involves the preferential accumulation and concentration of the solute in the solvent phase. Such solvent extraction can be performed quickly on a large scale (3).

Extraction of lipophilic substances with water-immiscible organic solvents is a well-established process. In a typical antibiotic extraction, such as extraction of penicillin, the solute is transferred from a clarified fermentation broth into an organic phase; the concentrated product is then reextracted into an aqueous buffer. Final product recovery occurs through precipitation, crystallization, or evaporation. Although aqueous-organic two-phase systems have been used extensively in antibiotic recovery, a newer two-phase aqueous-aqueous extraction process has shown great promise in protein recovery.

14.2 THEORY

As noted previously, evaporation, drying, and extraction are all enrichment processes involving the separation of one component from another. They are based on the principle of equilibration, as opposed to rate-based separation techniques such as diffusional separation. Equilibration processes necessarily involve more than one phase. In evaporation, the solute distributes itself between a liquid and a vapor phase. In drying, it distributes between a solid and a liquid phase. In liquid extraction, the solute distributes between the feed and the solvent used for extraction.

During the enrichment of a solution, either by the vaporization of the solvent in the feed phase or by the accumulation of the solute in a more favorable solvent, the ratio of the mole fractions of solvent to solute in the two product phases must differ if any separation is to occur. Mathematically, we express this condition as follows:

$$\alpha_{ij} = \frac{x_{i1}/x_{j1}}{x_{i2}/x_{j2}} \tag{14.1}$$

In Equation 14.1, α is the separation factor between components i and j, and x denotes the mole fraction of components i and j in product phases 1 and 2. By convention, the components i and j are chosen such that the separation factor exceeds 1. Thus we can define the separation factor as the ratio of the mole fractions of the components in both product streams. The separation factor for a process is a theoretical factor that indicates whether that process can effect a separation between two components. For example, when α is 1, we derive the following relationships from Equation 14.1:

$$\frac{x_{i1}/x_{j1}}{x_{i2}/x_{j2}} = 1 \tag{14.2}$$

and

$$\frac{x_{i1}}{x_{j1}} = \frac{x_{i2}}{x_{j2}} \tag{14.3}$$

Equation 14.3 indicates that the ratio of the component mole fractions becomes equal in the two product streams when the separation factor is unity, and no separation occurs. When $\alpha > 1$, component i accumulates in product stream 1 and component j accumulates in product stream 2, resulting in separation. The inverse case occurs when $\alpha < 1$. Even when the separation factor is not unity and separation is theoretically possible, rate or cost limitations may make it impractical. In general, the separation attained under practical conditions falls short of the theoretically possible level, with the actual separation being given as a percentage of the ideal separation. In cases where complete separation of two phases occurs, as in the evaporation of water from seawater to leave dry salt, the separation factor becomes infinity. The separation factors for two-phase systems (such as vapor-liquid, liquid-liquid, and liquid-solid systems) are derived in the next few sections.

14.2.1 Vapor-liquid systems

Vapor-liquid systems are usually encountered in the fermentation industry, where organic solutions or aqueous-organic solutions are common. Because most organic solvents, as well as water, have significant vapor pressures, these applications typically employ evaporation to concentrate the final product.

If more than one component of a solution is volatile, the vapor pressures of the individual components can be calculated as follows (assuming ideal gases):

$$p_i = y_i P = \gamma_i x_i P_i \tag{14.4}$$

In this equation, p_i is the partial pressure of component i, P is the total pressure (equal to the sum of the partial pressures), y_i is the mole fraction of component i in the vapor phase, x_i is the mole fraction of component i in the liquid phase, and P_i is the vapor pressure of pure liquid i. For ideal solutions, the liquid-phase activity coefficient, γ_i, is equal to unity. Thus Equation 14.4 can be rewritten as

$$\frac{y_i}{x_i} = \frac{P_i}{P} \tag{14.5}$$

for ideal solutions. The ratio of the mole fractions in the vapor and liquid phases is equal to the fraction of the total pressure accounted for by the vapor of component i in the vapor phase. Hence the separation factor for two components i and j at equilibrium will be given by the following equation:

$$\alpha_{ij} = \frac{y_i/y_j}{x_i/x_j} = \frac{y_i/x_i}{y_j/x_j} = \frac{P_i/P}{P_j/P} = \frac{P_i}{P_j} \tag{14.6}$$

The theoretical separation factor for ideal components following Dalton's and Raoult's laws, therefore, is given as the ratio of their vapor pressures. In this case, the ratio of the vapor pressures (the separation factor) is also called the relative volatility. Because the vapor pressure increases with temperature, the relative volatility shows sensitivity to temperature. Over short ranges of temperature, however, we can assume that it is constant because it consists of the ratio of the vapor pressures. If the components do not behave as ideal gases, we must include correction factors in the previously derived equations.

In a binary liquid-vapor system in which component 1 is the more volatile, the relationship between the mole fractions of the components can be expressed as follows:

$$x_1 + x_2 = 1 \tag{14.7}$$

$$y_1 + y_2 = 1 \tag{14.8}$$

We can rewrite the equation for relative volatility for a binary system:

$$\alpha_{12} = \frac{y_1/y_2}{x_1/x_2} = \frac{y_1/(1-y_1)}{x_1/(1-x_1)} \tag{14.9}$$

This equation can be rearranged as follows:

$$y_1 = \frac{\alpha_{12}x_1}{1+(\alpha_{12}-1)x_1} \tag{14.10}$$

Figure 14.1 shows the vapor and liquid equilibrium compositions for different values of relative volatility, derived from Equation 14.10. Clearly, as the relative volatility increases, the mole fraction of the more volatile component increases in the vapor phase.

Now assume that a container holds a binary liquid-vapor mixture at a certain temperature. Next, assume that ideal conditions apply, so that Equation 14.10 gives the vapor compositions in the system. A small amount of the vapor is slowly pumped from the system, with the temperature of the solution held constant. As the vapor and the liquid are no longer in equilibrium, some of the liquid evaporates to reestablish equilibrium conditions. Meanwhile, the liquid composition itself has changed, as some evaporation has taken place. Thus, the evacuation process changes the composition of both the liquid phase and the vapor phase, with the liquid becoming richer in the less volatile component (see Figure 14.2). In an ideal solution, continuous vaporization would enrich the liquid in the less volatile component and increase the concentration of the more volatile component in the vapor phase.

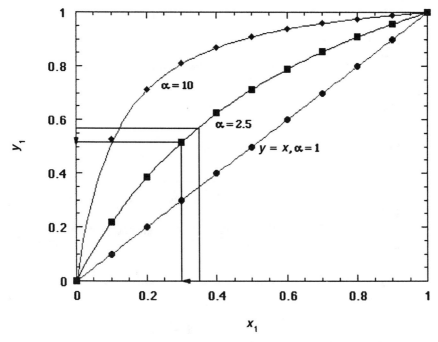

Figure 14.1 Vapor-liquid compositions at various values of α for ideal solutions.

The vaporization process can occur in alternative ways as well. For example, if the temperature changes while the system remains at constant pressure, the partial pressures of the components will change. The relative volatilities of the components will also change, prompting the molar compositions in the liquid and vapor phases to establish a new relationship. Figure 14.2 shows how this process works for a benzene-toluene system. As the figure shows, the initial equilibrium composition (x_a, y_a) changes to the new equilibrium composition (x_b, y_b) or (x_c, y_c) when the temperature increases or decreases from its initial value (4).

In the case of nonideal solutions, azeotropic mixtures could potentially form. At the azeotrope composition, equal concentrations of the components are found in both phases, and the separation factor becomes unity. At this point, any further vaporization will not alter the composition of either phase. An azeotrope forms, for example, in a chloroform-acetone system. Figure 14.3 shows the vapor-liquid equilibrium composition curve for such a system. As shown in the figure, at the azeotropic composition, the vapor-liquid equilibrium curve intersects the $y = x$ line. Hence continuous vaporization of a solution with an initial composition of the more volatile component placed above the azeotrope composition will tend to form the azeotrope composition; no further separation will then be possible.

Mathematically, we can explain the appearance of the type of vapor-liquid equilibrium curve shown in Figure 14.3 by noting that the separation factor is less than unity at lower concentrations of x_1. It gradually increases and finally becomes greater than unity at higher concentrations of x_1. Thus, in these cases, the separation factor or relative volatility changes in tandem with the solution's composition.

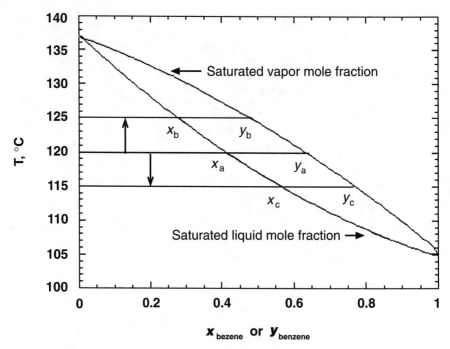

Figure 14.2 Temperature-composition plot for a benzene-toluene system.

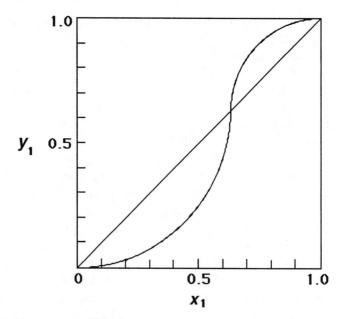

Figure 14.3 Liquid-vapor compositions for an azeotrope.

14.2.2 Liquid-liquid systems

In a two-phase liquid-liquid extraction process, the vapor of component i is in equilibrium with component i in both liquid phases:

$$p_i = y_i P = \gamma_{i1} x_{i1} \overline{P}_i = \gamma_{i2} x_{i2} \overline{P}_i \tag{14.11}$$

This relationship can be rewritten as follows:

$$\gamma_{i1} x_{i1} = \gamma_{i2} x_{i2} \tag{14.12}$$

$$\frac{x_{i1}}{x_{i2}} = \frac{\gamma_{i2}}{\gamma_{i1}} \tag{14.13}$$

The separation factor is

$$\alpha_{ij} = \frac{x_{i1}/x_{j1}}{x_{i2}/x_{j2}} = \frac{\gamma_{i2}\gamma_{j1}}{\gamma_{i1}\gamma_{j2}} \tag{14.14}$$

As Equation 14.14 makes clear, liquid-liquid extraction for volatile solutes takes place only in the case of nonideal solutions. For ideal solutions, the liquid-phase activity coefficients (γ) are equal to 1. The separation factor then becomes unity, which means that separation cannot occur for volatile solutes.

A better way to analyze the liquid extraction process is by using the distribution coefficient, K (see Chapter 11). It is defined as the molar (or mass) fraction of the solute in the raffinate and extract phases at equilibrium. The solution to be extracted is called the feed (with component A and solute C); the liquid used to extract the solute is called the solvent (component B). After contact between the feed and the solvent, the solute transfers to the solvent. The solvent-rich stream is called the extract (solution with B and C), and the feed stream now stripped of the solute is called the raffinate (mostly A and some C). Mathematically, K can be defined as follows:

$$K = \frac{y^*}{x} = \frac{\text{mole fraction of solute in extract}}{\text{mole fraction of solute in raffinate}} \tag{14.15}$$

Larger values of K are usually desirable because extraction can proceed with less solvent. Table 14.1 lists several distribution coefficients. Another important parameter is the separation factor, or selectivity, which indicates how effectively solvent B works in separating the feed solution into A and C. The separation factor can be rephrased in extraction terminology:

$$K_A = \frac{y_A^*}{x_A}; \; K_C = \frac{y_C^*}{x_C} \tag{14.16}$$

$$\alpha_{AC} = \frac{K_A}{K_C} \tag{14.17}$$

As mentioned earlier, no separation can take place if the separation factor is unity.

The partition coefficient in extraction is analogous to the distribution coefficient. The coefficient of a solute partitioned between two phases is defined as follows (Chapter 11):

$$K_p = \frac{y}{x} = \frac{\text{solute mole fraction in the top phase}}{\text{solute mole fraction in the bottom phase}} \tag{14.18}$$

Table 14.1 Partition Coefficients for Various Biological Compounds

Compound	Solute	Solvent	K
Amino acids	Glycine	n-Butanol	0.01
	Alanine	n-Butanol	0.02
	Lysine	n-Butanol	0.2
	Glutamic acid	n-Butanol	0.07
	α-Aminobutyric acid	n-Butanol	0.02
	α-Aminocaproic acid	n-Butanol	0.3
Antibiotics	Celesticetin	n-Butanol	110
	Cycloheximide	Methylene chloride	23
	Erythromycin	Amyl acetate	120
	Lincomycin	n-Butanol	0.17
	Gramicidin	Benzene	0.6
	Novobiocin	Butyl acetate	100
	Penicillin F	Amyl acetate	32
	Penicillin K	Amyl acetate	12
Proteins	Glucose isomerase	PEG 1550/K$_3$PO$_4$	3
	Fumarase	PEG 1550/K$_3$PO$_4$	0.2
	Catalase	PEG/crude dextran	3

Source: Adapted from Atkinson B, Mavituna F. Biochemical engineering and biotechnology handbook, 2nd ed. New York: Stockton Press, 1991:1016.

The solutes encountered in biotechnology applications usually take the form of weak acids or weak bases, and changes in pH influence the effectiveness of their extraction. For example, in the processing of penicillin, the penicillin product is extracted from the broth by an organic solvent, amyl acetate. The penicillin is then back-extracted into an aqueous solution of a different pH. The dependence of the partition coefficient on solution pH is, for a weak acid, given by the following relation (see Chapter 11):

$$K_p = \frac{K_i}{1 + (K_a/[H^+]_H)} \tag{14.19}$$

In Equation 14.19, K_i is the intrinsic partition coefficient:

$$K_i = \frac{\text{(concentration of weak acid)}_L}{\text{(concentration of weak acid)}_H} \tag{14.20}$$

In Equations 14.19 and 14.20, the subscripts L and H refer to the light phase and the heavy phase, respectively, and K_a is the dissociation constant of the weak acid in water:

$$RCOOH \rightleftharpoons RCOO^- + H^+ \tag{14.21}$$

$$K_a = \frac{[RCOO^-]_H [H^+]_H}{[RCOOH]_H} \tag{14.22}$$

Rearranging Equation 14.22 to solve for K_P, and substituting pH for $\log(1/H^+)$ and pK_a for $\log(1/pK_a)$, we have

$$\log_{10}\left[\left(\frac{K_i}{K_p}\right) - 1\right] = pH - pK_a \tag{14.23}$$

For weak bases, a similar equation can be derived:

$$\log_{10}\left[\left(\frac{K_i}{K_p}\right)-1\right]=pK_b-pH \tag{14.24}$$

14.2.3 Liquid-solid systems

In most liquid-solid systems of interest in biology, the degradation of the solid component at higher temperatures is the most important factor. To remove the product from liquid-solid systems, one generally employs crystallization from a supersaturated solution or the complete drying of the solid.

Crystallization may involve the solvent in crystal formation. If the crystals do not incorporate solvent, or if total drying is used with nonvolatile solids, the separation factor becomes infinity, and perfect separation is theoretically attainable. In real-world operation, however, equipment design limitations can affect the separation factor. For example, in the evaporation of seawater, small droplets of salt solution may become entrained in the water vapor, resulting in a less than perfect yield.

Depending on the product, solution to be processed, separation factors desired, and economic considerations, a liquid-solid separation process can be designed to operate in either batch or continuous fashion. The introduction of the feed and separating agent can take place in a batchwise or continuous manner, as can the extraction of the product. In addition, the separating agent and the feed can be introduced in a cocurrent, countercurrent, or a crosscurrent fashion (see Chapter 11). This variability opens up a wide variety of designs, each of which may prove suitable for a particular process.

In almost all evaporation and drying processes, heat serves as the separating agent. This heat needs to be continuously supplied to ensure effective separation. For this reason, we will not discuss batch feeding of the separating agent further here. Instead, we will describe a few of the processes commonly employed in liquid-solid systems, along with the relevant mass and energy balance equations.

In a batch technique, the feed enters the equipment, a separating agent (heat) is supplied, and the different phases are allowed to equilibrate. Only then are the feed and the product phases removed. The material balance for any component i and the overall material balance in such a system are described as follows:

$$x_{ij}F = x_{i1}P_1 + x_{i2}P_2 \tag{14.25}$$

$$F = P_1 + P_2 \tag{14.26}$$

In these equations, F is the number of moles (or mass) of feed charged, P_1 and P_2 are the number of moles (or mass) of the two product phases removed, and x represents the mole or mass fraction of component i. In the evaporation of a liquid, for example, the concentrations of a component in the liquid and vapor phases at equilibrium can be expressed as

$$C_{i1} = KC_{i2} \tag{14.27}$$

where K is the equilibrium constant.

Heat energy can be supplied to a system by a hot gas in drying or steam condensation in evaporation, although radiation drying and evaporation are not uncommon. Of the total heat supplied to a system, q, only the fraction q_d is utilized in the drying or evaporation process and the remainder q_1 is lost to the surroundings:

$$q = q_d + q_1 \tag{14.28}$$

In batch drying, the application of energy raises the temperature of the wet solids to the vaporization point, heats the dry solids and the vapor, and vaporizes the liquid:

$$q_d = m_s C_{ps}(T_f - T_i) + m_l C_{pl}(T_v - T_i) + m_l \lambda + m_l(T_f - T_v) \tag{14.29}$$

In Equation 14.29, the subscripts s, l, and v stand for solid, liquid, and vapor, respectively, the subscripts i and f indicate the initial and final conditions, m is the mass, T is the temperature, C_p is the specific heat, and λ is the latent heat of vaporization. All of the liquid is assumed to evaporate. If the solids retain some liquid, then Equation 14.29 must be modified accordingly.

The heat required for drying can come from the cooling of a hot gas or a constant temperature heat source, such as a heating plate. If a hot gas is used, we can calculate the heat transferred to the system as follows:

$$\frac{q}{t} = m_g(1 + H_i)C_{gi}(T_{hi} - T_{hf}) \tag{14.30}$$

where m_g is the mass flow rate of the dry gas, H_i is the humidity of the gas at inlet, C_{gi} is the humid heat of gas at inlet, t is the time of drying, and T_{hi} and T_{hf} are the inlet and outlet temperatures of the gas, respectively.

If a constant temperature source is used to dry the solid, then the heat supplied will be

$$\frac{q}{t} = UA\Delta T \tag{14.31}$$

where U is the overall heat transfer coefficient, A is the area through which the heat transfer occurs, and ΔT is the temperature difference between the heat source and the solids. Because the temperature of the solids varies continuously, we can use a log mean temperature difference or determine the temperature by using integration techniques.

In continuous drying, the energy and mass balances resemble those for batch dryers, except that we use rates of mass and heat transfer instead of absolute mass and heat. Note, however, that batch evaporation is rarely performed. Semibatch operations, such as Raleigh distillation, are more common. For evaporators, the condensation of the steam supplies the heat:

$$q_s = m_s(H_s - H_c) \tag{14.32}$$

In Equation 14.32, q_s is the rate of heat release due to steam condensation, m_s is the rate at which the steam flows through the steam jacket, H_s is the specific enthalpy of the steam, and H_c is the specific enthalpy of the condensate. Table 14.2 gives the enthalpies of saturated liquid and vapor for water at various temperatures. If we

Table 14.2 Enthalpies of Saturated Liquid and Saturated Vapor for Water

Temperature (°C)	Pressure (MPa)	Enthalpy (kJ/kg) Liquid	Enthalpy (kJ/kg) Vapor
0.01	0.00061	0	2500.5
5	0.00087	21.0	2509.7
10	0.00123	42.0	2518.9
20	0.00234	83.8	2537.2
30	0.00425	125.7	2555.3
40	0.00738	167.5	2573.4
50	0.01234	209.3	2591.2
60	0.01993	251.2	2608.8
70	0.03118	293.0	2626.1
80	0.04737	334.9	2643.1
90	0.07012	376.9	2659.6
100	0.10132	419.1	2675.7
110	0.14324	461.3	2691.3
120	0.19848	503.8	2706.2
130	0.27002	546.4	2720.4
140	0.36119	589.2	2733.8
150	0.47572	632.3	2746.4

Source: Adapted from ASHRAE. Fundamentals handbook, S.I. 1993:17.46–17.49.

neglect the superheating of steam and the subcooling of condensate, then q_s will equal the latent heat of condensation of steam. In addition, energy enters the system as feed liquid:

$$q_f = m_f H_f \tag{14.33}$$

The heat from steam condensation vaporizes a portion of the feed. Thus the heat absorbed by the system will be

$$q_c = m_c H_c + (m_f - m_c) H_v \tag{14.34}$$

where m_c and H_c are the mass flow rate and specific enthalpy of the condensate, respectively, and H_v is the specific enthalpy of the vapor. An overall energy balance for the equipment, assuming negligible heat losses, can be calculated as follows:

$$q = q_s + q_f \tag{14.35}$$

We can also calculate the heat transferred from the steam to the vaporizing liquid in the tubes by using overall heat transfer coefficients. Equations exist for determining the heat transfer coefficients for the liquid film on the shell side and tube side. In conjunction with the thermal conductivity of the wall and the fluid temperatures, the values obtained by these equations can be used to calculate the heat transfer rate through the wall.

14.3 RAYLEIGH DISTILLATION

In Rayleigh distillation (also known as batch distillation), one of the phases is fed or withdrawn in a batchwise manner, while the other phase is alternatively fed or withdrawn continuously. The batch-mode phase is assumed to be well mixed, and

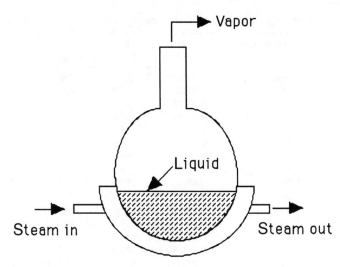

Figure 14.4 Raleigh evaporator.

the fraction of the continuous phase at any instant exists in equilibrium with the batch fluid at the same instant.

Consider, for example, the equilibrium vaporization process illustrated in Figure 14.4. In this process, the feed liquid is initially introduced in a batchwise fashion to a still vessel, with heat being added continuously. As the well-mixed liquid in the vessel evaporates, the mole fraction of the volatile component in the solution decreases in tandem. At any instant, the composition of the liquid lies on the vapor-liquid equilibrium curve, while the vapor formed at that instant exists in equilibrium with the liquid composition. The evaporation continues until the liquid composition reaches the target composition.

Because the evaporation is a continuous process in which the liquid composition and the equilibrium vapor composition change continuously, we must write differential mass balances. If dL represents the moles (or pounds) of liquid of composition x_i evaporated in a time dt, and dV is the number of moles (or pounds) of vapor formed, then

$$y_i dV = -d(x_i L) = -L dx_i - x_i dL \qquad (14.36)$$

The moles (or mass) of liquid evaporated equal the amount of vapor formed. Hence the overall differential mass balance is given by

$$dV = -dL \qquad (14.37)$$

Substituting Equation 14.37 in 14.36, we have

$$-y_i dL = -L dx_i - x_i dL \qquad (14.38)$$

Rearranging Equation 14.38 gives

$$L dx_i = (y_i - x_i) dL \qquad (14.39)$$

Equation 14.39 can be integrated starting with the limits L_0, the initial amount of liquid fed to the vessel, and x_{i0}, the initial liquid composition, to obtain the values at the end of the distillation process:

$$\int_{L_0}^{L} \frac{dL}{L} = \int_{x_{i,0}}^{x_i} \frac{dx_i}{(y_i - x_i)} \qquad (14.40)$$

We can integrate Equation 14.40 graphically by plotting x_i versus $1/(y_i - x_i)$ and determining the area under the graph; alternatively, we can employ numerical techniques. For the simplest case of a binary vapor-liquid system and an assumption that the relative volatilities are constant, Equation 14.9 can be used to solve Equation 14.40. In such a case, we get

$$\ln\left(\frac{L}{L_0}\right) = \int_{x_{1,0}}^{x_1} \frac{dx_1}{\left(\dfrac{\alpha_{12}}{\dfrac{1}{x_1} + \alpha_{12} - 1} - x_1\right)} \qquad (14.41)$$

$$\ln\left(\frac{L}{L_0}\right) = \frac{1}{\alpha_{12} - 1} \ln\left(\frac{x_1(1 - x_{1,0})}{x_{1,0}(1 - x_1)}\right) + \ln\left(\frac{1 - x_{1,0}}{1 - x_1}\right) \qquad (14.42)$$

14.4 EQUIPMENT

Because of the thermolability of bioproducts, a bioseparation designer must take temperature limits and exposure times into consideration when selecting equipment and processes. The temperature dependence of product degradation or loss of activity of biological compounds can usually be described by an Arrhenius expression:

$$C = C_0 e^{-kt} \qquad (14.43)$$

$$k = A e^{-E/RT} \qquad (14.44)$$

In these equations, C is the concentration of active material after exposure to temperature T for a time t, C_0 is the concentration of the active material at the beginning of exposure, k is the rate constant for the degradation process, E is the activation energy for the degradation process, and R is the gas constant.

For distillation in rotary vacuum evaporators and stills, which involve longer residence times, relatively low temperatures are required. The possibility that the hydrostatic pressure encountered in such equipment could overheat the liquid must also be considered. To concentrate a product while keeping a short residence time, evaporation can be carried out with a thin-film apparatus. A critically significant factor in the food and drug industries is the need for sterilization and extreme purity.

In this section, we will briefly describe the equipment commonly used in the biochemical industry for evaporation and drying operations.

14.4.1 Evaporation

In most industrial-scale evaporators, steam heats the solution through a conducting metal sheet. These systems are known as indirect-contact evaporators. In some cases, no heating surface is required and the heat is provided directly to the liquid, as in submerged combustion evaporators and flash evaporators.

Evaporators can be operated in either single- or multiple-effect evaporation modes. In single-effect evaporation, the hot vapor resulting from the evaporation of the liquid is discarded. In multiple-effect evaporation, however, the vapor from one unit heats the liquid in the next unit. Because the vaporization of 1 kg of water requires approximately 1.5 kg of steam, it may be necessary to add external steam to properly heat the downstream evaporators. In some cases, the vapor from the last stage is used to preheat the liquids entering the various evaporator stages. Except for these minor differences, single- and multiple-effect evaporators have similar designs (5).

Evaporators can be also classified as either single-pass or circulation systems. In single-pass evaporators, the entire feed liquid passes through the evaporator once and leaves as a concentrated solution. To obtain a more concentrated product, a series of single-pass evaporators can be operated in a multiple-effect evaporation mode. Such evaporators are especially well suited for dealing with heat-sensitive solutions, as their use of low-pressure evaporation allows the temperature of the liquid to remain low. Also, the liquid can be immediately cooled after a single, rapid pass through the evaporator.

In contrast, in circulation evaporators, only a part of the liquid evaporates during each pass. The remaining liquid mixes with fresh feed and recirculates through the unit. This mixture is usually held as a pool from which liquid is drawn for each circulation and to which the unevaporated liquid returns after each pass. Circulation evaporators can use either natural or forced circulation, depending on whether the liquid circulates via natural convectional currents or a pump, respectively.

Depending on their construction, evaporators can be classified into vertical or horizontal types. Vertical evaporators can be further subclassified into short-tube, long-tube, agitated-film, falling-film, or rising-film units.

14.4.1.1 Flash evaporators

Flash vaporization, a single-stage distillation technique, is ideally suited for single-pass operation (Figure 14.5). In this type of equipment, the feed liquid is heated to the required temperature via a heat exchanger. The pressure of this heated liquid is then suddenly reduced by passing it through a throttle valve, and vapor forms

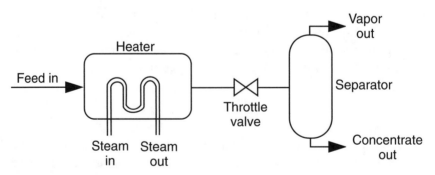

Figure 14.5 Flash vaporizer.

because of adiabatic expansion. The vapor and the liquid, which exist in equilibrium, are then separated.

14.4.1.2 Short-tube vertical evaporators

A short-tube vertical evaporator consists of a cylindrical tank with heating sheets in the center of its body (Figure 14.6). The heating sheets include a number of tube openings through which the feed liquid can rise as it is heated. The tubes are typically 25 mm in diameter and 2 meters in length. The feed enters at the bottom and rises through these tubes, with the liquid-vapor mixture separating at the top. The liquid can be removed from the top, as in single-pass types, or it can be returned to mix with the feed liquid, as in natural-circulation types. In the latter case, a large opening in the center of the sheet area permits the ingress of the return liquid.

14.4.1.3 Long-tube vertical evaporators

The construction of a long-tube vertical evaporator is similar to that of a vertical shell and tube heat exchanger. The equipment consists of tubes, generally 3 to 10 m long, that are encased in a shell. Steam passes through the shell side in a cross-flow fashion. At the same time, the liquid to be concentrated either rises or falls through the tubes, depending on whether the equipment comprises a falling-film or rising-film evaporator. In falling-film types, the tubes have a large diameter (50 to 250 mm); in rising-film evaporators, they are very narrow (diameter of 25 to 50 mm).

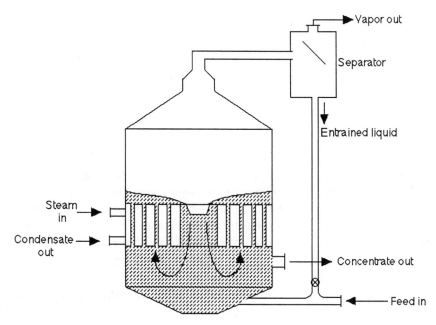

Figure 14.6 Short-tube vertical evaporator.

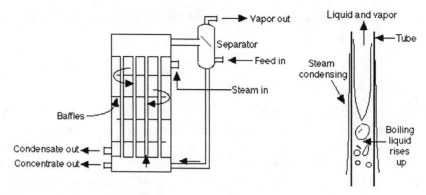

Figure 14.7 Long-tube vertical, rising-film evaporator.

In rising-film evaporators, the mixture of feed and concentrate enters at the bottom of the tubes and is heated as it rises through them (Figure 14.7). The hot liquid-vapor mixture from the top enters a separator, and the concentrate also mixes with the fresh incoming liquid. The separator include baffles, where the vapor-liquid mixture can enter to increase the effectiveness of the separation, especially in cases where the liquid tends to foam. Although the concentrate may be withdrawn before it mixes with fresh feed, a more typical setup involves its removal from the bottom of the tubes after mixing.

Falling-film evaporators, on the other hand, usually operate in single-pass mode and are used to concentrate highly heat-sensitive materials. One problem with this type of unit relates to its distribution of the liquid as a thin film. To achieve this effect, slots through which the liquid can enter and flow down the tube walls are cut in the tubes; alternatively, mechanical distributors such as sprinklers may be used. Most of the vapor flows down the tubes along with the liquid, and a separator operates at the bottom of the apparatus. Vapor can also pass through the center of the tubes, necessitating the installation of a vapor exit at the top of the equipment.

14.4.1.4 Agitated-film evaporators

When dealing with highly viscous solutions or in cases where high transfer coefficients are required, it is essential that a thin film form properly on the walls of the tube. The agitated-film evaporator can meet this need. This type of evaporator consists of a long, jacketed tube that holds a rotor with vertical blades (Figure 14.8). Because the gap between the rotor blades and the tube wall is very small, the blades spread the liquid flowing down the tubes as a thin film. The vapor rises to the top and exits after disengaging from any entrained liquid, and the concentrate is removed from the bottom of the equipment.

The mechanical action of the blades ensures that the film is continuously agitated, leading to very high heat transfer rates between the wall and the liquid film. Because high transfer rates can be accomplished without excessive heating, this type

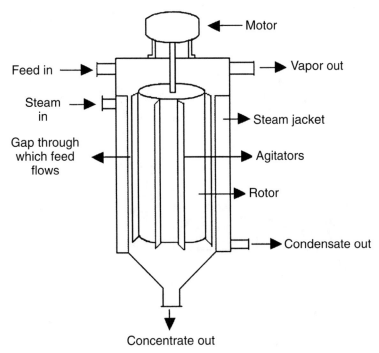

Figure 14.8 Agitated-film evaporator.

of evaporator can accommodate heat-sensitive materials. It is expensive, however, and its treatment capacity is low because it incorporates only a single tube.

14.4.2 Drying

14.4.2.1 Fluidized bed dryers

Air passed through a distributor can be used to fluidize granular solids, ranging from several hundred microns to a few millimeters in size. Fluidization occurs when the velocity of a gas flowing through the bed of solids suffices to keep the particles in a state of suspension. This suspension of solids in the gas, known as fluidized solids, behaves as a dense fluid. In fact, it can be drained from the bed through pipes and valves much as a liquid would be removed. Figure 14.9 depicts a cylindrical fluidized bed.

It is more economical to heat a gas to a higher temperature than to pump the same gas to achieve a higher velocity, and it is easier to handle smaller gas volumes than larger ones. Reflecting these facts, fluidized bed dryers keep the gas velocity as close to the minimum fluidization velocity as possible.

If drying rates must be varied at different stages and if a cooling zone should be created before the final solids discharge, then a horizontal fluidized bed dryer can be used (Figure 14.10). In this type of dryer, all particles have an identical residence time; in contrast, residence times are randomly distributed in a cylindrical bed. If the mixture includes aggregated solids or excessive moisture, then initial stages of fluidization may include some sort of mechanical vibration (6).

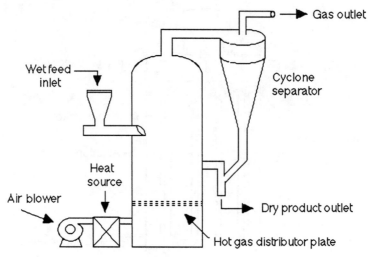

Figure 14.9 Cylindrical fluidized bed dryer.

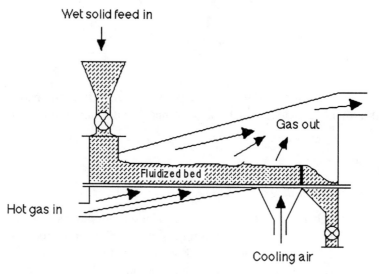

Figure 14.10 Horizontal fluidized bed dryer.

14.4.2.2 Flash drying

When the gas velocity in a fluidized bed increases significantly, the solids become entrained in the gas stream—a principle used to advantage in flash dryers. The construction of a flash dryer is similar to that of a fluidized bed dryer. The mixture of hot gases and solids flows through the dryer for a certain length, followed by collection of the solids. Although the temperature of the gas stream in the dryer usually remains high, the residence time of the solids is small enough (4 to 5 seconds) that these components do not experience a large-scale rise in their temperature. If the solids take the form of cakes, the dryer system will incorporate a pulverizer-conveyor system as well.

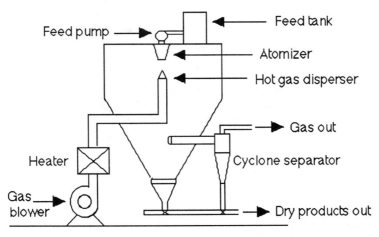

Figure 14.11 Spray dryer.

14.4.2.3 Spray drying

A spray dryer can handle solutions and slurries as well as wet solids. In this technique, a spray nozzle atomizes the feed solution or slurry. The material is then dispersed into the hot gas stream as fine droplets. Evaporation of the liquid occurs rapidly, with the residual solids in the gas being collected in a bin at the bottom of the dryer. The residence time of the solids is very short, and no significant rise in temperature occurs. The flow of the gases and the liquids can be arranged so that the dryer operates in either a cocurrent, countercurrent or crossflow mode.

Figure 14.11 depicts a typical spray dryer. Such dryers have found application with enzymes, antibiotics, yeast extracts, and high-solubility pharmaceuticals.

14.4.2.4 Drum drying

Like the spray dryer, the drum dryer can handle slurries and solutions. In this system, however, heat transfer occurs through conduction rather than convection. A drum dryer typically consists of rotating twin drums, which are steam-heated from the inside and half immersed in a pool of the solution to be evaporated (Figure 14.12). As the drum rotates at low speed, it becomes covered with a thin layer of liquid. This liquid dries as that particular section of the drum emerges from the pool. The dried material can then be scraped off the surface of the drum by a knife edge and removed via a conveyor. Solids remain in contact with the hot surface of the drum for only 5 to 15 seconds, which prevents the decomposition of heat-sensitive products.

14.4.2.5 Freeze drying

Freeze drying (also known as lyophilization) is an indirect drying technique suitable for use with extremely heat-sensitive materials. In this technique, the wet solids are first frozen by reducing the temperature below the solvent's triple point. Next,

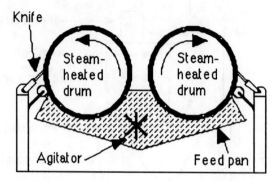

Figure 14.12 Double-drum dryer.

the temperature is raised and the pressure reduced simultaneously, prompting the sublimation of the frozen solvent. In sublimation, a solid goes directly to the vapor phase, bypassing the liquid stage and thus boiling of the product. The freeze drying system usually operates with a very low vacuum to prevent any oxidization of the products.

Heat is transferred to the frozen solids either by direct-contact heating plates or by radiation. Because approximately 700 kcal of heat is needed to sublime 1 kg of water at −40 °C, this technique is relatively expensive. For this reason, freeze drying is usually reserved for use with high-value products such as antibiotics, enzymes, vaccines, and blood fractions. Although continuous freeze drying is possible, batch mode is usually preferred because it offers lower operating costs.

14.5 SUMMARY

Evaporation and drying are popular techniques for solvent removal. Evaporation can take place in a semibatch process such as Rayleigh distillation, though it is more commonly performed in a continuous fashion. Flash evaporators, short- and long-tube vertical evaporators, and agitated-film evaporators are widely used in industry. Drying can take place via fluidized bed dryers, flash dryers, spray dryers, drum dryers, and freeze dryers. All solvent removal processes must take care to limit the exposure of thermolabile compounds to excessive temperatures.

14.6 PROBLEMS

14.1 You are evaporating 1000 kg of water at 25 °C in an open vessel.

(a) Calculate the heat energy required, neglecting the heat losses to surroundings. Determine how much saturated steam at 5 atm is required to carry out the evaporation.

(b) Calculate the amount of moisture that must be removed if 2000 kg of wet feed with 85% moisture content is to be dried to a final product with only 3% moisture content.

14.2 A horizontal fluidized bed dryer is used to dry a pelleted, wet solid from an initial moisture content of 70% to a final moisture content of 5%. The pellets take the form of small spheres, 0.1 mm in diameter with a density of 0.3 g/cm³. Dry air at 5 atm and 120 °C is supplied at a flow rate of 500 kg/h. The bed is operated at the minimum fluidization velocity, given by

$$V_M = \frac{g(\rho_p - \rho)}{150\mu} \frac{\varepsilon}{1-\varepsilon} \left(\frac{6v_p}{s_p} \right) \tag{14.45}$$

In this equation, g is the acceleration due to gravity (980 cm²/s), ρ_p is the particle density, ρ is the gas density, μ is the gas viscosity (0.02 cP), ε is the void fraction in the bed, v_p is the volume of each pellet, and s_p is the surface area of each pellet. If the height of the bed is 0.6 m, the cross-sectional area is 2000 cm², the exit temperature of the gas is 100 °C, and the rise in the temperature of the solids should not exceed 15 °C, calculate the following:

    (a) The minimum fluidization velocity
    (b) The porosity of the bed
    (c) The mass of the bed
    (d) The processing rate of the bed, in terms of kg/h of feed and kg/h of product
    (e) The residence time of the solids in the bed

Assume that the specific heat of the solids is 0.52 cal/g °C, the specific heat of water is 1 cal/g °C, drying does not influence the integrity or density of the solids, and the heat loss to the surroundings is negligible.

14.3 A single drum dryer, steam-heated from inside and 1.5 m in diameter and 3 m long, is used to dry a dilute slurry. The solids make up 30 wt% of the slurry, which is at a temperature of 27 °C. The drum picks up a thin (1.5 mm) film of slurry, which dries completely on the drum surface. If one-third of the drum is immersed in the slurry and the temperature of the solids cannot exceed 40 °C, calculate the following:

    (a) The speed at which the drum must rotate
    (b) The rate at which the dry solid is processed
    (c) The flow rate of steam needed for the drying if the dryer uses saturated steam at 125 °C

For these calculations, assume the density of the solids to be 1.37 g/cm³, the overall heat transfer coefficient to be 300 cal/(m² °C s), a scraper knife placed just above the slurry, and no heat loss.

14.4 An agitated film evaporator is used to concentrate a viscous, heat-sensitive antibiotic solution. The evaporator has an internal diameter of 15 cm and a length of 2 m. The walls of the evaporator are constructed of stainless steel (0.4 cm thick) with a thermal conductivity of 45 W/(m °C). The steam side heat transfer coefficient of the unit is 10,000 W/(m² °C) and the heat transfer coefficient on the viscous fluid side is 2200 W/(m² °C). If the feed, which is 8% antibiotic by weight, has a flow rate of 1500 kg/h, and if a concentrate of 60% antibiotic is desired as the final product, calculate the following:

    (a) The overall heat transfer coefficient (U_i) of the system, given the formula

$$U_i = \frac{1}{\dfrac{1}{h_i} + \dfrac{x_w}{k_m}\left(\dfrac{D_i}{D_L}\right) + \dfrac{1}{h_0}\left(\dfrac{D_i}{D_0}\right)} \tag{14.46}$$

in which the subscripts *i* and *o* stand for inside and outside, the subscript *L* indicates the log mean diameter, *D* is the diameter of the tube, *x* is the thickness of the wall, *K* is the thermal conductivity of the wall, and *h* is the heat transfer coefficient

(b) The mass rate of steam needed if saturated steam is supplied at 150 °C

14.7 REFERENCES

1. King JC. Separation processes. New York: McGraw-Hill, 1980.
2. Kakac S, ed. Boilers, evaporators, and condensers. New York: John Wiley and Sons, 1991.
3. Goldberg E, ed. Handbook of downstream processing. London: Chapman and Hall, 1997.
4. McCabe WL, Smith JC, Harriott P. Unit operations of chemical engineering. New York: McGraw-Hill, 1993.
5. Blanch HW, Clark DS. Biochemical engineering. New York: Marcel Dekker, 1996.
6. Treybal RE. Mass-transfer operations. New York: McGraw-Hill, 1980.

15

Cell Disruption

Many biological products of interest are secreted from cells. A large number of undesired cell constituents and products, however, are either present inside cells or produced intracellularly. Consequently, many bioseparations must include a cell disruption step for recovery of the desired product (including those listed in Table 15.1). We can classify the various cell disruption techniques as mechanical or chemical, with the effectiveness of a particular method being measured either in terms of the extent of cell disruption or by the amount of functional product recovered.

Some of the various cell disruption techniques are suited for large-scale processes; others are laboratory-scale procedures. The latter techniques generally stress the purity of the recovered product and the method's ease of use, paying little or no attention to the overall effectiveness of recovery and its associated costs. Industrial processes, on the other hand, offer high yields that minimize the cost of production, cannot significantly affect upstream and downstream processes, should rely on short residence times, and must be easily automated. The equipment itself also represents an important issue when selecting a cell disruption technique for large-scale processing; it must be tightly sealed and easily sterilized. Even if the proper disruption technique and equipment are chosen, large-scale procedures may place limits on both downstream and upstream processes.

Generally, techniques such as French pressing and sonication are reserved for laboratory-scale procedures, and mechanical disruption techniques such as high-pressure homogenization and grinding in ball mills are applied to industrial-scale processes. Freeze-thawing, although effective as a cell disruption option, is more applicable to laboratory procedures because of the high cost of freezing large cell suspensions and the technique's discontinuous nature (1).

15.1 CELLS AND CELL MEMBRANES

Most microbial and higher animal cells share a similar construction, so that many of the biochemical reactions are the same for all of these cell processes. Acknowledging this broad similarity, we will provide only a brief description of cell construction and composition here.

A cell mostly consists of a jelly-like cytoplasm (70% to 80% water by weight) in which the cell constituents are suspended. The remainder of the cell (20% to 30% by weight) is composed of cell walls and membranes, proteins, lipids, and nucleic acids. Approximately 50% of a cell's dry weight usually takes the form of protein, and nucleic acids account for 5% to 30% of the dry weight, depending on the

337

Table 15.1 Product for Which Cell Disruption Is Required

Product Type	Examples
Vaccines	Tetanus, meningitis
Enzymes	L-Asparginase, glucokinase, glycerokinase, invertase, mandelate dehydrogenase, phenylalanine dehydrogenase, sarcosine dehydrogenase
Spore release	Spore preparations
Spore breakage	Spore enzymes
Toxins	Enterotoxin A
Subcellular constituents	Mitochondria, chloroplasts
Other	Recombinant insulin, recombinant growth hormone, protein A, protein G

Source: Adapted from Foster D. Cell disruption: breaking up is hard to do. Biotechnology 1992;10:1539–1541.

genome size (2). Lipids, although essential for the formation of semipermeable membranes and cell walls, rarely serve as a major component of the cell's dry weight. Their proportion of this weight remains nearly constant. In contrast, proteins and nucleic acids weigh fractions of the cell's dry weight, it varies depending on the surrounding environment and the cell's current growth phase.

A plasma membrane surrounds the cell. This physical boundary separates the cytoplasm and its constituents from the surroundings, thereby regulating the movement of material into and out of the cell. Approximately 10 nm thick, the plasma membrane consists of a fluid bilipid layer in which proteins and glycoproteins involved in the transport of various ions and chemicals are embedded. Although the lipids in the plasma membrane usually take the form of phospholipids, some glycolipids and cholesterol are present as well. Typically, cell membranes account for 5% to 10% of the dry weight of a cell (3).

The cells of plants, bacteria, and fungi include thick cell walls (10 to 20 nm thick) that give these cells their characteristic shapes and strengthen the structure formed by the cells (Figure 15.1). These cell walls are quite rigid, but possess pores in their structures through which water and other dissolved materials can pass. Although most animal cells do not possess these types of cell walls, cell wall-like structures sometimes appear in some such cells. For example, the proteins spectrin and ankyrin form a cytoskeleton that gives red blood cells their characteristic doughnut shape (4).

A plant cell wall is usually composed of cellulose, pectin, and lignin. Cellulose, the major component, can undergo degradation by the enzyme cellulase.

The cell walls of bacteria, on the other hand, consist of peptidoglycans; this network of linear heteropolysaccharides cross-linked by peptides usually makes up 20% of the cell by dry weight. Interestingly, the peptidoglycan layer in *E. coli* is not a single layer. Exponentially growing cells may possess two or three peptidoglycan layers measuring 6.6 ± 1.5 nm thick, and researchers have identified four to five layers measuring 8.8 ± 1.8 nm in stationary cells (5). Gram-negative bacteria, for example, exhibit an outer membrane made of lipids, proteins, and lipopolysaccharides surrounding a thinner cell wall.

The peptidoglycan layer imparts strength to the cell wall, with this strength depending on the extent of cross-linking and the thickness of this layer. Hence cells at different phases and rates of growth demonstrate differing cell strengths, as the

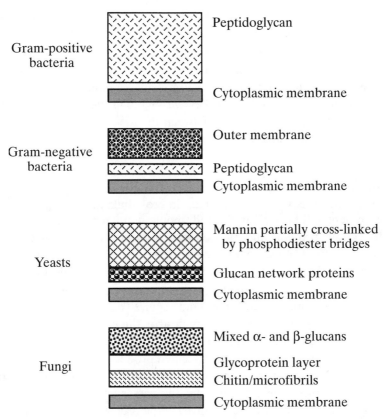

Figure 15.1 Simplified illustration cell envelopes for various microorganisms. (Adapted from Rehm H-J, Reed G. Biotechnology, vol. 2. Weinheim: VCH Publishers, 1985:737.)

degree of cross-linkage and the thickness of the peptidoglycan layer increases as cells move from the exponential growth phase to the stationary phase (6). In addition to the peptidoglycan layer, the outer cell membrane may help to maintain the cell shape under some circumstances and serve as the skeleton of the cell wall (7, 8).

In contrast to the structure of bacteria, fungal cell walls are 100 to 250 nm thick and generally composed of the polysaccharide chitin (although some fungal cells incorporate cellulose in the wall). Yeast cell walls consist of mannans and glucans.

15.2 CELL DISRUPTION TECHNIQUES

Because of their pharmaceutical and biochemical applications, bacterial and fungal microorganisms have been studied in more detail than animal cells. Generally, rupture of microbial cells proves more difficult than rupture of most animal cells. Some preliminary grinding, however, must be carried out on animal cells to open these tissues. Disruption of microbial cells that have entered a growth phase and are actively dividing is somewhat easier to accomplish because the constricted division site acts as a stress concentration point (6). The percentage of constricted cells in a population depends on the cells' growth rate, with higher percentages of these cells being present at higher growth rates. Cells in their exponential growth phase are also larger, and possess peptidoglycan layers that are less highly cross-linked than the comparable layers in stationary-phase cells (6). As a result, fast-growing cells are easier to disrupt than their slower-growing counterparts.

Based on disruption studies of *E. coli* by high-pressure homogenization, researchers have found a correlation between the mean effective cell strength-to-cell length ratio and the degree of cross-linking:

$$S_n = 34.2X - 8.61L_n + 48.2 \qquad (15.1)$$

In this equation, S_n is the mean effective cell strength of nonseptated bacteria in a given cell sample, X is the fractional murein cross-linkage, and L_n is the average length of nonseptated bacteria in the cell sample in micrometers.

Of the various proteins and biochemicals found in the cell, some exist in the cytoplasm and others appear in specific organelles and compartments. Consequently, it may be necessary to breach not only the cell as a whole, but also specific compartments. On the other hand, the recovery of some products—especially those present in the periplasmic space—might not require the complete disruption of the cell. Indeed, excessive cell disruption may destroy the product.

The selection of a specific disruption technique and the time of disruption or the number of passes needed to attain the specified product purity should be based on the optimum product recovery. In addition, the process designer must account for the particle size distribution attained after the passes and make note of the presence of any chemical or enzymatic additives in the process. These factors will affect the complexity of the downstream purification steps and determine the product recovery economics. For example, the disruption of yeast in a bead mill is preferred to disruption in a Gaulin-type homogenizer, even though the bead mill yields a difficult-to-process debris suspension. In comparison with the homogenizer, however, it offers a simpler single-pass design and can operate continuously (9).

Table 15.2 lists the various mechanical and chemical techniques available for cell disruption. These procedures are described in more detail in the following sections.

15.2.1 Mechanical cell disruption

15.2.1.1 Freeze-thaw

The freeze-thaw technique employs slow freezing and thawing of the cell paste. The ice crystals formed during the freezing step disrupt the cell. The cell paste should

Table 15.2 Cell Disruption Techniques

Mechanical	Chemical
Freeze-thaw	Osmotic shock
Homogenization	Solubilization with detergents
Ultrasonic cavitation	Organic solvents
Mechanical grinding	Alkali treatment
Shearing	Enzymatic digestion

not be exposed to high temperatures during the thawing process, as it might destroy the useful products. This method is slow and inefficient, and the freezing and thawing steps are associated with high costs. As a result, this technique has gained little commercial acceptance.

Instead of alternating the temperature between freeze and thaw conditions, a more efficient technique is to vary the pressure while keeping the cell paste frozen at −25 °C (10). As the pressure varies between 2000 and 4000 kg/cm², a phase change occurs and ice translates from one crystalline form to the other. This phase change, which is accompanied by a volume change, causes the cells to rupture.

15.2.1.2 Homogenization

When a solution of cells passes through a restricted orifice under high pressure, the high shear stresses placed on the cells result in cell disintegration. The most important component of the equipment used for this purpose, which is called a homogenizer, is the valve. The homogenizer also uses an air- or motor-driven pump to attain pressures as high as 2000 atmospheres, with a valve adjustment mechanism permitting the fine-tuning of the shear stress placed on the cells.

The most common industrial-scale homogenizer is the Gaulin-Manton homogenizer. In this design, a reciprocating positive-displacement piston pump forces the liquid cell suspension through a restricted valve under high pressure. The maximum pressures depend on the size of the homogenizer, with larger units offering smaller maximum operational pressures. With this type of equipment, it may be necessary to pass the cell suspension through the homogenizer several times to achieve the desired disruption. Larger-scale equipment can attain pressures of 55 MPa, while smaller homogenizers can provide pressures ranging from 200 to 400 MPa (11).

Gaulin-Manton homogenizers employ one of three cell-processing techniques (12): single pass, batch recycle, and continuous recycle and bleed. Figure 15.2 depicts simplified flow diagrams for each of these methods. In the figure, I is the bleed rate, k is the fraction of cells surviving each pass, n is the number of passes through the machine, t is the processing time, V is the volume of suspension in the reservoir, W is the flow rate through the homogenization valve, C_0 is the initial concentration of unbroken cells, and C is the input concentration of cells.

In the single-pass technique, the proportion of cells surviving each pass through the homogenizer valve is given by

$$\frac{C_{out}}{C_{in}} = K \tag{15.2}$$

In Equation 15.2, C_{out} is the concentration of the unbroken cells in the outflow of the homogenizer, C_{in} is the concentration of cells in the inflow, and K is the fraction of cells that survive each pass through the homogenizer (12). If the process is repeated n times, then

$$\frac{C_f}{C_0} = K^n \tag{15.3}$$

where C_f is the final outflow concentration after n passes and C_0 is the initial inflow concentration. The overall cell processing rate in the homogenizer is given by the following equation:

Single-Pass Technique

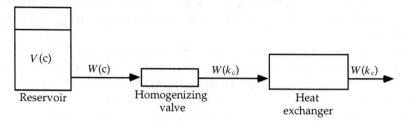

Batch Recycle Technique

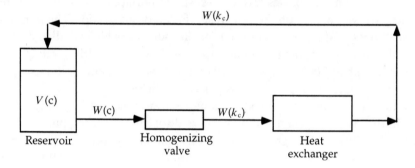

Continuous Recycle Technique

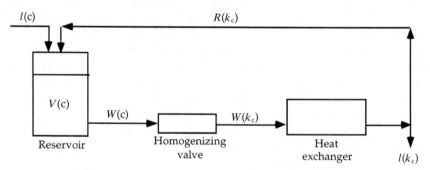

Figure 15.2 Gaulin-Manton homogenizer arrangements. (Adapted from: Jakoby WB. Methods in Enzymology. 1971;22:486.)

$$\text{Processing rate } (P) = \frac{\text{flow rate to homogenizer } (W)}{\text{number of passes } (n)} \tag{15.4}$$

In the case of a batch recycling system, the fraction of unbroken cells in the outflow to the feed is calculated as follows:

$$\frac{C_f}{C_0} = \exp\left(-\frac{Wt}{v(1-K)}\right) \tag{15.5}$$

Here W is the flow rate of the cell suspension through the homogenizer and t is the processing time. The overall processing rate in such a case is then

$$P = \frac{V}{t} = -\frac{W(1-K)}{\ln\left(\dfrac{C}{C_0}\right)} \tag{15.6}$$

In the case of a continuous recycle and bleed process, the ratio of the outflow to inflow concentrations of unbroken cells is described by

$$\frac{C_f}{C_0} = \frac{1}{\dfrac{W}{I}\left(\dfrac{1}{K}-1\right)+1} \tag{15.7}$$

where I is the bleed rate. We calculate the processing rate as follows:

$$P = I = \frac{W(1-K)}{\left(\dfrac{C}{C_0}-1\right)K} \tag{15.8}$$

Researchers have found that the protein release rate in yeast follows first-order kinetics:

$$\ln\left(\frac{R_m}{R_m - R}\right) = K_R n \tag{15.9}$$

In Equation 15.9, R_m is the theoretical maximum possible protein released and R is the actual protein released in n passes. The specific protein release rate, K_R, depends on temperature, the number of passes, and the operational back-pressure of the homogenizer. Investigators have developed the following expression to describe the dependence on back-pressure:

$$K_R = K_0(\Delta P)^\alpha \tag{15.10}$$

In this equation, α lies between 2 and 3. Equations for single-pass, batch recovery, and continuous recycle and bleed processes can be written in a similar fashion.

15.2.1.3 Ultrasonic cavitation

When sound passes through a liquid at sufficiently high intensities, cavities form because of the alternating pressure changes in the liquid. These cavities grow in size and then collapse violently, creating local pressures as high as 1000 atmospheres and temperatures of as much as 10,000 K (13). These conditions cause the cells to disintegrate. In water, the threshold for cavitation occurs at frequencies of 8 to 20 kHz.

Using an M.S.E.-Mullard Ultrasonic disintegrator, researchers have shown that the design of the system should take into account the volume of the cell suspension. Otherwise, the system may provide insufficient energy or it may include excessive energy that produces high liquid turbulence and low efficiency (14). Similarly, formation of bubbles in the liquid during cavitation, a phenomenon known as cavitation unloading, must be avoided as it dissipates the ultrasonic energy in the form of heat.

In general, rod-shaped microorganisms are more easily disrupted than cocci, and gram-negative cells disintegrate more readily than gram-positive cells (14, 15). Usually, the tougher cell walls of gram-positive microorganisms rule out the use of ultrasonic disruption (16). Although yeast cells are more easily disrupted than gram-positive cells, they do not dissociate as readily as gram-negative cells. All of the cell components in yeast, however, may not be released simultaneously. Experiments with the ultrasonication of yeast, for example, have demonstrated that enzymes such as alcohol dehydrogenase and some nucleic acids are released early in the sonication process, before the cells incur any visible damage (14).

When working with cell suspensions, higher viscosities, higher salt concentrations, and lower surface tensions all reduce cavitation and hence reduce disintegration of the cells. Fine powders (10 to 50 μm) can act as cavitation nuclei, promoting cell disintegration. For the same reason, vessels open to air exhibit higher disintegration efficiencies than closed vessels.

Another phenomenon observed during ultrasonication involves the formation of free radicals. Although they have no significance in cell disintegration, free radicals can deactivate some sensitive enzymes. If this deactivation occurs, compounds must be added to the solution to neutralize the free radicals.

Enzyme release during sonication can be described by a first-order equation:

$$\frac{R}{R_m} = 1 - \exp(K_r t) \tag{15.11}$$

In this equation, R is the protein released up until time t, R_m is the maximum amount of protein that can be released, K_r is the specific protein release rate, and t is the sonication time. Events such as free-radical formation and heat generation can also lead to enzyme degradation, which can be described by a first-order equation as well:

$$\frac{S}{S_0} = \exp(-K_d t) \tag{15.12}$$

Here S is the specific activity of the enzyme of interest, S_0 is the specific activity of the enzyme at time zero, K_d is the specific degradation rate, and t is the sonication time. The total active enzyme recovered by sonication is calculated as

$$E = E_0(1 - \exp(-K_r t))\exp(-K_d t) \tag{15.13}$$

where E_0 represents the total active enzyme.

The administration of shock waves may also promote cell disruption (17). Shock waves consist of single pressure pulses of microsecond duration, with a peak pressure of several hundred atmospheres. They have been employed in the medical treatment of urinary calculi.

The energy input to a lithotripter (that is, the equipment used to generate the shock waves) is calculated as follows:

$$W = \frac{1}{2}CU^2D \tag{15.14}$$

In Equation 15.14, W is the electrical energy input, C is the capacitance, U is the operating voltage, and D is the discharge number. Cell disruption increases with increasing energy input, through either an increase in the discharge number or a boost in the voltage. Given the same energy input into the system, however, better cell disruption will occur with a high voltage and small discharge number than with a small voltage and high discharge number. Even if complete disruption of the cells is not achieved, a considerable fraction of the cells will lose membrane integrity (shown by their inability to exclude tryptan blue stain). Cavitation appears to play a major role in the mechanism of shock wave disruption, as an increase in pressure—which suppresses cavitation—decreases cell disruption (17).

Ultrasonication and shock waves not only disrupt the cells, but also hinder the proliferation of the viable cells for the ensuing 24 hours; after this period, the normal growth rate resumes. Use of these techniques, therefore, might lead to the presence of fewer cells in the downstream separation steps and reduce the hazards of contamination in those stages.

Designers continue to develop newer and more efficient cavitation devices for cell disruption (18). For example, one hydrodynamic cavitation setup forces the cell suspension through a throttle valve at high velocity, producing a pressure drop on the other side of the valve. If the pressure on the other side of the throttle valve is lower than the liquid's vapor pressure, cavitation results. Further downstream, the cavities collapse, generating a pressure impulse in the immediate surroundings and disrupting the cells.

During normal operation of a cavitation system, heat is generated through a combination of cavitation collapse, cavitation unloading, and absorption of sonic energy by the liquid. For a 30 mL cell suspension, a 50 W disintegrator can raise the temperature from room temperature to 60 °C in approximately 10 minutes. Because this type of system requires cooling steps, ultrasonic disintegration is not suited for industrial-scale processes—large-scale cooling is a costly proposition and any industrial-scale ultrasonic apparatus would have an unduly large energy requirement.

In highly viscous solutions (such as concentrated cell suspensions), the temperature in the immediate vicinity of the ultrasonic probe increases to 80 °C even if the bulk of the solution exists at very low temperatures. The distintegration process, therefore, must include provisions for effective mixing. Apart from this design consideration, it is difficult to compare the results from different designs and studies, as no physical parameter for the measurement of ultrasound dosage has won universal acceptance.

15.2.1.4 Mechanical grinding

Many designs have been developed for grinding cell pastes, frozen cells, and cell suspensions on both laboratory and industrial scales. These designs vary in complexity, ranging from instruments as simple as a mortar and pestle to automated, motor-driven systems using glass or metal beads. One major advantage associated with grinding systems is the ease with which the equipment can be scaled from the laboratory to industrial scale.

A bead mill, for example, incorporates a cylindrical container with a central shaft on which discs of various shapes and sizes are mounted. The cell suspension and beads are added to the container, with the beads usually accounting for 50% to 60% of the total mixture volume. The agitator increases the beads' velocity, and grinding between the beads disrupts the cells. This type of equipment can be run in either batch or continuous mode. In the continuous design, the system must include a retainer or filter at the outflow to retain the beads.

Cell disruption in a bead mill follows first-order kinetics:

$$\frac{C_f}{C_0} = \exp(-Kt) \tag{15.15}$$

In this equation, C_0 and C_f are the concentrations of viable cells before and after milling, respectively, t is the residence time in the mill, and K is the rate constant. The protein release in the mill can also be described by a first-order rate equation (19):

$$\frac{dR}{dt} = K(R_m - R) \tag{15.16}$$

Here R_m is the maximum protein release possible, R is the protein released at time t, t is the processing time, and K is the specific release rate. The specific release rate depends on a multitude of factors, including the size, weight, and loading of the beads, the agitator speed, the blade design, the cell concentration, and the temperature. In addition, the geometry of the chamber and the design of the agitator are important considerations (20).

Bead mills have longer residence times than do high-pressure homogenizers, and they generate considerable heat. As a result, such systems must include efficient cooling steps to remove the excess heat. Increasing the bead load not only increases the efficiency of disruption, but also increases the temperature of the suspension because of additional friction generated by the poorly conducting beads.

In one small-scale application, researchers have studied bead mill optimization in the recovery of cytochrome b5 (21). In this case, the bead mill consisted of a vibratory chamber holding stainless steel or Teflon tubes of 20 mL capacity. These tubes held 10 mL of cellular suspension and glass beads. The chamber was vibrated at a frequency of 75 Hz. Using an experimental factorial design technique, the investigators concluded that the concentration of the cellular suspension does not have a significant effect on the recovery of cytoplasmic protein. The bead load and time of mill operation, however, are significant. The following expression can be used to describe the release of cytochrome b5 protein from *E. coli* TB1 cells:

$$Y = 14.93 + 0.71\left(\frac{t-6}{4}\right) + 0.36\left(\frac{l-15}{5}\right) + 0.56\left(\frac{t-6}{4}\right)\left(\frac{l-15}{5}\right)$$
$$- 0.48\left(\frac{t-6}{4}\right)^2 - 0.54\left(\frac{l-15}{5}\right)^2 \tag{15.17}$$

Here Y is the cytochrome b5 yield (mg b5/g dry cell weight), t is the operation time in minutes, and l is the bead load in grams. Using Equation 15.17, we can find the optimum operation conditions. Indeed, experimenters have actually achieved the optimum product recovery predicted by Equation 15.17.

Studies in a dyno mill indicate that a critical velocity exists for the operation of a bead mill, below which only insignificant cell disruption occurs; this critical velocity apparently increases with the bead size (22). Although cell disruption is a first-order rate process, the enzyme release from the disrupted cell need not be immediate. In fact, it strongly depends on the location in the cell (23). Using this information, researchers have proposed a two-step model for the enzyme release from the cell:

$$\text{Undisrupted cell} \xrightarrow{k_1} \text{disrupted cell} \xrightarrow{k_2} \text{enzyme release} \tag{15.18}$$

In this reaction, k_1 and k_2 are the rate constants for the disruption and enzyme release steps, respectively. In the case of cytoplasmic and periplasmic enzymes, which are released immediately after cell breakage, k_2 is large and the overall rate depends on k_1. For membrane proteins, k_2 is significant and the enzyme release from the cell represents a second-order process.

The enzyme release effectiveness depends on both the agitation in the bead mill chamber and the cell disruption itself. Taking these considerations into account, designers have developed an improved 180 mL grinding vessel, made of nylon (24). This vessel can prepare 50 to 150 mL of enzyme extracts.

15.2.1.5 Pressure shearing

The basic principle of the X-press shearing device (25) is the same as that of the Hughes press—that is, forcing a frozen cell paste through a narrow opening under high pressure promotes the disintegration of cells. The single-stage operation of the Hughes press has one disadvantage that the X-press lacks: The X-press's receiver compartment is essentially the same as the charge compartment, allowing the cell paste to cycle back and forth through the aperture and thereby permitting a high degree of cell disruption.

The X-press consists of two identical cylindrical compartments connected to an aperture; this opening is very small compared with the diameter of the compartments, but much larger than the cell diameters. Applied pressure forces the frozen cell paste to flow from one compartment to the other. Cell disruption occurs via changes in the crystal structure of the ice that accompanies a phase change and through the shear forces generated when the cells flow through the opening.

Studies in a small X-press have demonstrated that neither the size nor the area of the aperture affects the disintegration of *E. coli* (25). On the other hand, experiments with *B. megaterium* in a large X-press have shown that the abundance and size of the apertures influence the disintegration process. In these studies, as the size of the openings decreased, the maximum pressure needed to cause the frozen cell paste to flow increased, thereby enabling greater disintegration to occur. The differing results achieved with these sets of studies, however, may reflect the fact that *B. megaterium* is more resistant to cell disruption than is *E. coli.*

In another shearing device, known as the French press, a liquid cell suspension is subjected to pressures as high as 20,000 psi. A sudden drop in pressure then disrupts the cells. One problem with this technique, as with other mechanical techniques, is that the energy imparted to the cell suspension becomes converted to heat. As a result, the system must provide cooling to avoid damaging the product. Normally, each pass results in a temperature rise of approximately 10 °C.

The French press is generally reserved for small-scale processes, but it can be easily scaled to larger processes. Its operational efficiencies, however, make it less attractive than other large-scale disruption processes.

In a modification to the French press technique developed by Constant Systems, the pressure is released through an orifice instead of a valve; the mechanical energy is converted into kinetic energy in the fluid stream, which cools as it moves over a surface and its velocity diminishes (26). Alternatively, the kinetic energy of the stream can be directed to operate a centrifuge that separates the cell debris from the disruptate.

15.2.2 Chemical cell disruption

15.2.2.1 Osmotic shock

The osmotic shock technique is based on a relatively simple principle: When two systems are connected by a membrane that is permeable to the solvent, the solvent will flow from the side of higher chemical potential to the side of lower potential until the system attains chemical equilibrium. The addition of solutes to pure water decreases the solvent water's chemical potential. On the other hand, this chemical potential increases with increasing pressure:

$$\mu_{H_2O} = \mu_{H_2O}^0 + \overline{V}_{H_2O}P + RT\ln(x) \tag{15.19}$$

In Equation 15.19, $\mu_{H_2O}^0$ is the chemical potential of water at standard conditions, P, $\overline{V}$, and T are the pressure, partial molar volume, and temperature of the water, respectively, and x is the mole fraction of water.

When cells are suspended in pure water, the concentration of solutes inside the cells exceeds that outside the cells. Consequently, the chemical potential of water is greater inside the cell. Water flows into the cell, increasing the intracellular pressure. This osmotic flow continues until the decrease in the potential inside the cell due to the presence of solutes is balanced by an increase in the pressure due to the flow. Mathematically, we can express this chemical equilibrium as follows:

$$\mu_{H_2O}(\text{outside}) = \mu_{H_2O}(\text{inside}) \tag{15.20}$$

In this equation, μ_{H_2O} is the chemical potential of water. Equation 15.20 can be rewritten in terms of pressure and molarity:

$$\mu_{H_2O}^0 + \overline{V}_{H_2O}P_{out} = \mu_{H_2O}^0 + \overline{V}_{H_2O}P_{in} + RT\ln(1 - x_1) \tag{15.21}$$

The terms on the left and right sides of the new equation identify the variables on the outside and inside of the cell, respectively. The mole fraction of solvent on the outside of the cell is unity, because pure water exists in the environment. The mole fraction of the solvent water inside the cell is $(1 - x_1)$, where x_1 is the mole fraction of the solute in the cell. Assuming that the molar fraction of solute x_1 is small and that the partial molar volume for water equals its molar volume, we can rewrite the equations as follows:

$$P_{out} - P_{in} = \left(\frac{RT}{\bar{V}_{H_2O}} \right) \ln(1 - x_1) \tag{15.22}$$

$$P_{out} - P_{in} = \left(\frac{RT}{\tilde{V}_{H_2O}} \right) (-x_1) = -RTc_1 \tag{15.23}$$

Equations 15.22 and 15.23 indicate that, at equilibrium, the pressure inside the cell exceeds the pressure outside the cells, usually by several atmospheres. Because they lack a cell wall, animal cells are easily lysed under such conditions. In contrast, microbial cells are more difficult to disrupt, requiring the use of higher osmotic pressures. To overcome this problem, these cells may be suspended in a sucrose buffer to increase the solute concentration in them. When the cells have absorbed enough sucrose, they are washed, centrifuged, and resuspended in pure water.

This technique, although highly efficient, is not well suited for large-scale processes that handle large amounts of fluids. Instead, it is used in conjunction with other techniques that dissolve the cell walls, with the final disruption of cells occurring through osmosis. Osmosis is also used to release some hydrolytic enzymes and binding proteins from *E. coli* and other gram-negative bacteria. These enzymes, which are located in the periplasmic space between the cell wall and the cytoplasmic membrane, are released into the solution. The cells remain viable. As this process releases only 4% to 7% of the total cell proteins, it can offer very high levels of purification (12).

15.2.2.2 Solubilization with detergents

Under appropriate conditions of pH, ionic strength, and temperature, detergents can lyse cells efficiently. Generally, cationic detergents lyse both gram-positive and gram-negative bacteria. Anionic detergents lyse primarily gram-positive bacteria. In contrast, nonionic detergents have little effect on either kind of bacteria.

Researchers suggest that detergents act on the lipid membranes of the cell wall, where they combine with the lipoproteins to form micelles and then solubilize the membranes (11). This process renders the cell permeable to the passage of certain constituents.

Even in cases where detergents do not provide total disruption, they can be combined with other techniques to achieve greater effectiveness—for example, this type of combination method can solubilize membrane-bound enzymes that are otherwise difficult to release (11). Using detergents, however, can complicate downstream purification steps.

15.2.2.3 Lipid dissolution

Organic solvents, such as isopropyl alcohol and ethanol, have been used to disrupt some yeast strains such as *Kluyveromyces* (27). Other solvents, such as toluene, benzene, and ethyl acetate, have also been applied in cell disruption. Some of these solvents are hazardous and must be handled with care.

15.2.2.4 Alkali treatment

One of the easiest and least expensive methods of cell disruption involves treatment of bacteria with alkali. This option is adaptable to both laboratory and industrial scales. It has one major disadvantage: Many cell components are susceptible to damage and suffer a loss of conformation and activity at high pH values. To respond to alkali treatment, the product must remain stable at high pH for at least a brief duration. This technique has been used to produce l-asparaginase from *Erwinia* (28, 29) and human growth hormone from *E. coli* cells (30).

15.2.2.5 Enzyme digestion

The application of lytic enzymes, such as egg lysozyme, some microbial glucanases, and proteases, can be used to degrade cells. In the gentle technique of enzyme digestion, the enzymes attack and degrade specific components of the cell wall, leaving the products unaffected in most cases.

Egg lysozyme, in particular, hydrolyzes the β-1,4-glycosidic linkage in bacterial cell walls. Gram-positive bacterial cell walls, for example, contain large amounts of peptidoglycans. They are susceptible to attack by egg lysozyme because β-1,4 bonds link the alternate residues of glucosamine and muramic acid. In contrast, gram-negative bacteria have a lower proportion of peptidoglycans in their cell walls, plus an outer cell membrane that covers the peptidoglycans. This structure renders them more resistant to lysozyme. The addition of EDTA to gram-negative cell suspensions can chelate metal ions on the cell surface, releasing lipopolysaccharides from the cell wall and enhancing disruption (11). For both gram-positive and gram-negative bacteria, the final disruption occurs via osmotic pressure generated from the ionic strength of the suspension buffer and the weakening of the cell wall.

Thanks to its gentle nature, enzymatic disruption releases unbroken strands of nucleic acids, which increases the solution's viscosity. To avoid this effect, the enzyme deoxyribonuclease can be added to the suspension to digest the DNA strands. Magnesium ions must be administered to activate the deoxyribonuclease, which must be added to the solution in sufficient quantities, especially when the system relies on the action of EDTA. The cell walls of some yeast cells contain β-1,3-glucans, which are susceptible to digestion by microbial glucanases. In other cases, cell disruption has relied on the use of proteases. Despite the gentleness of the technique and its efficiency, it has not seen wide use at the industrial scale because of the high cost of the enzymes.

15.3 SUMMARY

Bioseparation processes often require the extraction of a product from cells. The choice of a disruption procedure depends heavily on the membrane construction of the particular cell. Mechanical cell disruption techniques include the freeze-thaw method, homogenization, ultrasonic cavitation, mechanical grinding, and pressure shearing. Chemical techniques include osmotic shock, detergent solubilization, alkali treatment, and enzyme digestion. Enzyme digestion, a very gentle technique, delivers high product yields but is highly expensive. Mechanical methods, such as pressure shearing, are more economical at the industrial scale.

15.4 PROBLEMS

15.1 *Saccharomyces cerevisiae* was disrupted in a dyno mill. Cell disruption and enzyme release were monitored with time, giving the data presented in Tables 15.3 and 15.4.

(a) Plot the cell disruption with time and determine k_1, the cell disruption rate constant.

(b) Using the appropriate axis labels, plot the enzyme release with time for both invertase and alkaline phosphatase. From the shape of the graphs, what can you conclude about the kinetic order of the enzyme release, and why?

15.2 Certain cellular enzymes are not released immediately after cell disruption, and their release may be a time-dependent process. To accommodate these characteristics, designers generally use the two-step reaction given in Equation 15.18.

(a) Using this model, write rate equations for the first and second steps.

(b) Combine the equations to determine the enzyme release in terms of k_1, k_2, and time.

(c) Using calculations from Problem 15.1 and the equations found in parts (a) and (b) of this problem, determine k_2.

Table 15.3 Experimental Data

Time (s)	Viable Cells
0	10^6
30	0.48×10^6
60	0.27×10^6
120	0.15×10^6
180	0.05×10^6

Table 15.4 Experimental Data

Time (s)	R/R_m Invertase	R/R_m Alkaline Phosphatase
0	0	0
30	0.42	0.18
60	0.51	0.42
120	0.88	0.7
180	0.95	0.85
240	0.99	0.92

Table 15.5 Bacterial Cell Composition

Constituent	Percent Dry Weight	Average Molecular Weight
Proteins	50	45,000
Nucleic acids	25	1,000,000
Lipids	20	500
Salts	5	50

15.3 A normal bacterial cell has a diameter of approximately $1\,\mu m$ and the constitution shown in Table 15.5. Assuming that all of these components are soluble and that their specific volume is negligible, estimate the osmotic pressure inside such a cell. Assume that the cells are at $4\,°C$ and suspended in pure water at atmospheric pressure.

15.4 A Manton-Gaulin homogenizer is operated in single-pass, batch, and continuous modes. Given that $K = 0.5$, $C_0 = 200\,g/L$, a flow rate through the valve of $100\,L/min$, and a target of 80% cell disruption, determine the following:

(a) What is the time t or the number of passes required (as applicable) needed to achieve the target cell disruption for $5000\,L$ of cell suspension? Assume that the bleed rate I for the continuous mode is 50%.

(b) How does changing the bleed rate change the operation of the continuous mode as compared with the other two modes of operation? Plot the different processing rates for all modes as a function of K.

15.5 The power consumption in a Gaulin-Manton homogenizer is given as $0.35\,kW/100$ bar pressure applied. Assuming a linear relationship between power consumption and pressure applied, $K = K_0(\Delta P)^{2.9}$, and $K = 0.5$ at $7000\,psi$, plot the cost of operating the homogenizer under the three modes. Use $I = 25\%$, 50%, and 75% bleed rates for the continuous mode. Use all necessary data from Problem 15.4.

15.5 REFERENCES

1. Tannenbaum SR, Wang DIC. Single-cell protein II. Massachusetts: MIT Press, 1975.
2. Davis P, Solomon EP. The world of biology, 4th ed. Philadelphia: Saunders College Publishing, 1990.
3. Zubay GL. Biochemistry, 3rd ed. Dubuque: William C. Brown, 1993.
4. Aiba S, Humphrey AE, Millis NF. Biochemical engineering, 2nd ed. New York: Academic Press, 1973.
5. Leduc M, Frehel C, Siegel E, van Heijenoort J. Multilayer distribution of peptidoglycan in the periplasmic space of *Echerichia coli*. J Gen Microbiol 1989;135:1243–1254.
6. Middelberg AJ, O'Neill BK, Bogle IL, Gully NJ, Rogers AH, Thomas CJ. Modeling bioprocess interactions for optimal design and operating strategies. Trans Inst Chem Eng 1992;70C:213.
7. Hennig U. Determination of cell shape in bacteria. Ann Rev Microbiol 1975;29:45–60.
8. Zorzopulos J, de Long S, Chapman V, Kozloff LM. Evidence for a net-like organization of lipopolysaccharide particles in the *Escherichia coli* outer membrane. FEMS Microbiol Lett 1989;61:23–26.
9. Rito-Palomares M, Lyddiatt A. Impact of cell disruption and polymer recycling upon aqueous two-phase processes for protein recovery. J Chromatogr B 1996;680:81–89.
10. Edebo L, Heden CG. Disruption of frozen bacteria as a consequence of changes in the crystal structure of ice. J Biochem Microbiol Tech Eng 1960;2:113–120.
11. Rehm H-J, Reed G. Biotechnology, vol. 7a. Weinheim: VCH Publishers, 1987.
12. Charm SE, Matteo CC. Scale-up of protein isolation. Meth Enzymol 1971;22:476–556.

13. Noltingk BE, Neppiras EA. Cavitation produced by ultrasonics: theoretical conditions for the onset of cavitation. Proc Phys Soc 1951;64B:1032–1038.
14. Hughes DE. The disintegration of bacteria and other microorganisms by the M.S.E.-Mullard ultrasonic disintegrator. J Biochem Mic Tech Eng 1961;3:405–433.
15. Neppiras EA, Hughes DE. Some experiments on the disintegration of yeast by high intensity ultrasound. Biotech Bioeng 1964;6:247–270.
16. Wiseman A. Enzymes for the breakage of microorganisms. Process Biochem 1969;4:63–65.
17. Wagai T, Omoto R, McCready VR, eds. Ultrasound in medicine and biology: proceedings of the second meeting of the World Federation for Ultrasound in Medicine and Biology. New York: Elsevier, 1980.
18. Save SS, Pandit AB, Joshi JB. Microbial cell disruption: the role of cavitation. Chem Eng J 1994;55:B67–B72.
19. Currie AJ, Dunnill P, Lilly MD. Release of protein from baker's yeast (*Saccharomyces cerevisia*) by disruption in an industrial agitator mill. Biotech Bioeng 1972;14:725–736.
20. Rehm H-J, Reed G. Biotechnology, vol. 2. Weinheim: VCH Publishers, 1985.
21. Belo I, Santos JL, Cabral JS, Mota M. Optimization study of *E. coli* cell disruption for cytochrome b5 recovery in a small-scale bead mill. Biotechnol Prog 1996;12:201–204.
22. Melendres AV, Unno H, Shiragami N, Honda H. A concept of critical velocity for cell disruption by bead mill. J Chem Eng Japan 1992;25:354–356.
23. Melendres AV, Honda H, Shiragami N, Unno H. Enzyme release kinetics in a cell disruption chamber of a bead mill. J Chem Eng Japan 1993;26:148–152.
24. Morin A, Gera R, Leblanc D. Nylon-made 180 mL grinding vessel for laboratory-scale cell disruption. J Mic Meth 1993;17:233–237.
25. Edebo L. A new press for the disruption of micro-ogranisms and other cells. J Biochem Mic Tech Eng 1960;2:453–479.
26. Foster D. Cell disruption: breaking up is hard to do. Biotechnology 1992;10:1539–1541.
27. Fenton DM. Solvent treatment for β-D-galactosidase release from yeast cells. Enzyme Microb Technol 1982;2:229–232.
28. Wade HE. U.K. patent application 40344/68. 1968.
29. Le MS, Spark LB, Ward PS, Ladwa N. Microbial asparagine recovery by membrane processes. J Membrane Sci 1984;21:307–319.
30. Sherwood RF, Court J, Mothershaw A, Keevil W, Ellwood D, Jack G, et al. From gene to protein: translation into biotechnology. New York: Academic Press, 1982.

PART IV

BIOPROCESS SYNTHESIS

16

Integration of Individual Separation Steps

Early in the development of a bioseparation process, the designer must consider how the individual units will act in concert with one another. It is generally easier to think in a linear fashion than to contemplate a complete set of connected unit operations that interact in complex, nonlinear ways. Failure to consider the ramifications of removal of an impurity in an extraction step that eventually hampers crystallization, for example, can prove quite damaging during scale-up, however.

In practice, the design team should conduct at least a full run-through at the bench scale to ensure the integrity of the process scheme as a whole before proceeding to the pilot scale. Analysis of a set of processes (also referred to as a process train) is also an essential step because it can identify potential problem areas or areas of opportunity for process improvement. It is also useful to work with simple solutions at times, along with complex or "real" bioreactor solutions, to identify any components that hinder the effectiveness of a particular separation step.

16.1 BIOSEPARATION PROCESS HEURISTICS

Process design would be greatly simplified if we could enter all manufacturing inputs and product specifications into a computer program and the software would then draw a flow diagram specifying all equipment and operating conditions for an optimized process. Unfortunately, although computer-aided bioprocess design is an invaluable tool, bioproduct stream composition and biological molecular properties are quite poorly defined. As a result, bioseparation process design remains a labor- and thought-intensive procedure.

As we saw in Chapter 2, however, many diverse bioseparation processes share common patterns in process sequencing and integration. Recognizing these similarities has inspired the development of a useful set of heuristics ("rules of thumb") for the designer, some of which appear in Table 16.1. These heuristics mostly arise from two facts: biological products exist in highly diluted aqueous solutions in the bioreactor, and many current commercial products are manufactured in batches. Regardless of how sophisticated property estimation and process modeling software becomes, process heuristics will remain valuable time savers because they help to quickly define logical choices and sequences in processing.

Note, however, these heuristics are akin to English-language spelling in one important way: They work in general, but many exceptions exist. The following sections discuss both some of the more popular bioseparations process heuristics, which follow these rules of thumb, as well as some exceptions to them.

Table 16.1 Summary of Bioseparation Heuristics

Heuristic	Rationale
Reduce volume early in the process sequence	Lowers the size of process vessels and capital and operating costs
Save the most expensive (usually high-resolution) bioseparation step for last	Complies with customer and regulatory requirements; reduces cost by using the lowest process flow rate
Follow the KISS principle	Simplification reduces downtime and cost by using the fewest number of process steps
Resolve components well as early as possible	Minimizes product losses caused by enzymatic degradation; improves downstream separation processes by removing impurities
Increase the final product concentration in the bioreactor by minimizing inhibition mechanisms	Higher concentrations improves the yield and simplifies separation steps

16.1.1 Reduce volume early in the process sequence

The two most important variables that should be minimized during manufacturing are processing cost and time. Initially, the designer identifies key separation steps as being necessary to meet the customer's specifications. After that point, however, processing-related costs are closely tied to the volume or flow rate that must undergo processing from the bioreaction step. In dilute solutions, in which most of the material moving through the manufacturing facility must be discarded, rapid volume reduction should isolate the product of interest before proceeding to the purification and formulation steps. In essence, this heuristic calls for the removal of water. For products of limited solubility, water removal should not exceed the solubility limit if a precipitate form of the product is unstable or if the volume of impurities present in the precipitate exceeds the customer's specifications.

Water removal can prove quite energy-intensive when the system employs vaporization to eliminate this component of the mixture. Some bioseparation processes avoid vaporization, yet still apply this heuristic for solutes that decompose or have a higher boiling point than water. In these systems, the desired product moves into another phase via precipitation, extraction, sorption, or affinity binding.

Example 16.1

The chemical backbone structure of milbemycin, a relatively hydrophobic biopharmaceutical, is shown in Figure 16.1. In the recovery of such products, solvent extraction greatly reduces the volume and concentrates the end product. This option makes high-resolution, expensive steps such as HPLC more economical. In the solvent extraction, 1-butanol is added directly to the *Streptomyces* broth. After several extraction and centrifugation steps (Figure 16.2), the volume can be reduced more than 100-fold, from 2339 L of aqueous broth to 19 L of oil.

16.1.2 Save the most expensive step for last

This heuristic is tied to the volume-reduction rule of thumb in that product concentration and process flow rates are generally reduced by the end of the process.

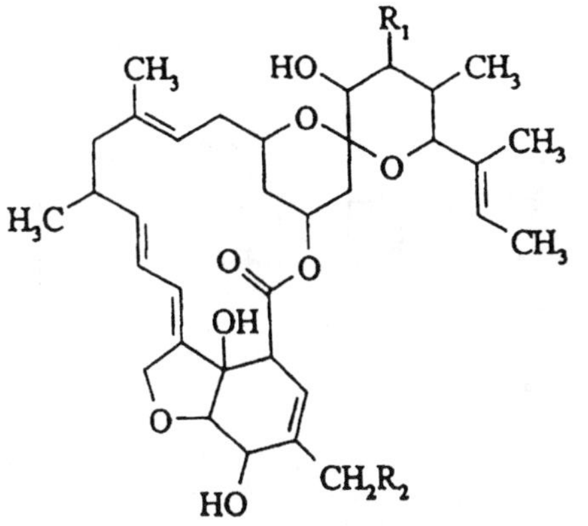

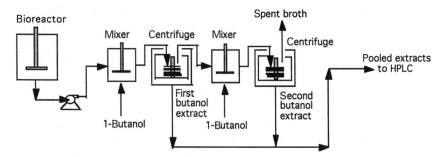

Figure 16.1 Backbone structure of milbemycin analogues.

Figure 16.2 Extraction sequence for milbemycin.

The more costly separation step is employed near the end of the manufacturing process because it is also usually the highest-resolution step.

Two important reasons explain why this highest-resolution step occurs so late in the process. First, the capital and operating costs for this step are linked to the process flow rate. Because of the relatively high costs associated with high-resolution bioseparation steps, flow rates should clearly be minimized at this point in the procedure. Second, the implementation of a high-resolution step at the end of the process ensures that validation of product quality takes place immediately before product formulation and packaging. Customers and plant operators can rely on information provided by a high-resolution separation to ensure that this step adequately removes impurities.

Example 16.2
The recovery of therapeutic proteins, such as interferon, typically implements the high-resolution, expensive separation steps toward the end of the process (Figure 16.3). After cell disruption via homogenization, ammonium sulfate is administered to precipitate the protein fraction; this precipitation leaves behind cell debris. The pellet is then dissolved. After dialysis, an immunosorbent column containing monoclonal antibodies is used to bind the interferon. After washing the column to remove entrained material, elution from the anti-interferon column is performed at pH 2.5. As a further chromatographic method, cation exchange with elution using

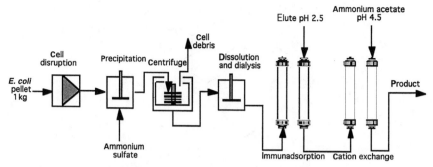

Figure 16.3 Human leukocyte purification process. (Adapted from Staehelin T, Hobbs DS, Kung H-F, Lai CY, Pestka S. Purification and characterization of recombinant human leukocyte interferon (IFLrA) with monoclonal antibodies. J Biol Chem 1981;256:9750–9754.)

ammonium acetate at pH 4.5 is carried out to ensure that the final product has high purity (1). This separation scheme has provided reliable results for more than 600 cycles.

16.1.3 Follow the KISS principle

It is difficult to define exactly what constitutes a simplified bioseparation process design. This simple-minded heuristic—Keep It Simple, Stupid (KISS)—reminds us that a streamlined, elegant process is very much desirable if only because it facilitates monitoring and troubleshooting in a manufacturing setting. Many engineers refer to the "robustness" of a process—in a robust process, small fluctuations in operating conditions will not cause the system to completely fail to meet product specifications. Looking at a process from the KISS perspective suggests that a desirable process sequence contains as few steps as possible. Any additional step introduced into a simple process sequence should therefore offer substantial benefits in terms of product quality and process operability before it is accepted into the system.

An extreme application of the KISS principle involves the use of a single affinity technique to obtain an extracellular product immediately after the bioreactor step. Elution of the affinity-recovered product followed by dilution or precipitation would complete the bioseparation process. In fact, this scheme offers many benefits in biopharmaceutical manufacturing. A single affinity technique can prove expensive, however, and may lead to lower yields.

The KISS principle's opposite—that reliance on a single step can shut down manufacturing if it fails—has important ramifications for process design as well. According to this philosophy, simplifying the process diagram does not lead to a robust process, but instead creates a risky scenario.

Example 16.3

Aspartic acid has a relatively low solubility (less than 1 g/dL at room temperature) compared with other amino acids. We can use an immobilized cell column to create a continuous production process for aspartic acid (Figure 16.4). Heating the effluent

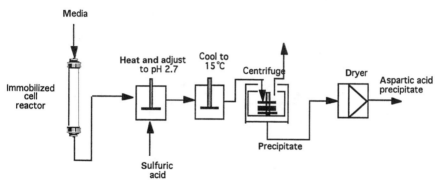

Figure 16.4 Aspartic acid production.

stream from the column permits the rapid denaturation of proteins and other secreted substances. To minimize solubility, we simply adjust the pH to the isoelectric point of aspartic acid (pH 2.7) with sulfuric acid. Cooling to 15 °C yields aspartic acid in precipitated form. A second precipitation step and drying gives the final product. Thus, in this process, the fortuitously low solubility of this amino acid leads to an extreme example of early volume reduction.

16.1.4 Resolve components well as early as possible

Early resolution of components simplifies the bioseparation as it minimizes the number of steps needed. It may also enhance product quality, because enzymatic degradation or product denaturation may occur in the presence of impurities. An important caveat must be noted in the implementation of this guideline, however: High-resolution steps are generally more expensive, and reducing process volume first may prove more economical.

Both crystallization and precipitation have been used to implement this heuristic. A primary crystallization (or precipitation) step may not provide the high purity required by some customers. It can, however, more economically resolve components that can then be purified into the final product form via dissolution, purification, and recrystallization.

Example 16.4

After cell separation by either centrifugation or membrane filtration, cation exchange at low pH can bind lysine to the ion-exchange packing while simultaneously permitting the removal of most of the negatively charged or neutral species (Figure 16.5). Columns rarely appear early in a process, especially when the goal is the separation of a low-value product such as lysine. Because lysine has a net charge of +2 at low pH, however, a column offers a high capacity. The high affinity of lysine for the cation-exchange functional group ensures that the column retains few other species. Ammonium hydroxide is introduced to elute the lysine from the column at high concentration. Following this step with a reverse-osmosis membrane allows ammonia gas and water to be removed, leaving behind lysine in its free-base form.

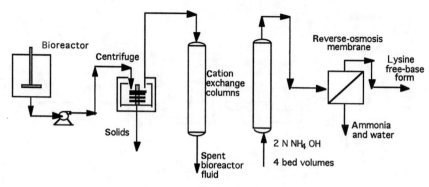

Figure 16.5 Initial lysine bioseparation steps with cation-exchange product isolation.

16.1.5 Minimize inhibition mechanisms in the bioreactor

High product concentrations before the bioseparation process sequence may eliminate the need to implement a water removal step. In some instances, high product or impurity concentrations in the bioreactor can significantly inhibit cellular productivity rates. Integrating a bioseparation step with the bioreactor step can yield higher product concentrations and production rates. Unfortunately, this integration can prove difficult to control during both process start-up and ongoing operation. On the other hand, the potential gains in profit margins and manufacturing rates must be weighed against potential operability problems—such problems may lead to frequent or lengthy shutdowns. (This idea is more fully explored in Section 16.4.)

16.2 ISSUES IN CONCURRENT BIOSEPARATION AND BIOREACTOR PROCESS DEVELOPMENT

One especially challenging aspect of bioprocess engineering arises when we attempt to manufacture products that have never been made at scales larger than the several nanograms or micrograms obtained by biochemists and microbiologists. Without large amounts of the product or bioreactor fluid, how can bioseparation process development take place? Many development groups face the daunting task of designing a separation process even as the upstream bioreactor is being brought to industrial scale. From a marketing and economic perspective, it is rarely feasible to wait until the bioreactor development group finishes before beginning the bioseparation process design. As a result, alternative approaches have been developed.

16.2.1 Take the lab-scale process and scale it directly with no changes

This extremely conservative approach is employed only when time is of the essence, the scaling factor is not very large, and the final product will carry a high price. It does not represent a scientifically grounded or valid engineering approach in the long run unless the benchtop scientist has carefully screened alternative separation schemes before choosing a particular sequence of separation steps.

Most often, production-scale design is inherently different from the lab-scale design simply because it must satisfy different objectives. At the lab scale, the bench-top scientist attempts to verify that the process yields the expected product. Thus product recovery and purification revolves around positive results generated through analytical techniques such as chromatography and bioactivity assays such as ELISA. Because so many other issues need to be addressed at the bench scale, product recovery yield is rarely important. In stark contrast, although product quality remains important, yield takes center stage in production-scale work. After all, the effort (and cost) in making the product in the bioreactor step counts for naught if product recovery and purification yield are low.

16.2.2 Design a bioseparation process based on the closest existing commercial product

The phrase "don't reinvent the wheel" is often heard in engineering. That is, a good starting point or base case for a process design may be to use a known commercial process for a product similar to the one under development. When the final product must satisfy regulatory criteria (such as a food additive or a pharmaceutical), an attractive strategy may be to work from an approved methodology to create a new product that will be used in a similar way as an existing one. Categorization of the type of product and following existing processes (such as those described in Chapter 2) can therefore be a useful exercise.

One inherent danger in this approach, however, is that the properties of the new product may differ dramatically from those of existing commercial products. As a result, adapting existing commercial processes to the new product may not yield an efficient and robust process. In addition, following existing commercial processes will not allow development groups to try out new ideas that could potentially improve product quality, yield, or process economics.

In any event, the adaptation of well-documented commercial processes will leave much room for making important design specifications, such as membrane selection, chromatographic packing selection, extractant selection, and crystallization conditions (2).

16.2.3 Pilot-scale experimentation with "spiked" bioreactor fluid

In a practical approach to dealing with the problem of insufficient bioreactor fluid volume or low product concentration during process development, one can "spike" a typical bioreactor fluid stream with the desired amount of bioproduct. This simple idea can help determine the effect of common impurities, such as antifoaming agents and media components, on potential separation processes. Of course, this method assumes that we have enough pure product to spike the bioreactor fluid and create a simulated bioreactor stream.

The spiking method is most often used in the industrial development of bioprocessing routes applicable to products manufactured by existing organic chemical processes. It may also be employed to develop industrial-scale processes for products obtained directly from plants and animal matter. Finally, it has benefits in bioseparation process development where existing bioreactor technology can produce only very small amounts of the product, but further study of metabolic pathways and inhibitory mechanisms might permit the manufacture of the product at higher concentrations.

16.3 EXPERT SYSTEMS IN PROCESS SYNTHESIS

In the past, we have been forced to rely on designer experience or investigate process alternatives solely through human decision making. In recent years, however, computer-based expert systems have become available. Process engineers in the oil industry, for example, have long used economic optimization routines based on linear programming techniques to select from among several separation process alternatives.

Unfortunately, the complexity of natural products such as proteins greatly complicates the generation of explicit equations and rules for predicting product characteristics, transport parameters, and the properties of impurities. As a result, traditional computer-aided process synthesis cannot be easily applied to most biotechnological separations.

Some have suggested that, instead of using explicit equations, expert systems might rely on many logical rules to guide process synthesis (3). In essence, expert systems elevate the role of heuristics to a more exact science. They utilize databases and require extensive interaction with the human designer (4). Several expert system shells are available through the World Wide Web—one particularly useful shell, Jess, is written in Sun's Java language (5).

When using any computer program in process design, caution is warranted. Extra care should be taken when incorporating expert systems into the design strategy, however—after all, the system is only as good as the rules on which it relies. Expert systems require knowledgeable individuals to run the routines. In any event, the candidate flow diagrams must be tested before the designer places too much faith in the computer-generated choices. Figure 16.6 provides a flow diagram depicting how an expert system might be employed to enhance process synthesis.

16.4 INTEGRATION OF BIOREACTION AND BIOSEPARATION STEPS

Integrating a bioseparation step with the bioreactor step offers potentially significant economic and productivity advantages. For example, it might permit the continuous removal of a product that is toxic to cells, or it might repress feedback significantly by reducing gene expression. Such integration might be used to advantage when continuous product removal mimics cellular or tissue behavior in vivo, as in the production of monoclonal antibodies. This strategy could also minimize processing time by continuously removing a product during its formation in the

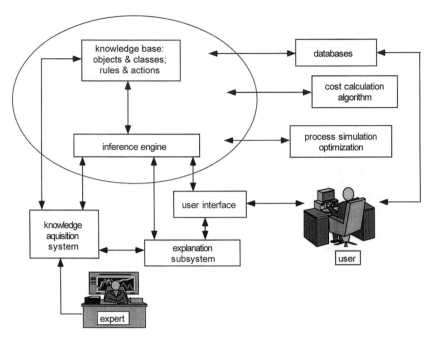

Figure 16.6 Expert system architecture in process synthesis.

bioreactor—otherwise, enzymatic degradation or unwanted biopolymer aggregation can sometimes lead to significant product losses.

The main disadvantage of coupling separation with product formation involves the loss of some control over the individual unit operations. A problem with the separation medium such as a membrane module or adsorption unit could potentially bring about the shutdown of the bioreactor, and vice versa. This type of integration also complicates the startup of the unit, as little product exists (and therefore is available for removal) until the bioreactor reaches a requisite cell density. Among the more serious and practical problems of integration are the difficulty in maintaining aseptic operation and the undesirable removal of metabolites or cell fragments along with the product.

Essentially two methods exist for keeping cells in the bioreactor during integration: cell recycling and immobilization of cells.

In cell recycle (depicted in Figure 16.7), the bioreactor fluid is usually pumped to a membrane that retains cells and permits permeation of the product. Other cell recycle systems use liquid extraction or adsorbents to remove the product, leaving cells behind (6). An ingenious approach to the production of volatile products such as ethanol is to pump the bioreactor fluid to a vessel that is maintained under a vacuum. The product becomes volatilized, leaving cells, biopolymers, and most of the metabolites behind.

The second method of continuous product removal involves the immobilization or trapping cells onto a support (Figure 16.8). Supports can consist of highly porous particles or membranes. Cell lines that grow only when attached to some surface are particularly well suited for integration in this manner.

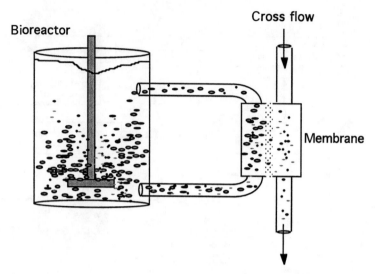

Figure 16.7 Cell recycle with a membrane for continuous product recovery.

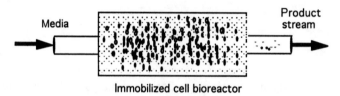

Figure 16.8 Immobilized cell reactor with continuous product removal.

16.5 MAKING THE BIOREACTOR STEP BIOSEPARATION-FRIENDLY

No designer should ever develop steps to carry out a specific task in a manufacturing process without thinking about the subsequent steps needed to create the final product. The bioreaction step, for example, offers plenty of room to improve and simplify the later bioseparation process steps. The bioseparation process designer must therefore cooperate with the bioreactor designer in areas such as the following (3):

- Finding growth media compositions that do not cause fouling or create the need for a separation step
- Researching alternatives to cell disruption, such as controlled protein secretion or permeabilization of cell matrices, for recovering intracellular products
- Minimizing the addition of salts or antifoaming agents needed to control bioreactor conditions
- Implementing fusion technology to add affinity tails to proteins such as polyamino acids, carbohydrates, and biotin

The bioreactor engineer places a high value on media design, because growth and metabolic regulation are highly dependent on the nutrients available to the cell. Media costs, on the other hand, are very important when the process will yield lower-priced products such as carboxylic and amino acids. Unfortunately, many media components are ill defined for two reasons: They may consist of natural product mixtures such as molasses, corn syrup, and serum, and they may contain impurities that add unwanted color or components to the process stream. A fully defined medium is one in which each component is identified explicitly. Such media may not be economically attained today; at worst, they may be practically impossible to achieve, as in the case of animal cell media. Nevertheless, use of a fully defined medium is very desirable simply because all ingredients used in the process then have a known purpose in cell growth and regulation.

Inclusion bodies are depots in cells that form when high concentrations of proteins accumulate because they cannot pass through the cell wall or membrane. The standard method for obtaining products from inclusion bodies (or any other intracellular product) is cell disruption. Unfortunately, as noted in Chapter 15, cell disruption releases debris that may complicate downstream bioseparation processes; it also frees proteolytic enzymes that can degrade the desired product. One alternative to cell disruption relies on the permeabilization of the cell matrix; intracellular products can then escape without creating cell membrane fragments. This technique has been demonstrated in yeast cells. Yeast cell walls consist of 85% glucan, and glucanases can easily permeabilize the cells (7). In another alternative to cell disruption, scientists have used (or modified) cell hosts to facilitate secretion of the product. Organisms such as *E. coli* and modified *B. subtilis*, for example, can allow for compartmentalization or facilitate the secretion of proteins (8).

Creating a new marker by fusing an affinity tail to a protein product is among the more creative ways of facilitating bioseparations. Enzymes, polypeptide binding proteins, antigenic epitopes, carbohydrate-binding domains, polyamino acids, and biotin-binding domains have all been used for this purpose (3). After removal of the product, the tail is cleaved, either chemically or enzymatically. This technology also offers potential utility in assaying and monitoring the process.

16.6 CONSIDERATIONS IN FINAL PRODUCT FORMULATION AND ENVIRONMENTAL IMPACT

One or more steps that deal explicitly with product formulation can be inserted between the bioseparation steps and the acquisition of the final product. In many cases, a formulation may better satisfy the customer's needs, as providing the product in a particular form can sometimes enhance its activity and value. For example, detergent enzymes remove stains from clothes by undergoing proteolysis in a surfactant solution above room temperature. Additives may therefore be introduced in the final formulation to enhance enzyme activity or stabilize the enzyme. As polyethylene glycol (PEG) can help to maintain protein stability, two-phase aqueous extraction of detergent enzymes (such as subtilisin A) has been proposed as a bifunctional step. The enzyme partitions into the PEG phase and the enzyme/PEG solution could potentially improve the shelf life of the product and maintain enzyme activity in harsh environments.

A shorter shelf life and lower enzyme activity loss may result when freeze-drying proteins from salt-rich solutions. To maintain enzyme activity, additives that do not freeze can be incorporated into the formulation. In addition, to eliminate a buffer exchange or additive step, it may be wise to investigate buffer salts and other additives that are compatible with enzyme activity and integrate these compounds into the bioseparation step.

Around the world, waste minimization has emerged as a critical concern in separations of all sorts. From a processing standpoint, the easiest method for dealing with waste streams is to eliminate their production altogether. Because they are usually water-based, biotechnology processes are generally regarded as being cleaner than traditional chemical processes. Nevertheless, their generation of biohazardous wastes, concentrated salt solutions, spent organic solvents from extraction steps, and spent chromatographic packings and membranes call attention to the need for diligence in designing clean bioprocesses.

Several simple steps can minimize the need for waste treatment:

- Reducing the number of buffer exchange steps in protein production
- Emphasizing the use of regenerable or recyclable solvents, chromatographic packings, and membrane materials
- Wherever possible, using continuous production systems rather than batch bioreactors, as the latter generate waste cell matter that must be disposed after each run
- Minimizing the number of processing steps
- Finding alternative uses for cell matter and waste streams such as animal feed or fuel, even if it requires further processing of this undesired material

The industrial processes used to manufacture the equipment, materials, and energy used in bioprocessing should also be considered when assessing the environmental impact of a bioprocess design. Facilities that supply clean air and water to the process are, of course, an essential part of the overall design. These considerations should steer the designer toward specifying the use of durable equipment, which can be of a modular design and therefore applied in a variety of different processes. In addition, processes with low energy consumption are more desirable. Clean-room air handlers and clean-water production facilities should also be low energy consumers.

16.7 SUMMARY

Heuristics play an important role in guiding bioseparation process synthesis and integration of separation processes with the other manufacturing steps. Expert systems can be used to archive these heuristics and aid in process selection and sequencing. Integration of the separation process steps can occur via genetic engineering of the cell, bioreactor design, separation process sequence selection, and design of the final product formulation. Careful analysis of the cost/benefit ratio for any development aimed at close process integration should be performed.

Faced with a short time frame for commercialization, it may be tempting to skimp on process integration. Planning and a cross-disciplinary approach to integrating the manufacturing process, however, can yield benefits in the long run.

16.8 PROBLEMS

16.1 Citric acid production has been commercialized using two different bioseparation strategies. In the Pfizer citric acid process (Figure 16.9), the fungal bioreactor step is followed by the addition of calcium to precipitate calcium citrate. This precipitation step removes many of the impurities, but the calcium citrate must be redissolved with sulfuric acid to eventually produce the acid form (which is used as an acidulant in foods). Gypsum ($CaSO_4$), a by-product of this process, must be either sold or disposed of in other ways.

Another process for citric acid production uses a long-chain tertiary amine extractant in an organic solvent diluent (Figure 16.10). (Note that the solvent extraction process and back-extraction processes are multivessel countercurrent cascades.) Citric acid is extracted in its acid form and subsequently back-extracted into a fresh aqueous phase. Although the extractant and diluent are recycled, small amounts of both are lost after each process run.

(a) Discuss the relative merits of each process with respect to the heuristics discussed in this chapter.

(b) What information do you need to decide which process has the potential to produce a milder environmental impact?

16.2 In facilitated protein processing, an affinity tail codon and a signal peptide codon are added to the gene for the target protein. This approach yields a target protein that is made available in the periplasm, or fully excreted into the aqueous media; its tail can be used to purify the protein because it imparts radically new properties to the protein. Affinity or ion-exchange chromatography, and even two-phase aqueous extraction, can be used to purify the modified protein on this basis.

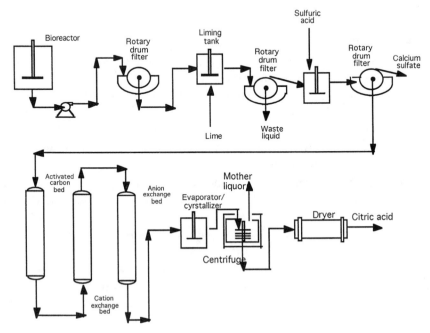

Figure 16.9 Pfizer process.

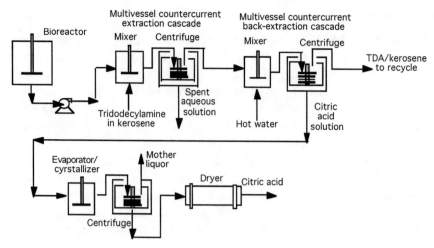

Figure 16.10 Alternative solvent extraction process.

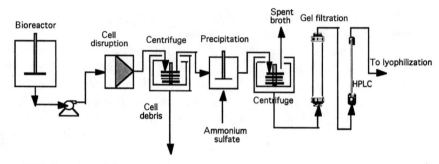

Figure 16.11 Generic protein separation process.

After elution of the protein from the affinity column, the affinity tail is cleaved enzymatically or chemically. A typical protein purification process (shown in Figure 16.11) uses one or more chromatographic steps.

(a) What are the potential strengths and drawbacks of using facilitated protein processing with fusion technology compared with the standard technique?

(b) How do these two processes compare with the use of an antibody column manufactured by hybridoma culture? (The underlying concept of this type of affinity chromatography is that antibodies will form a perfect fit with a target antigen, and this complex can then be easily reversed with a mild change in pH or salt concentration.)

16.3 Acetic acid can be recovered by complexation extraction using alamine 336 in an organic diluent phase. It can also be recovered with the aid of nonionic adsorbents. Referring to the data given in Tables 16.2 and 16.3, discuss the merits and benefits of recovering acetic acid if it can be produced by bacterial fermentation at concentrations between 10 and 25 wt%. Assume that the acid is being produced for food use.

16.4 Given the range of properties exhibited by carboxylic acids (see Chapter 4),

Table 16.2 Acetic Acid Adsorption Data

Adsorbent	Surface Area (m^2/g)	Wet Density (g/mL)	Equilibrium Concentration (wt%)	Uptake (g/g)
Amberlite XAD-4	760	0.22	1	0.05
			5	0.14
Ambersorb XEN-563	630	0.45	1	0.11
			5	0.17

Table 16.3 Acetic Acid Solvent Extraction Data

Adsorbent	Organic/Water (Volume Ratio)	K_d or K_b	Equilibrium Concentration (wt%)
Tridodecylamine in hydrocarbon solvent (25°C)	0.7	$K_d = 2.5$	0.1
Tridodecylamine in hydrocarbon solvent (60°C)	0.7	$K_b = 3.2$	60

develop a set of rules not discussed in this chapter to guide process selection and integration for carboxylic acids and their derivatives.

16.5 Imagine that lactic acid, l-lysine, and ethanol can be produced by a single organism in concentrations between 10 and 20 wt%. Develop a separation process that will yield food-grade products.

16.6 Discuss the relative merits and drawbacks of building three biotechnology facilities to produce three different low-molecular-weight products, such as lactic acid, l-lysine, ethanol, versus having a recombinant organism produce all three at reasonably high concentrations and then developing a separation process to resolve the three products.

16.9 REFERENCES

1. Staehelin T, Hobbs DS, Kung H-F, Lai CY, Pestka S. Purification and characterization of recombinant human leukocyte interferon (IFLrA) with monoclonal antibodies. J Biol Chem 1981;256:9750–9754.
2. Schuler M, ed. Chemical engineering problems in biotechnology, vol. 1. New York: American Institute of Chemical Engineers, 1989.
3. Verrall MS, ed. Downstream processing of natural products: a practical handbook. Cichester: Wiley and Sons, 1996.
4. Leser EW. Prot ex: an expert system for selecting system for selecting the sequence of processes for the downstream purification of proteins. University of Reading: Ph.D. thesis, 1995.
5. http://herzberg.ca.sandia.gov/jess/
6. Atkinson B, Mavituna F, eds. Biochemical engineering and biotechnology handbook. New York: Macmillan, 1983.
7. Kocková-Kratochvílová A. Yeast and yeast-like organisms. Germany: VCH Verlagsgesellschaft, 1990.
8. Ford CF, Souminen I, Glatz CE. Recovery of a charged-fusion protein from cell extracts. Prot Exp Pur 1991;2:95.

17

Product Formulation

Formulation is the processing step in which both chemical and biological pharmaceutical products become converted to the final state in which they are packaged, stored, and dispensed. A successful formulation process will offer both efficiency and safety. Additionally, the resulting product must remain stable through product constitution, analysis, processing, storage, and, in some cases, reconstitution.

17.1 FORMULATION CHARACTERISTICS

An efficient formulation should not only retain the full bioactivity of the product, but also release it in the proper organ or tissue, at the desired time and rate. The desired final product would be effective, aesthetically pleasing, convenient to use, and economical. For example, the ideal healing agent formulation would be easily applied to the wound site, have a bearable odor, be viscous enough that the drug stays in place at the point of application, retain its bioactivity after the preparation has dried, and diffuse at a reasonable rate into the wound.

After product constitution, it should still be possible to determine the agent's properties, such as its concentration and activity, through convenient analytical techniques. Unfortunately, changes in properties such as viscosity or turbidity of the drug after formulation often require the design of new or modified analytical techniques to make even routine measurements of these characteristics.

The formulation should also withstand any further processing steps, such as freeze drying, without losing any of its bioactivity. Many times, we may need to reconstitute a freeze-dried product; in these cases, the product should regain its original characteristics without forming aggregates or losing bioactivity. Note, however, that any additional precautions necessitated by the use of a complicated manufacturing process will add to the cost of the final product. A highly robust formulation will not be affected by the type, size, or speed of the equipment, or small changes in composition. Altering the process scale can change the effectiveness of some formulations, however, as can modifying the final desired product dose.

Another important formulation criteria relates to its physical and chemical stability during storage. The agent should retain its activity and aesthetic qualities during storage, without any loss of color, odor, taste, and macroscopic integrity. It must also avoid loss of conformation because of unfolding and aggregation at the molecular level. Chemically, the drug might undergo oxidation or other chemical processes—all of which can diminish its activity. Some products, such as TGF-β, have

372

an affinity for glass, and the protein Relaxin is degraded by light. As a result, care must be taken when choosing the storage container for these compounds.

In many cases, rapid absorption and excretion of a pharmaceutical can lead to dangerously high levels of the drug in the bloodstream for short periods of time. Consequently, the formulation should ensure that all the active drug is not released into the body in a single large dose. Controlled release of the drug is also warranted when the compound can diffuse rapidly into other regions after being locally administered. A more desirable formulation of the drug will allow for its local release at a slow and steady rate. To achieve this slow rate of release, the active drug can be enclosed in a polymer matrix or its structure changed chemically so as to reduce its solubility. In designing a formulation for controlled release, we must consider the region in which the drug will be administered, the desired rate of release, and the duration of that release.

17.2 EXCIPIENTS

Excipients are chemicals that are added to the biologically active drug as part of the formulation to attain the previously discussed objectives. The excipients themselves must meet certain criteria, such as being safe for both users and the immediate environment. The important role played by these chemicals becomes obvious when we consider that many promising drugs fail to reach the marketplace because developers cannot create safe and efficacious formulations for them. The safety of most formulations is closely linked to dosage and the method of administration, so their manufacturers must prepare specific instructions on the drugs' use to safeguard consumers' safety.

Excipients commonly added to formulations include antimicrobial compounds, chelating agents, and antioxidants. These compounds preserve and stabilize the active drug from changes in pH, moisture content, and physical and chemical degradation. In addition, excipients are used as solid or liquid solvents. In such cases, the formulation process must account for density differences between the drug and the solvent, adsorption characteristics, chemical and physiological influences of the excipients on the drug, and the release rate of the drug from the solvents.

Many times, we may need multiple formulations of the same drug, with each version customized to a certain mode of administration. For example, a drug can be incorporated in microcapsules to be taken by the spoonful or encased in an outer shell; it might also take the form of a powder that becomes a gel when mixed with water for pediatric patients or be taken with a liquid by adult patients. In all of these cases, the particular formulation needs to be appropriate for its intended use, and drug potency must remain consistent.

17.2.1 Thickeners and binders

Thickeners and binders may increase the physical stability of the end product or enhance the convenience of administration of some fluids. When these compounds are used in the formulation process, care must be taken to ensure uniformity of the thickened product across various batches. Examples of thickeners and binders include acacia, agar, starch, sodium alignate, gelatin, methyl cellulose, bentonite, and silica.

17.2.2 Surface-active agents

Surface-active agents, such as sodium lauryl sulfate, are occasionally added to formulations. Although they may not be suitable for internal consumption because of toxicity concerns or unpleasant taste, these compounds are nevertheless appropriate for external application. As a result, they are incorporated into many lotions and creams. Examples include talc, silica, stearic acid, and hydrogenated vegetable oils.

17.2.3 Colors and flavors

Colors are used to distinguish different drugs from one another, and to give a uniform appearance to different batches of the same product. Flavors represent an important addition to many pediatric formulations, masking the bitter taste associated with many drug compounds.

17.2.4 Preservatives

The addition of preservatives to a formulation can serve to stabilize the product during storage. These additives protect the drug against microbial, chemical, and physical degradation. Physical degradation, such as a loss of color or change in consistency, crystallinity, or porosity, usually occurs because of an absorption of moisture. Water is considered an undesirable addition to a formulation when its inclusion will diminish the product's stability. Indeed, many pharmaceutical products have very short shelf lives in water. In such cases, either the manufacturer must use a more stable solvent or the product must be freeze-dried for storage, then reconstituted immediately prior to use. To minimize such physical degradation, the manufacturer should couple the proper freeze drying technique, which takes into consideration the product's glass transition characteristics, with appropriate packaging.

Chemical degradation is normally related to the product's exposure to moisture and light. It often involves reactions such as oxidation, hydrolysis, or cleavage of disulfide links. To prevent such changes, a manufacturer can add antioxidants such as thymol, tocopherols, or BHT to the formulation. Similarly, the manufacturer must prevent microbial growth in the product, which can occur if the formulation includes compounds such as carbohydrates. This composition makes some drug formulations very susceptible to microbial attack; other formulations may be protected by the inherent antimicrobial characteristics of the drug itself.

17.3 DOSAGE FORMS

The various dosage forms can be broadly classified as either solids or liquids (although other minor groups, such as suspensions, emulsions, aerosols, and semi-solids, also exist). Selection of a particular dosage form is based on its safety and the most effective route of administration.

Liquids are typically used when large doses of the drug must be administered orally or when the formulation includes a liquid solvent. These types of formulations are also essential when the optimal route of administration involves injection, when the application of a lotion is needed to ensure dermal penetration, when the drug will be used as eye or ear drops, and when the agent will be administered in the form of a pressurized aerosol. Because liquid solutions are prone to microbial contamination, the ideal formulation will remain stable at high temperatures, which permits autoclaving of the solution. The drug should not precipitate out of the solution, nor should it settle. Pastes are commonly used for drugs that will be applied to unbroken or damaged skin.

Solid formulations are more stable than liquids and suspensions, can be provided in unit doses, and offer the convenience of a compact form. Tablets account for the majority of solid formulations. These formulations are designed for oral consumption or, in some cases, as implants for controlled drug release. Injectable drugs that are difficult to resolubilize, however, will take the form of solution ampoules, rather than solids.

In the processing of a solid formulation, various solids are usually mixed, sometimes accompanied by the application of heat. Thus the manufacturer must determine how the processing conditions will affect the active drug. The dose or amount of active ingredient in the tablet also influences the choice of processing conditions. For example, if the drug is very potent and the tablet should include only very small quantities of this compound, then the major unit operation will focus on uniform mixing of this drug in the diluent and other additives. On the other hand, if the drug accounts for a large part of the formulation, then mixing becomes easier but the drug's physical properties will greatly influence the granulation and compression operations. If the solubilities of the drug and the other constituents in the formulation differ, then absorption of moisture may result in recrystallization, leading to nonuniform distribution. In any event, after proper mixing of the formulation ingredients, the drug and associated excipients are placed in a die, where they are compressed into a tablet form.

Controlled release of the drug can be achieved through high compression of the drug into tablet form, ensuring a slow rate of disintegration for the tablet. Alternatively, the pharmaceutical can be dispersed in an inert, porous polymer matrix. It can also be granulated, and the granules coated with a resin—a design known as an encapsulation.

17.4 ENCAPSULATION

In encapsulation, a material is placed within or coated by another material. The encapsulated material, which is known as the core or internal phase, can consist of a solid, a liquid, or even a gaseous substance. The outer shell forming the coating is called the wall material or membrane. A polymeric wall material must, of course, be nontoxic. A wide selection of polymers is currently available, including naturally occurring polymers (such as gelatin, gum arabic, and beeswax), semi-synthetic polymers (such as methyl cellulose, stearic acid, and hydrogenated castor oil), and synthetic polymers (such as polyvinyl alcohol, various polyamides, and polyesters).

Encapsulation is usually performed to control the rate of release of the core drug. It may also be used to protect the drug from the environment to increase stability, target the site of release, and mask undesirable properties (such as a bitter taste). Although encapsulation can increase the drug's tolerability, it does not make the agent less toxic, except in the case of certain special formulations. An encapsulated drug must also remain bioavailable.

One popular encapsulation method is spray drying. In this technique, the wall material is hydrated and added to the core material. Homogenization then disperses the core in the wall material solution. The mixture is fed to a spray dryer, where the atomized droplets come into contact with hot air and form the encapsulated particles. Spray drying is a well-established technique, and the equipment used in this process is inexpensive and readily available. The final product is usually a fine powder needing further agglomeration, however, and the drying process can cause the core material to deteriorate. The use of a low-temperature spray drying technique may circumvent these limitations. In this method, the emulsion of the core and wall material is atomized with ethanol, a solvent that can be easily evaporated at low temperatures and pressures.

Encapsulation can also be performed through extrusion. In this technique, the core material is homogenized in a molten carbohydrate mass, and the mixture is forced through an extrusion dye into a dehydrating liquid such as isopropyl alcohol. After evaporation of the solvent, the formulation system can physically cut the remaining solid into individual doses.

A third encapsulation method involves coacervation. In this technique, the core material is suspended in a solution of the polymeric wall material. The polymer-rich phase is then separated from the solution by the application of a nonsolvating polymer or a change in temperature or pH. The choice of the encapsulating material is important, as it determines the dissolution rate of the formulation and, therefore, the release rate of the drug (1).

Numerous other methods are used for pharmaceutical encapsulation, including some that can alter the activity of the compound. In liposome entrapment, for example, a liposome is formed through the addition of phospholipids to an aqueous solution. After transfer to an organic solvent, the addition of excess water disrupts the lipid bilayer, thereby forming heterogeneous multilamellar liposomes in which the active drug is encased. Ultracentrifugation, molecular-sieve chromatography, or dialysis can be used to remove the liposomes and obtain formulations with uniform size distributions. When injected intravenously, such liposomes preferentially accumulate in the liver and spleen (2).

If the encapsulation procedure should simply coat a material, then the manufacturer may employ fluidized bed coating. In this procedure, the core material is suspended in a fluidized bed. The coating material is atomized and sprayed into this bed, where it forms a single layer on the particles of core material. Drugs can also be formulated as gelatin or alginate beads (3). These beads are created in a solution of the bioactive agent, separated by filtration, and dried. When rehydrated, the beds will release the pharmaceutical compound at a rate reflecting the drug's solubility.

The most common method of encapsulation involves the dispersion of the drug in a polymeric matrix to form microspheres or microcapsules. Microspheres can be as small as a few nanometers in diameter. More easily produced than microcapsules, they can be administered through intravenous injection or oral consumption. Microcapsules are small hollow bodies, several hundred microns in diameter, that surround the drug compound. The properties of the microcapsules depend on the type and thickness of the wall material, the targeted organ or design purpose, the core material properties, and the production technique used. The microcapsule material should, of course, be nontoxic and biocompatible. If the microcapsules will use an oral mode of administration, then the excipients used in the capsule construction must be compatible with the gastrointestinal tract.

Because microcapsules are so small, normal engineering fabrication procedures—such as extrusion or molding—are useless in their manufacture. Instead, their production involves a biochemical "self-assembly" process, such as micelle formation or multiple emulsions. Micelles, although employed in certain applications, have several limitations, including a low drug-to-excipient ratio. Thus microcapsule manufacture more commonly relies on multiple emulsions. Two types of multiple emulsions exist: 1) water-in-oil-in-water (W/O/W), which is best suited for the encapsulation of water-soluble drugs, and 2) oil-in-water-in-oil (O/W/O), which is preferred for the encapsulation of fat-soluble drugs.

Drug loading in microcapsules formed by the W/O/W and O/W/O techniques can sometimes be less than 10% on a weight basis; even more of the pharmaceutical compound may be lost during microcapsule isolation and dying. In a recent innovation known as the core-binding technique, the drug is complexed to the polyelectrolyte core of the microcapsule to increase microcapsule loading (4). To date, it has been used to encapsulate two antimalarial drugs, quinine and chloroquine.

17.5 FREEZE DRYING

In the food and pharmaceutical industries, freeze drying is a popular technique for removing water from the desired product. It is suitable for only certain products, however, and it may require a pretreatment step to avoid the nonhomogenous formation of ice crystals during the freezing process. Similarly, homogenization of immiscible mixtures may be required before the freezing step takes place. The specific conditions during freeze drying are extremely important, as they determine the crystalline structure formed and thus the physical properties of the final product after sublimation.

Preliminary studies of the product's temperature behavior can determine the conditions under which the product will attain the desired crystalline network structure. In addition, the manufacturer should investigate whether the product's bioactivity will be altered by a change in physical or chemical structure during the freeze drying process. If a loss of bioactivity occurs, then protective agents can be added to the product solution before freeze drying begins.

To study the structure of water-protein mixtures before freezing, computer simulations are recommended. These simulations must use models that account for the flexibility of chemical bonds, as protein crystals in an aqueous environment invariably have water associated with their molecular structures. Naturally, one must study the phase transitions in the system to avoid situations that may change the product's structural or chemical properties.

Although it is difficult to make generalizations about how freeze drying will work with any given product, and development of a process often requires extensive experimentation, we can nevertheless derive a limited theoretical model. In the following discussion, we assume that water is the solvent; if the process will use a different solvent, then we must include appropriate equations or corrections.

17.5.1 Theory

As the temperature of a solution decreases, the solution becomes progressively cooler until freezing begins. When the temperature falls even further, undercooling of the solvent ceases and a spontaneous ice nucleation process begins. Although nucleation occurs at $-40\,^{\circ}$C in absolutely pure water, the presence of particulate matter in a solution serves to increase the nucleation temperature. Each of the nuclei formed grows into an ice crystal, with the number and size of the crystals ultimately being determined by the temperature and rate of cooling. Through the removal of the solvent water in the form of ice, the remaining solution becomes concentrated with the solute. As the ice front grows, it moves away from the cooling surface.

Most liquid polymers and polymeric solutions can exist in various states, such as liquid, crystalline, amorphous rubber, and glassy states. In a crystal, molecules are immobilized in a regular three-dimensional pattern. In the liquid form, however, molecules are mobile. In the amorphous rubber and glassy states, they have constrained mobility. For all polymeric solutions, as the temperature decreases or the concentration increases, the polymer reaches a point at which it changes from a viscous, rubbery state to a hard and relatively brittle glassy state (Figure 17.1). This transition resembles a phase change from a solid to a liquid, although it is technically a state change.

As noted earlier, freezing leads to ice formation, accompanied by the increasing concentration of the solution. Eventually ice formation in the residual solution becomes negligible, and we can consider the water content to be constant. The temperature at which this development occurs is called the glass transition temperature, T_g. The glass state is very stable, being characterized by only a small amount of molecular motion. At temperatures exceeding T_g, the greater mobility of the mol-

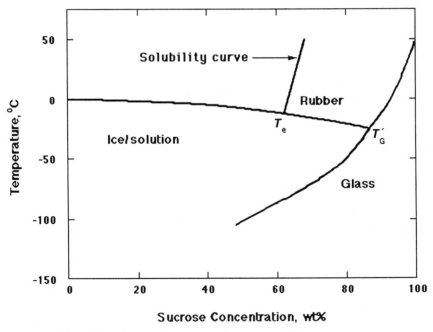

Figure 17.1 Solid-liquid state diagram for a sucrose-water system (5).

ecules can cause products to deform, degrade, or have a short shelf life. For this reason, the glass transition temperature is sometimes called the collapse temperature. It can be determined via techniques such as differential thermal analysis and electrical resistance measurement.

To calculate the amount of water in a solution that can be frozen, we use Raoult's law (6). During the freezing process, if the solid and liquid phases are in equilibrium, then the partial pressures of the water in the two phases are equal ($p_i = p_{\text{solution}}$). According to Raoult's law, the vapor pressure of the solution is related to the mole fraction of the solvent in the solution as follows:

$$\frac{p_{\text{sol}}}{p_{\text{w}}} = \frac{W}{W + S\left(\dfrac{M_{\text{w}}}{M_{\text{S}}}\right)} \tag{17.1}$$

In this equation, W is the freezable water in the sample, S is the amount of solute, and M is the molecular weight of the water or solute. The activity of water a_{w} is defined as the ratio of the vapor pressure of ice to that of the pure solvent ($a_{\text{w}} = p_i/p_{\text{w}}$). Hence we can express the activity of water at the freezing point as follows:

$$a'_{\text{w}} = \left(\frac{p'_i}{p'_{\text{w}}}\right)_{T_{\text{F}}} = \frac{W}{W + S\left(\dfrac{M_{\text{w}}}{M_{\text{S}}}\right)} \tag{17.2}$$

As the temperature drops below the freezing point, a portion of the water turns to ice. If x is the amount of ice formed, then the activity at any point where $T < T_{\text{F}}$ can be expressed as follows:

$$a_w = \left(\frac{p_i}{p_w}\right)_T = \frac{W - x}{W - x + S\left(\dfrac{M_w}{M_S}\right)} \tag{17.3}$$

Equations 17.2 and 17.3 can be combined to eliminate the term $S(M_w/M_S)$. We can then define the amount of ice formed in terms of activity:

$$\frac{x}{W} = \frac{1 - \left(\dfrac{a_w}{a'_w}\right)}{1 - a_w} \tag{17.4}$$

Equation 17.4 can also be expressed in terms of the vapor pressures:

$$\frac{W - x}{W} = \frac{\left(\dfrac{p'_w}{p'_i}\right) - 1}{\left(\dfrac{p_w}{p_i}\right) - 1} \tag{17.5}$$

Thus we can determine the amount of frozen water by measuring the vapor pressure of the ice. For temperatures not far below the freezing point, however, we can predict the amount of water frozen by simply calculating the activity of water:

$$a = \exp\left(-A + \frac{B}{T} + C\ln(T)\right) \tag{17.6}$$

In Equation 17.6, $A = 27.306$, $B = 513.848$ K, and $C = 10.435$.

During the drying step, the rate of sublimation of the ice from the frozen solid is equal to the rate of evaporation minus the rate of condensation. It can be determined by using the Knudsen equations (6):

When $l < \lambda$, $\alpha = 1$,

$$G_s = (p_s - p'_c)\left(\frac{M}{2\pi RT}\right)^{1/2} \tag{17.7}$$

When $l > \lambda$,

$$G_s = \alpha(p_s - p'_c)\left(\frac{M}{2\pi RT}\right)^{1/2} \tag{17.8}$$

In these equations, G_s is the rate of sublimation, p_s is the saturated vapor pressure of ice, p'_c is the partial vapor pressure of water outside the condenser, M is the molecular weight of water vapor, R is the universal gas constant, T is the absolute temperature, l is the distance between the evaporator and the condenser, and λ is the mean free path of the vapor molecules. The coefficient of evaporation α is defined as the ratio of the actual sublimation rate to the rate calculated from kinetic theory under conditions of zero resistance to evaporation; this coefficient lies between 0 and 1.

The kinetic rate constants for processes such as product deformation and deterioration in the region of T_g are sensitive to temperature. They can be accurately predicted by the Williams-Landel-Ferry (WLF) equation (7):

$$k(T) = \frac{(C_1(T - T_g))}{(C_2 + (T - T_g))} \qquad (17.9)$$

The rate of product degradation is given by

$$R = R_g \exp\left(\frac{(C_1(T - T_g))}{(C_2 + (T - T_g))}\right) \qquad (17.10)$$

where R is the product degradation at the storage temperature T, and R_g is the degradation rate at the glass temperature T_g.

17.5.2 Technique

The first step in the freeze drying process is, of course, the freezing of the sample. The vial containing the sample solution is placed on a shelf where the temperature is controlled for optimum ice crystal formation. As the freezing proceeds, ice forms and the concentration of the solutes in the unfrozen solvent increases. This greater concentration of all water-soluble constituents serves as the main source of any subsequent chemical changes. For example, the presence of even dilute amounts of salt or other extraneous chemicals in the solution may evolve into a serious problem, leading to the loss of activity of many biological products. In addition, the increase in the constituents' concentrations could significantly increase the rates of reactions that would have only negligible consequences at room temperature. For example, salt mixtures that remain stable at room temperature have reacted rapidly and irreversibly during freezing (8). To prevent the undesired concentration of salts in the solution, the manufacturer can add protectants such as sugars or sugar alcohols to the solution (9).

After freezing the product to the required temperature, the product must then be dried. Drying is usually performed by evacuating the chamber that contains the vial and then raising the shelf temperature. The heat energy supplied evaporates the solvent and can potentially alter the product's conformation. Because the product must retain its microscopic and macroscopic structure during the drying operation, it may be necessary to add stabilizers to prevent any modifications to its conformation. The water vapor formed in the drying process also acts as an insulator, upsetting the heat transfer and possibly creating nonhomogeneous regions in the product or denaturation through excessive local heating. Drying equipment must therefore be designed to avoid such problems. In many cases, temperatures and vacuum conditions may vary according to the stage in the drying process. Drying usually involves two stages: sublimation and desorption (9).

Sublimation is carried out at low temperatures and moderate vacuum. During this stage, the solvent evaporates from the frozen or crystalline state (that is, it sublimes). Sublimation usually takes place at or below T_G, the glass temperature of a solid, so as to prevent the formation of liquid water. T_G is a unique point on the T_g curve—that is, the intersection of the liquid curve and the glass (T_g) curve.

After sublimation of the crystalline material through primary drying, desorption removes the remaining liquid solvent at slightly higher temperatures under high vacuum. At this stage, the liquid water acts as a plasticizer. Consequently, the desorption process is controlled by the diffusion of water from the amorphous solid product. The diffusion of water depends on the local texture, so it may not be uniform for the entire product. Variations may create local regions of high water content that could potentially decrease the product's shelf life. For many products, the final formulation should retain some solvent—overdrying can sometimes prove as deleterious to product quality and shelf life as underdrying. The final processed product is then packaged and stored so as to minimize contact with air or sources of moisture.

Processes that favor amorphous glass formation, as well as storage conditions that retain such structures, provide optimum product preservation. A high-quality and stable product will exist in the glassy state during primary drying. With some products, increasing the moisture content of the sample will cause crystallization of excipients. The presence of any moisture in the product decreases T_g—a relationship that storage temperatures must take into account. The preferred temperature for primary drying and storage is slightly lower than T_G, which is considered the maximum safe storage temperature for the product (10).

For some products, such as human growth hormone (hGH), freeze drying does not degrade the final formulation but nevertheless yields an unstable solid. To increase the final product stability, the manufacturer may add a combination of excipients such as mannitol and glycine. These excipients, however, can actually destabilize the product when excessive moisture or oxygen is present in the system (8).

New freeze drying techniques have been developed to deal with products that cannot withstand the rigors of the normal process. If a product is sensitive to low temperatures or does not form a stable suspension with the solvent during freezing, its manufacturer may select a "soft ice" technique (9). When freeze drying extremely dilute solutions of some biological products, the active product may sometimes be removed with the vapor during sublimation. To prevent the loss of expensive biological products, for example, a supporting matrix may adsorb the active product during sublimation. This adsorbed product can then be reconstituted with an appropriate solvent. Another important development is the continuous freeze drying process (9).

Example 17.1
Basic fibroblast growth factor (bFGF) is a protein drug used to promote wound healing and bone growth (11). Like other proteins, this compound forms aggregates during the freeze drying process. With most proteins, if hydrophobic interactions cause aggregation, then the manufacturer can generally disrupt this process by adding surfactants such as SDS. With bFGF, however, SDS does not work in the same way. The primary reason for aggregation in bFGF is scrambled disulfide linkages. Glycosaminoglycans (GAGs) such as heparin, although they stabilize the bFGF protein conformation, nevertheless lead to aggregation and cannot extend the product's shelf life. On the other hand, the addition of salts or carbohydrates, such as sucrose, lactose, and glucose, can prevent this aggregation. These additives also preserve bFGF's conformation during freeze drying and reconstitution (12).

The bFGF product can be used in solution form as long as the solution also includes stabilizers such as heparin, GAG, β-1,3-glycansulfate, dextran sulfate, or cyclodextrin sulfate. The solution can be administered by a spray pump or dropper, absorbed into a gel foam or a collagen sponge, or consumed in tablet form. In the latter case, the freeze-dried product consists of a mixture of the aluminum salt of β-cyclodextrin sulfate (ACDS) and bFGF.

bFGF can also be formulated in microspheres, which are used primarily to promote wound or burn healing. To create these microspheres, the manufacturer dissolves bFGF and dithiothreitol in water at pH 6.0 and then adds 20 μm starch particles to the solution. This well-mixed suspension is lyophilized to form a powder.

bFGF impregnated in carboxymethylcellulose gel is used to promote bone formation in a demineralized bone matrix. Similarly, fibrin gel containing bFGF has been used to stimulate bone fracture repair (11).

17.6 SUMMARY

Product formulation—the final step in any bioprocess scheme of much interest to the pharmaceutical industry—is used to create the form in which drugs are packaged, stored, and dispensed. A desirable formulation will deliver the bioactive drug in a manner appropriate for the treatment route. Chemicals known as excipients are added to the drug to influence the formulation's properties, including its stability, color, odor, and taste. Formulations are generally either solid or liquid forms, and are processed by freeze drying or encapsulation.

17.7 PROBLEMS

17.1 A freeze drying process was employed to remove water from a biological sample. The freezing point of the solution was determined to be $-2\,°C$, and the sample was frozen by flowing nitrogen around the sample. Assume that it is possible to freeze the sample at any temperature by controlling the flow rate of nitrogen.

(a) Determine the amount of water frozen from the sample at $-40\,°C$.

(b) Plot the amount of water frozen for temperatures $T < T_F$. Using the equations provided, predict the maximum amount of water that can be frozen. At what temperature is this point reached? What happens at lower temperatures?

(c) Redo part (b) assuming the freezing point to be $-8\,°C$ and $-17\,°C$.

17.2 Based on Problem 17.1, what can you conclude regarding the accuracy and suitability of the equations provided in this chapter?

17.8 REFERENCES

1. Speiser P. Microencapsulation by coacervation, spray encapsulation, and nanoencapsulation. In: Nixon JR, ed. Microencapsulation. New York: Marcel Dekker, 1976.
2. Gregoriadis G, McCormack B. Liposomes and polysialic acids as drug delivery systems. In: Karsa

DR, Stephenson RA, eds. Encapsulation and controlled release. Cambridge, UK: The Royal Society of Chemistry, 1993.

3. Schacht E. The use of gelatin and alignate for the immobilization of active agents. In: Karsa DR, Stephenson RA, eds. Encapsulation and controlled release. Cambridge, UK: The Royal Society of Chemistry, 1993.

4. Warburton B. Microcapsules from multiple emulsions. In: Karsa DR, Stephenson RA, eds. Encapsulation and controlled release. Cambridge, UK: The Royal Society of Chemistry, 1993.

5. Franks F. Freeze-drying: from empiricism to predictability. The significance of glass transitions. Dev Biol Standard 1992;74:9–19.

6. Mellor JD. Fundamentals of freeze drying. London: Academic Press, 1978.

7. Levine H, Slade L. Water as plasticizer: physico-chemical aspects of low moisture polymeric systems. Water Sci Rev 1988;5:79–185.

8. Pikal MJ, Dellerman K, Roy ML. Formulation and stability of freeze dried proteins: effects of moisture and oxygen on the stability of freeze dried formulations of human growth hormone. Dev Biol Standard 1992;74:21–38.

9. May JC, Brown F. Developments in biological standardization. International Association of Biological Standardization, Ed. 1992.

10. Hatley RM. The effective use of differential scanning calorimetry in the optimization of freeze drying processes and formulations. Dev Biol Standard 1992;74:105–122.

11. Wang YJ, Shahrokh Z, Vemuri S, Eberein G, Beylin I, Busch M. Characterization, stability, and formulations of basic fibroblast growth factor. In: Pearlman R, Wang JY, eds. Formulation, characterization, and stability of protein drugs. New York: Plenum Press, 1996.

12. Prestrelski SJ, Tedischi N, Arakawa T, Carpenter JF. Dehydration-induced conformational transitions in proteins and their inhibitions by stabilizers. Biophys J 1993;65:661–671.

18

Bioprocess Economics

Accurate operating and capital cost estimation is crucial in making process decisions, as is establishing a minimum price for the resulting product. Once these estimations are made, the manufacturer can undertake careful economic analyses using various profitability measures as part of the decision-making process. Rather than discussing this important topic in great detail, this chapter will provide references and information about currently available process development and cost estimation software. During real-world process design and economic decision making, equipment vendors and software developers will be the manufacturer's indispensable allies. Our goals in this chapter are to explain the concepts involved in equipment cost estimation and to introduce the methods used to calculate the capital and operating expenses in bioproduct manufacturing.

18.1 RESOURCES AVAILABLE FOR COST ESTIMATION

Modern software tools allow the designer to concentrate on the creative aspects of process design, rather than dealing with the tedium of calculations. Well-known software titles for bioseparation process design and economics include BioAspen (1) and SuperPro Designer (2). These and similar software design packages perform several valuable tasks:

- They simplify the creation of a process flow sheet.
- They track the flow of components through individual vessels and identify final product stream composition and amounts.
- They track the energy and chemical consumption in the process.
- They help determine the appropriate size of equipment.
- They help estimate capital and operating costs.
- They provide economic analyses.

To take advantage of these powerful tools, the user must be skillful enough to set up the information needed to design the process and to interpret the information provided by the software.

To specify energy requirements and to track the amounts of different components entering and exiting the process vessels, the user must enter the relevant compounds and their physicochemical properties into the computer program. The components of the process stream can be single molecules, such as water or carbon dioxide, or pseudocomponents, such as biomass and cell debris. Although most process design software includes an extensive database of components and their properties, the user may want to add his or her own pseudocomponents and

Table 18.1 Typical Physicochemical Properties for a Pseudocomponent and Two Common Compounds Used in Design Simulation (2)

Property	Biomass	Carbon Dioxide	Glucose
Formula	$C_{1.8}H_{0.5}O_{0.2}$	CO_2	$C_6H_{12}O_6$
Molecular weight (g/mol)	147.6	44.01	180.16
Particle size (μm)	1.5	0	0
Liquid/solid density (g/mL)	1050	1	1.562
Liquid/solid heat capacity (J/g mol K)	103	100	220
Boiling point (K)	800	194.76	419.2
Critical temperature (K)	647.3	304.18	647.3
Critical pressure (Pa)	50	75.27	50
Compressibility factor (Z)	0.229	0.274	0.229
Saturated vapor pressure			
(Antoine coefficient a)	8.1437	9.8106	8.1437
(Antoine coefficient b)	1746.15	1347.786	1746.15
(Antoine coefficient c)	0	–0.2	0
Henry's constant (atm m^3/g mol)	0	0.02188	0

molecules or perhaps change some of the properties provided in the software database. (References giving physicochemical properties appear at the end of this chapter.)

Table 18.1 gives a typical list of molecules and pseudochemical properties needed to determine a system's energy and material balances. The balances for each process vessel are analyzed to produce design specifications for the entire process.

Once the component list has been established, the designer can draw a flow sheet for the base case. A base case is the best initial design, though not necessarily the final design. After estimating the costs and profitability of this base case, the designer then explores sensitivity analyses and process variations to identify potential improvements to the base case design.

The performance characteristics of each unit must be specified individually to establish the vessel size, exit stream composition and flow rate, energy requirements, and size and composition of other streams needed to operate the unit. Table 18.2 outlines critical design parameters for selected bioseparation units. The designer relies on many different sources to gather this type of information, including pilot- and laboratory-scale experimental data, vendor data, theoretical design calculations (see previous chapters), and commercial-scale data from prior projects. Process designs may also require that the separation equipment vendor supply detailed information. If this information is not available during the preliminary design stage, "best guess" estimates can often be used as placeholder values without creating large errors in the overall process economics.

18.1.1 Capital cost estimation

The capital cost for a specific unit is directly related to equipment size, which is specified in the design. Equipment size is used to calculate the equipment purchase

Table 18.2 Typical Critical Specifications Needed for Ion Exchangers, Mixer-Settler Extractors, UF Membranes, and Disc-Stack Centrifuges

Unit	Design Parameters	Process Scheduling	Materials/Energy Costs
Ion-exchange column	Linear velocity Binding capacity Volume Length/diameter	Turnaround time Elution/wash steps Cycles/batch	Resin cost Resin life
Mixer-settler extractor	Partition coefficients Flow rate Residence time Number of stages		Power consumption
UF membrane	Concentration factor Average flux Rejection coefficients		Unit power consumption Membrane life
Disc-stack centrifuge	Sigma factor Dense phase particle concentration Product yield		

cost, which is added to other expenses to obtain the total capital cost of a particular process unit.

Once the overall plant design has been specified, the designer can include other capital costs associated with the construction of a working production plant. To calculate these costs, the designer might, for example, obtain pricing information for the specified construction materials from vendors or refer to detailed cost information from earlier designs. Note, however, that the use of nonstandard construction materials can affect pricing.

Process design software greatly simplifies this cost estimation process, as cost databases will automatically calculate purchase costs. The developer must realize, however, that the choice of vendor and other market factors can influence the project's final cost. The user can introduce such factors into the software databases to reflect these variations and obtain a more accurate cost estimation.

Capital costs for a process unit are based on the vessel size and the number of vessels needed in a particular design. When developed through the use of software or design curves, the estimation usually includes only the purchase price.

One simple, but effective estimation tool is a power law equation that scales the cost based on the needed size (3):

$$\text{Cost} = (\text{base cost}) \left[\frac{\text{capacity}}{\text{base capacity}} \right]^{n} \tag{18.1}$$

The exponent used in Equation 18.1 will depend on the user's prior design experience or vendor-supplied information. Process design software can provide an exponent or a more complex method for estimating capital costs based on this simple model.

Other costs that are factored into the calculation of the total capital costs include materials costs, installation costs, and the number of backup units required to maintain process schedules if a particular unit must be serviced on-line. To account for these costs, we need only introduce a multiplying factor after determining a cost based on vessel size. Materials selection is usually limited to different grades of steel, or possibly plastic or glass, depending on the type of vessel under consideration. Purchasing backup units can prove quite costly—if each vessel receives one backup unit, for example, then the capital cost will double. Normally, backup units are purchased only for lower-cost, mechanically intensive items such as pumps.

Once the individual units have been analyzed, we must assess the overall capital costs for plantwide items. These items include the following (4):
- Process piping to connect the equipment to the main process and utility lines
- Instrumentation to control and monitor the process
- Building infrastructure
- Exterior infrastructure improvements to provide needed fire safety, security, and shipping of raw materials and products
- Construction of any auxiliary facility such as a steam generation plant

To estimate these costs, we determine how much each item applies to the individual pieces of equipment in the plant.

Process developers and design teams refine the cost estimate in stages. Throughout many of these stages—from rough to more detailed designs and cost estimations—simple multiplying factors are employed to determine the total costs of building the facility based on the purchase prices of the equipment (5).

18.1.2 Operating cost estimation

The operation of a bioprocess unit requires some or all of the following utilities:
- Energy in the form of electrical power or heat (usually supplied by steam)
- Water, usually deionized and purified, for cooling and processing
- Filtered air
- Chemicals in aqueous solution or solvents
- Disposal or treatment of any effluent streams

These costs are usually determined by standard pricing per unit measure, although the prices for a given item may vary depending on the plant site. Effluent streams are usually managed for the plant as a whole, as similar treatment and disposal protocols may apply to different chemical components or several vessels may emit the same substances at various parts of the facility.

Annual operating costs also include equipment maintenance and expenses related to operator monitoring and control. These more detailed costs are typically calculated by applying a factor based on unit purchase price.

As a general rule of thumb, process development favors operations that are not energy-intensive and do not consume large amounts of chemicals. The use of chemicals, for example, raises several concerns:
- Does their use trigger special regulatory restrictions on worker exposure or disposal?
- Have they been targeted by the government for production reduction?
- Does their use pose special hazards, such as flammability or explosivity?

For example, the use of a flammable solvent in a process unit would require that its storage and operation area meet fire safety standards, thereby incurring a higher capital cost. Clearly, the bioseparation process designer must weigh both economic and operational risks in decision making and in selecting a particular bioseparation method.

18.2 ECONOMIC DECISION-MAKING MODELS

No single ideal methodology can determine the profitability of a new process investment and simultaneously account for the diverse needs of the investor and analyst. On the other hand, several methods collectively have proved useful in gauging the profitability of a new investment in the chemical process and bioprocess industries. We will discuss these methods, which are listed in Table 18.3, in terms of a sample design by Petrides and Sapidou (6).

To illustrate how each of these methods calculates costs, we first need to tabulate the cash flow of a project. We will use the results from a β-galactosidase design, shown in Table 18.4, to illustrate the implementation of these methods. After studying this introduction to profitability measures, the reader is encouraged to use available design software to perform these routine calculations; such software also allows the user to perform more sophisticated analyses that use continuous interest compounding and more complex depreciation strategies.

Table 18.3 Common Analyses Used in the Bioprocess Industry

Method	Basis	General Equation	Weakness
IRR	Uses capital investment to calculate project interest rate	$\dfrac{\text{average profits}}{\text{fixed investment} + \text{working capital} + \text{interest}} \times 100\%$	Does not account for project life
NPV	Value of investment based on present value of money	$\dfrac{\text{present worth of project}}{\text{fixed investment} + \text{working capital}}$	Does not address project magnitude
Return on investment	Maximum loan interest rate based on principal and interest of investment	$\dfrac{\text{average profit over earning life}}{\text{fixed investment} + \text{working capital}} \times 100\%$	Does not track cash flow timing
Payback period	Time required to recover original investment	$\dfrac{\text{depreciable fixed investment}}{\text{profit} + \text{average depreciation}}$	Does not consider future worth of investment

Table 18.4 Discrete Cash Flow* from a β-Galactosidase Process Design Created Using BioPro Designer (6)

Year	Capital Outlay	Sales	Operating Costs	Gross Profit	Loan Payment	Depreciation	Taxable Income	Taxes (40%)	Net After Taxes	Net Cash Flow
1	-32,609	0	0	0	0	0	0	0	0	-32,609
2	-43,478	0	0	0	0	0	0	0	0	-43,478
3	-33,445	16,509	14,963	1546	7916	10,326	-16,696	0	-6370	-39,815
4	0	33,018	16,638	16,380	7916	10,326	-1862	0	8464	8464
5	0	52,830	18,647	34,183	7916	10,326	15,941	6376	19,891	19,891
6	0	52,830	18,647	34,183	7916	10,326	15,941	6376	19,891	19,891
7	0	66,037	19,986	46,051	7916	10,326	27,809	11,124	27,011	27,011
8	0	66,037	19,986	46,051	7916	10,326	27,809	11,124	27,011	27,011
9	0	66,037	19,986	46,051	7695	10,326	28,030	11,212	27,144	27,144
10	0	66,037	19,986	46,051	7695	10,326	28,030	11,212	27,144	27,144
11	0	66,037	19,986	46,051	7695	10,326	28,030	11,212	27,144	27,144
12	0	66,037	19,986	46,051	7695	10,326	28,030	11,212	27,144	27,144
13	0	66,037	19,986	46,051	0	0	46,051	18,420	27,631	27,631
14	0	66,037	19,986	46,051	0	0	46,051	18,420	27,631	27,631
15	6271	66,037	19,986	46,051	0	0	46,051	18,420	27,631	33,902

* In thousands of dollars.

Table 18.5 Interest on Loans During the Life of the Project

	Fixed Capital	Working Capital	Total
Interest Paid	$33,471,000	$489,000	$33,960

18.2.1 Internal rate of return

As the first step in an economic analysis, we must determine the average yearly profit before and after taxes. In the β-galactosidase case given in Table 18.4, these profits are $28,106,000 and $19,122,000, respectively. The interest payments considered for this case is for a 10-year loan for 40% of the direct fixed capital ($108,696,000) at an interest rate of 12%, and a 6-year loan for 15% of the working capital ($836,000) at an interest rate of 12%. Table 18.5 gives the interest paid, based on amortization calculations.

Adding the total interest cost, the cost of working capital, and the average fixed capital investment gives a denominator of $143,492,000. With this method, the internal rate of return is 19.6% before taxes and 13.4% after taxes. Note that the internal rate of return (IRR) after taxes is actually based on the interest rate on the working capital. Such an analysis allows the investor or venture capitalist to determine whether a project offers an adequate return when compared with other investment vehicles, such as stocks. Its weakness is a failure to directly account for the lifetime of the project. This method also says little about the total magnitude of the project.

18.2.2 Payback period, including interest

Graphing the cumulative net cash flow over time can provide an estimate of how long it will take the project to pay off the initial investment. A more commonly used method for calculating the payback time is to divide the fixed, depreciable capital plus interest paid by the annual profit and depreciation. In the β-galactosidase example, this calculation becomes

$$\text{Payback} = \frac{(108,696 + 836) + 33,960}{27,144 + 10,326} = 3.8 \text{ years} \tag{18.2}$$

where the amounts are given in thousands of dollars.

18.2.3 Net present value

The time value of money is a useful measure when assessing projects with varying service lives. To calculate the net present value (NPV) of the β-galactosidase project, our analysis must incorporate the interest rate that the company can obtain in lieu of investing in the project. If stocks or other investments have a return of 8%, for example, then this interest rate is used as a discount factor to account for the time value of money. This discount factor is applied starting with the first year in which sales are generated. Table 18.6 shows the detailed calculations for the net profit after

Table 18.6 Calculations to Determine Net Present Worth for β-Galactosidase*

Year	After-Tax Cash Flow	Discount Factor (i = 8%)	Present Value	Discount Factor (i = 10%)	Present Value	Discount Factor (i = 12%)	Present Value
0							
1							
2	0						
3	−6370	0.925925926	−5898	0.909091	−5791	0.892857143	−5688
4	8464	0.85733882	7257	0.826446	6995	0.797193878	6747
5	19,891	0.793832241	15,790	0.751315	14,944	0.711780248	14,158
6	19,891	0.735029853	14,620	0.683013	13,586	0.635518078	12,641
7	27,011	0.680583197	18,384	0.620921	16,772	0.567426856	15,327
8	27,011	0.630169627	17,022	0.564474	15,247	0.506631121	13,685
9	27,144	0.583490395	15,838	0.513158	13,929	0.452349215	12,279
10	27,144	0.540268885	14,665	0.466507	12,663	0.403883228	10,963
11	27,144	0.500248967	13,579	0.424098	11,512	0.360610025	9788
12	27,144	0.463193488	12,573	0.385543	10,465	0.321973237	8740
13	27,631	0.428882859	11,850	0.350494	9684	0.287476104	7943
14	27,631	0.397113759	10,972	0.318631	8804	0.256675093	7092
15	27,631	0.367697925	10,160	0.289664	8004	0.22917419	6332
		Net Present Value	156,811		136,814		120,008
		Investment	109,532		109,532		109,532
		Net Present Worth	47,279		27,282		10,476

*Thousands of dollars, end-of-year cash flows and interest calculations used.

taxes (without depreciation, but including the loan payment) for the β-galactosidase example (6).

The NPV calculation is generally regarded as the most suitable method when the task at hand is decision making about process economics. Its strength lies in its ability to determine a project's value by taking the initial and working capital investment and transforming this investment into an interest rate; this interest rate shows the time value of money. The analysis in Table 18.6, for example, demonstrates that if we could obtain a 10% return by investing in something other than the production of β-galactosidase, then the present value of investing in the production of β-galactosidase is still $27.3 million. Thus the NPV method can help a company choose an investment that will maximize its future worth.

The main drawback of the NPV calculation is its failure to consider the magnitude of the project. For example, it does not account for the higher element of risk associated with a project that returns $26 million on a $110 million investment than that associated with a project that returns $20 million but requires only a $70 million investment.

18.2.4 Return on investment

We can also use a discounted cash flow to determine the profit value of a proposed project, compared with the original investment, by calculating the return on investment. This calculation involves a trial-and-error identification of the maximum interest rate at which money can be borrowed and still have the net cash flow over the project lifetime be equivalent to the principal and interest of the investment. The higher the return on investment, the more attractive the project. A typical manufacturing venture should have a return on investment before taxes that exceeds 20% given the relatively high risk involved.

The pertinent equations to determine the return on investment are as follows:

The future worth of the project due to the net cash flow (after taxes):

$$\begin{pmatrix} \text{Future} \\ \text{worth} \end{pmatrix} = \begin{pmatrix} \text{Year 1} \\ \text{cash flow} \end{pmatrix}(1+i)^{13} + \begin{pmatrix} \text{Year 2} \\ \text{cash flow} \end{pmatrix}(1+i)^{11} + \ldots + \begin{pmatrix} \text{Year 12} \\ \text{cash flow} \end{pmatrix}(1+i)$$

(18.3)

The future worth of the initial investment minus the working capital and salvage value of the facility:

$$\begin{pmatrix} \text{Future} \\ \text{worth} \end{pmatrix} = \begin{pmatrix} \text{Direct} \\ \text{fixed capital} \end{pmatrix}(1+i)^{13} - \begin{pmatrix} \text{Working} \\ \text{capital} \end{pmatrix} - \begin{pmatrix} \text{Salvage} \\ \text{value} \end{pmatrix}$$

(18.4)

Setting these two equations equal to one another permits us to undertake a trial-and-error determination of the return on investment. For the project described in Table 18.7, the return on investment is 13.7% after taxes. The more customary practice, however, is to consider the before-tax return on investment, which is calculated as 19.3% for the project detailed in Table 18.7. Based on standard practice, this

Table 18.7 Calculation of Return on Investment After Taxes*

Year	After-Tax Cash Flow	Growth Factor ($i = 13\%$)	Future Worth	Growth Factor ($i = 14\%$)	Present Value	Growth Factor ($i = 13.7\%$)	Present Value
0							
1							
2	0						
3	−6370	4.335	−27,611	4.818	−30,690	4.668	−29,735
4	8464	3.836	32,467	4.226	35,771	4.105	34,749
5	19,891	3.395	67,520	3.707	73,739	3.611	71,821
6	19,891	3.004	59,752	3.252	64,683	3.176	63,167
7	27,011	2.658	71,808	2.853	77,052	2.793	75,445
8	27,011	2.353	63,547	2.502	67,590	2.457	66,355
9	27,144	2.082	56,513	2.195	59,580	2.161	58,646
10	27,144	1.842	50,011	1.925	52,264	1.900	51,579
11	27,144	1.630	44,258	1.689	45,845	1.671	45,365
12	27,144	1.443	39,166	1.482	40,215	1.470	39,898
13	27,631	1.277	35,282	1.300	35,909	1.293	35,720
14	27,631	1.130	31,223	1.140	31,499	1.137	31,416
15	27,631	1.000	27,631	1.000	27,631	1.000	27,631
		Future Worth	551,566		581,088		572,057
		Future Worth of Investment	529,382		594,488		574,229
		Ratio	1.04		0.98		1.00

*Thousands of dollars, end-of-year cash flows and interest calculations used.

project would not be very desirable because it falls below the cutoff of 20% return before income taxes.

18.2.5 Choosing among projects and alternative investments

Using the four profitability measures described in the preceding sections, we could potentially obtain conflicting results when comparing different projects. For example, while one project might have the highest net present value, another project might yield a shorter payback period. When this situation arises, which project should we select?

When these methods yield conflicting information, the net present value calculation is often used as the tie-breaker because of its underlying philosophy—that is, its emphasis on the maximization of future worth. This analytical method incorporates the notion that alternative investments can provide an incremental rate of return. Consider a case in which one project requires a $100,000 investment and returns an annual profit of $25,000, and a second project requires a $200,000 investment and offers an annual profit of $35,000. With the net present value method, the project requiring a $100,000 investment would be preferred. The reason—the smaller investment yields a 25% return but the larger investment returns only 17.5%. Moreover, the incremental expense of $100,000 yields only $10,000 more per year, giving an incremental rate of return of 10%. It would be more prudent to take the extra $100,000 and invest it in another project offering a return greater than 10%. As this simple example demonstrates, we prefer that investments providing the required rate of return use the minimum amount of initial capital, unless the project has specific advantages that the economic analysis does not take into account.

18.3 SENSITIVITY ANALYSES

Many assumptions are made within the context of process economic calculations, especially in the early design stages. Sometimes these assumptions may lead to inaccurate forecasts. Although estimates of a project's overall profitability may depend on speculations about factors that are either not well known or beyond the designer's control, careful design and data collection can often overcome these problems. More specifically, they may allow for good decision making through sensitivity analyses.

A sensitivity analysis begins with a base case and then conducts a "What if?" scenario to analyze the complex interplay among process units in a bioseparation process. Importantly, it identifies economic trade-offs required in the manipulation of process variables. Sometimes minima or maxima in cost curves are difficult to predict a priori without a good design simulation of the entire process. A sensitivity analysis also helps to determine the economic impact of changing process variables by adding or subtracting individual units.

We can also produce substantial savings in batch processes with staggered schedules (7). When a new batch of cells is multiplying in the bioreactor, the downstream process vessels should not remain empty, simply waiting for the product to be generated. Instead, downstream process vessels and the bioreactor should operate nearly continuously, with minimum turnaround time. (Table 18.8 defines batch-processing terms.)

Example 18.1

In this example, we will examine the effect of ion-exchange column capacity on the economics of insulin recovery. Here we are concerned with an insulin process design from SuperPro Designer (2). The separation process includes one hydrophobic interaction column, one ion-exchange column, a reverse-phase HPLC column, and a gel filtration column. This column-intensive separation and purification process is a high-cost operation.

Figure 18.1 shows a portion of the process flow diagram of the hydrophobic interaction column step. This step is preceded by diafiltration and followed by an enzymatic step that cleaves proinsulin into insulin α and β. Elution and regeneration streams feed the C-102 HIC columns and are used in the cyclic batch operation of the column. Using the process and economic simulation information

Table 18.8 Scheduling Terms in Batch Processing

Term	Definition
Process time	Number of continuous hours in which the vessel is processing the product stream
Turnaround time	Number of continuous hours in which the vessel is being flushed, chromatography columns are being regenerated, and membrane modules are being cleaned in preparation for the process time period
Number of cycles	Whole number that gives the number of cycles that can be completed within one year

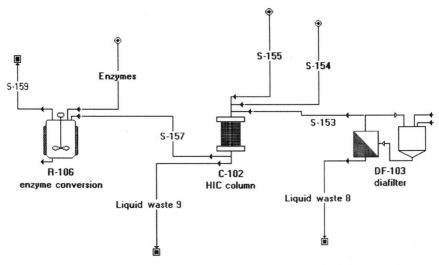

Figure 18.1 Section of the insulin flow diagram showing the HIC step and neighboring units.

provided, we can investigate how the capacity of the hydrophobic interaction column affects process costs.

Table 18.9 highlights the capacities studied and the resulting cost data generated by such a simulation. To design the column, we fixed its length and then varied its diameter, being careful not to exceed a specified limit. Fixing the length is a common ploy in column design, as it keeps the pressure drop constant and thereby prevents the column packing from being crushed by overpressurization. The number and size of columns is determined directly from the capacity available. In this sensitivity analysis, the column is run in a batchwise manner and constant column space velocities are maintained during the process operation, elution, and washing.

In addition to providing unit-specific data, the process economic simulation can generate cost and profitability information to help guide decision making. For instance, if increasing the capacity in this column has a substantial impact on the overall process economics, it may be worthwhile to direct research and development dollars into a search for higher-capacity column packings.

Consider the example in Figure 18.2, which shows how HIC column capacity

Table 18.9 Information Used in Developing Cost Curves for C-102 HIC Column

Proinsulin Capacity (mg/mL)	Number of Column Units	Capital Cost ($1000)	Packing Cost ($1000/year)
Base capacity: 20	8	2016	1736
10	15	3840	3472
15	10	2560	2314
25	6	1536	1389
30	5	1280	1157
40	4	1008	868

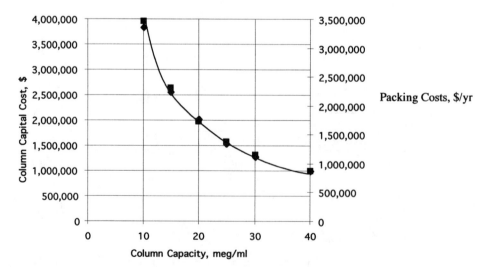

Figure 18.2 Effect of column capacity on the purchase and packing costs for a hydrophobic interaction column used to purify proinsulin.

affects the annual operating cost with depreciation for an insulin plant design that produces 1500 kg/year of insulin. To evaluate the effect of column capacity on the overall plant economics, the simulation should consider batch scheduling as a strategy for minimizing unit operator hours (which in turn minimizes labor costs) and providing economies of scale.

Example 18.2

In this example, we compare costs for liquid extraction and calcium precipitation of citric acid. These separation methods for recovering citric acid were discussed in earlier chapters.

The conventional method relies on precipitation of citrate using lime, followed by acidification to obtain the free acid form of citric acid. A by-product of the conventional separation is gypsum ($CaSO_4$). A portion of this process is illustrated in Figure 18.3.

An alternative method uses solvent extraction to partition the free acid form of citric acid, as shown in Figure 18.4. Back-extraction then recovers the purified acid, which is subjected to subsequent crystallization to obtain its final form.

Table 18.10 lists capital and operating costs associated with each of the two separation methods, starting with the tank used to hold the citric acid stream as it exits the fermenter. Although the solvent extraction method requires a smaller capital investment, it incurs a higher operating cost primarily because of its waste disposal costs.

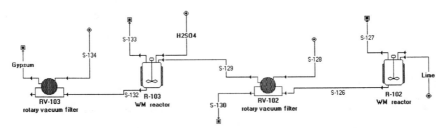

Figure 18.3 A portion of the conventional citric acid separation step showing the lime addition and gypsum formation steps.

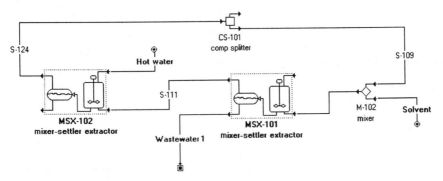

Figure 18.4 Solvent extraction steps for citric acid recovery.

Table 18.10 Capital and Operating Costs Associated with the Conventional and Alternative Methods for Recovery of Citric Acid

Cost	Conventional Method (Lime Precipitation)	Alternative Method (Solvent Extraction)	ΔCost ($ or $/year)
Capital equipment	$1,664,000	$1,079,000	$585,000
Operating labor	$62,726	$31,644	$31,082
Raw materials	$1,284,000	$818,000	$466,000
Waste treatment	$2,650,000	$4,438,000	−$1,788,000
Utilities	$172,000	$294,000	−$122,000
Annual costs	$4,168,726	$5,581,644	−$1,412,918

18.4 SUMMARY

Bioseparation process development is guided and evaluated in terms of both economic and social factors. Of the two measures, economics is the more straightforward. This chapter introduced basic concepts and equations used in economic decision making. Although no single economic model can suffice to guide decision making, some companies tend to rely more heavily on one model over others based on the past experience in a particular industry.

Software represents a welcome new addition to process developers' cost estimation toolkits. It can provide rapid flow-sheet creation and built-in cost curves and estimation methods. The reader is strongly encouraged to learn about these tools and, if possible, to use them to solve the examples and problems given in this chapter. More in-depth instructional materials and textbooks on cost estimation (including those listed in Section 18.6) will prove valuable in building an expertise in this important component of bioseparation process development.

18.5 PROBLEMS

18.1 Graph the cash flow for the β-galactosidase example provided in Table 18.4. Provide detailed descriptions of the major features shown in the graph.

18.2 Assume that in the production of β-galactosidase, the selling price of the enzyme increases in year 6 by 15% and remains at this higher price through year 15. How does this change affect the economic model predictions (i.e., internal rate of return, return on investment, net present value, and payback time)? Show detailed calculations and provide a table summarizing your results.

18.3 Producing 1000 kg/year of a new protein therapeutic requires a capital investment of $205 million. Assuming that the operating costs are approximately $53 million per year, what should be the selling price of this therapeutic to make this project worthwhile? Make assumptions about interest rates, capital outlay rates, depreciation, and any other values necessary based on the β-galactosidase example.

18.4 Your company must decide between two projects. One project involves the manufacture of a generic therapeutic agent, and the other deals with the manufacture of a new therapeutic that is currently in the last stages of clinical trials. The new therapeutic could potentially cure a specific subgroup of patients with a terminal

Table 18.11 Vessel Process and Turnaround Times

Vessel Number	Process Time (h)	Turnaround Time (h)
1	6	1
2	5	1.5
3	8	2

illness. What criteria do you suggest should be used in gauging the relative merits of these investments as compared with one another, and as compared with stocks and bonds? Tabulate the criteria and list methods for assessing their social and economic value. How would you factor your corporation's marketing strategy into the economic decision-making process?

18.5 Why does the solvent extraction process for separating citric acid in Example 18.2 (Table 18.10) incur such high waste treatment costs? What process modification(s) would you suggest to lower this cost?

18.6 Based on the data in Table 18.10, develop an economic model using the assumptions given in the β-galactosidase example that can determine which separation process is preferable. Assume that citric acid is sold for $3.55/kg and that 12,000,000 kg are made per year. The fermentation capital and operating costs are approximately $11 million and $7.5 million per year, respectively.

18.7 Given a process that contains three vessels with the process and turnaround times listed in Table 18.11, if each vessel has the same liquid volume capacity, how many vessels would you need to maintain continuous processing? If all of the vessels have the same cost, and the annual cost of each hour of downtime (when production is halted) is equal to 0.25 of the total vessel cost, what would you recommend for the process design? Now assume that vessel 2 costs twice as much as the other vessels and that its downtime cost is 1.4 times that for vessel 1 or 3. What would be the most cost-effective design?

18.6 REFERENCES

1. http://www.aspentech.com
2. http://www.intelligen.com
3. Peters MS, Timmerhaus KD. Plant design and economics for chemical engineers. New York: McGraw-Hill, 1991.
4. Perry RH, Green DW, eds. Perry's chemical engineers' handbook, 7th ed. New York: McGraw-Hill, 1997.
5. Ulrich GD. A guide to chemical engineering process design and economics. New York: Wiley, 1984.
6. Petrides DP, Sapidou ES, Calandranis J. Optimization of protein recovery using computer-aided process design tools. Biotech Bioeng 1995;48:529.
7. Schweitzer PA, ed. Handbook of separation techniques for chemical engineers. New York: McGraw-Hill, 1997.

Appendix A

The Laplace Transform

Without a doubt, the Laplace transform is the most useful tool for solving differential equations. It is a relatively simple technique, although it does require some rigor in its use of algebra.

In a nutshell, the Laplace transform involves three steps:

1. Applying an integral equation to each term in the differential equation.
2. Solving the "reduced" differential or algebraic equation for the independent variable
3. Finding the inverse Laplace transform of the resulting expression

This book does not provide a mathematical proof of this technique. Instead, Appendix A will discuss the mechanics of applying the Laplace transform method. Appendix B describes the calculation of the inverse transform so as to obtain the final answer with a spreadsheet.

To obtain a solid working knowledge of the Laplace transform method, it is best to take a detailed approach first. Later, you can use the shortcuts provided by the tabulation of common transforms and inverse transforms. To ensure that you become fully comfortable with the Laplace transform method, however, we strongly recommend repeated use of the detailed approach before resorting to the shortcut so that you will be fully adept at solving equations in any circumstance.

The crux of the Laplace transform method is the multiplication of an integral expression,

$$\int_0^\infty e^{-st} dt \tag{A.1}$$

to each term in a differential equation. We then evaluate the resulting integrals and apply the limits. This step creates an equation in "Laplacian space"—that is, the equation is reduced in complexity. Partial differential equations (P.D.E.) are reduced to ordinary differential equations (O.D.E.), or an O.D.E. is reduced to an algebraic equation. The equation in Laplacian space is subsequently solved and transformed back to normal space by a process known as obtaining the inverse transform. Often, we obtain the inverse simply by looking up each term in Laplace transform tables. On some occasion, however, the fastest way to obtain the inverse is to use a simple numerical inversion method, described in Appendix B.

Now that we have examined the Laplace method in general terms, we will study specific uses of this tool. A simple example for transforming a first-order O.D.E. is provided below.

For the first-order ordinary differential equation

$$\frac{dy}{dt} + 3t + 5 = 0 \tag{A.2}$$

we begin the Laplace transform method by multiplying each term in the equation by the integral expression in Equation A.1. Taking the first term and multiplying by the integral expression gives

$$\int_0^\infty \frac{dy}{dt} e^{-st} dt \tag{A.3}$$

To solve this integral, we apply the chain rule:

$$\int u \, dv = uv - \int v \, du \tag{A.4}$$

Letting $u = e^{-st}$ and $dv = \frac{dy}{dt}$, we can evaluate Equation A.3:

$$\int_0^\infty \frac{dy}{dt} e^{-st} dt = \left[y e^{-st} + s \int y e^{-st} dt \right]_0^\infty \tag{A.5}$$

The integral on the right in Equation A.5 defines the dependent variable in Laplacian space. By convention, this definition is

$$\overline{Y} = \int_0^\infty y e^{-st} dt \tag{A.6}$$

Applying the limits to the remaining term $y e^{-st}|_0^\infty$ yields

$$\int_0^\infty \frac{dy}{dt} e^{-st} dt = 0 - y(0) + s\overline{Y} \tag{A.7}$$

because $e^{-\infty} = 0$ and $e^0 = 1$, and the dependent variable y must be evaluated at $t = 0$.

The next step is to transform the term $3t$ using the chain rule:

$$\int_0^\infty 3t \, e^{-st} dt = -\frac{3}{s} \left[t e^{-st} + \frac{1}{s} e^{-st} \right]_0^\infty \tag{A.8}$$

You can test your understanding by deriving Equation A.8 yourself. Now L'Hopital's rule (1) is needed to evaluate the first term in the right-hand side because, when we apply the infinity limit, we do not obtain a definite answer:

$$\lim_{t \to \infty} t \, e^{-st} = \infty \cdot 0 \tag{A.9}$$

L'Hopital's rule is applied to this particular term in the following way:

$$\lim_{t \to \infty} \frac{t}{e^{st}} = \lim_{t \to \infty} \frac{\frac{d}{dt}(t)}{\frac{d}{dt}(e^{st})} = \lim_{t \to \infty} \frac{1}{s \, e^{st}} \tag{A.10}$$

This rule is useful in obtaining Laplace transforms the long way. Explanations and examples of this rule can be found in any calculus textbook.

Now we can evaluate t:

Table A.1 The Transforms of Terms in Equation A.2

Term	Laplace Transform
$\dfrac{dy}{dt}$	$s\overline{Y} - y(0)$
$3t$	$\dfrac{3}{s^2}$
5	$\dfrac{5}{s}$
0	0

$$\int_0^\infty 3t\, e^{-st} dt = \frac{3}{s^2} \tag{A.11}$$

The last term that needs to be transformed is the constant 5:

$$\int_0^\infty 5e^{-st} d = \frac{5}{s} \tag{A.12}$$

Table A.1 summarizes the results we obtained using the Laplace transform method. Note that the Laplace transform of zero is zero. Combining all of these transformed parts gives the following algebraic equation:

$$s\overline{Y} - y(0) + \frac{3}{s^2} + \frac{5}{s} = 0 \tag{A.13}$$

Equation A.13 can be solved for the Laplace transformed dependent variable $\overline{Y}$ after we have established an initial condition for $y(0)$. Using the initial condition that $y(0) = 0$,

$$\overline{Y} = -\left(\frac{3}{s^3} + \frac{5}{s^2}\right) \tag{A.14}$$

The only task left is to obtain the inverse transform so as to frame our answer in terms of y. Thankfully, many sources provide tables in which a function and its transform are given side by side (2, 3). These tables provide easy ways to transform and invert equations, term by term. Table A.2 should suffice for solving many problems. References providing more extensive tabulations appear at the end of this appendix.

Using Table A.2, we can solve Equation A.2:

$$y = -\frac{3}{2}t^2 - 5t \tag{A.15}$$

The reader should take time to study how we obtained this result and verify it by proving that it satisfies the equation and the initial condition $y(0) = 0$. Appendix B uses this example to illustrate how inverse transforms can be obtained numerically when an equation cannot be inverted after solving for its transform.

The last two entries in Table A.2 give Laplace transforms for the impulse func-

Table A.2 Selected Laplace Transforms

Function	Laplace Transform
t^n (for $n > -1$)	$\dfrac{n!}{s^{n+1}}$
1	$\dfrac{1}{s}$
$\dfrac{t^{n-1}}{(n-1)!}$ (for $n > 0$)	$\dfrac{1}{s^n}$
$\dfrac{t^{n-1}e^{-at}}{(n-1)!}$ (for $n > 0$)	$\dfrac{1}{(s+a)^n}$
$\dfrac{1}{(a-b)}\left(e^{-bt}-e^{-at}\right)$	$\dfrac{1}{(s+a)(s+b)}$
$\sin(at)$	$\dfrac{a}{(s^2+a^2)}$
$e^{-bt}\sin(at)$	$\dfrac{a}{(s+b)^2+a^2}$
$\delta(t)$	1
$u(t)$	0 for $t < 0$ 1 for $t > 0$

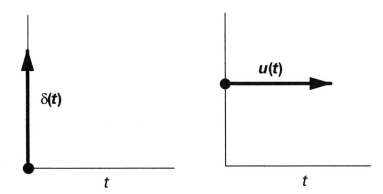

Figure A.1 Impulse function $\delta(t)$ and unit step function $u(t)$.

tion $\delta(t)$ and the unit step function $u(t)$. Figure A.1 graphs these two functions as a function of time. The impulse function demonstrates how an infinitesimally short disturbance of infinite height propagates through a model system. For example, this function can model the rapid injection of a solute in a column or a quick burst of sound. In contrast, a unit step function measures the propagation of a change in set point through a system. Such functions are used, for example, to determine how long an ion-exchange column can continue to remove ions before it requires regeneration, assuming that the feed concentration remains constant and the ion-exchange bed does not initially contain any of the ions to be removed from the feed.

REFERENCES

1. Swolkowski E. Calculus with analytical geometry. Boston: Prindle, Weber and Schmidt, 1977.
2. Oberhettinger F, Badii L. Tables of Laplace transforms. Berlin: Springer-Verlag, 1973.
3. Burrington RS. Handbook of mathematical tables and formulas, 5th ed. New York: McGraw-Hill, 1973.

Appendix B

Numerical Inversion, van der Laan's Theorem, and Huchel and Helmholtz-Smoluchowski Equations

Often the use of Laplace transforms for solving an equation leaves a solution that cannot be readily converted into "normal" space. In such a situation, a rather simple numerical inversion technique can be readily applied using a spreadsheet.

Although many other inversion techniques are available, none rivals the simplicity of the method first developed by Gaver (1), subsequently developed into a computer algorithm by Stehfest (2), and applied to the analysis of radial dispersion in a porous media by Moench and Ogata (3). This method involves the calculation of the following equation:

$$F(t) = \frac{\ln(2)}{t} \sum_{i=1}^{N} V_i \overline{Y} \left(\frac{i \ln(2)}{t} \right) \tag{B.1}$$

The coefficients V_i are determined using Equation B.2:

$$V_i = (-1)^{\left(\frac{N}{2} + i \right)} \sum_{k=(i+1)/2}^{\text{smaller}(i,N/2)} \left[\frac{k^{N/2}(2k)!}{(N/2 - k)! \, k! \, (k-1)! \, (i-k)! \, (2k-i)!} \right] \tag{B.2}$$

Although these equations are compact and it is easy to determine their values with a hand-held calculator or a spreadsheet, the symbols can sometimes be confusing. It is important that the properties of the parameters used in these two equations be appreciated. For the reader's convenience, they are summarized in Table B.1.

In the following discussion, we provide a numerical inversion based on the equation transformed in Appendix A. Although we can perform the inversion of this transform without resorting to numerical techniques, we will nevertheless use this equation to illustrate how Stehfest's formula is calculated and demonstrate the accuracy of this technique. From Appendix A, we are interested in calculating the inverse of:

$$\overline{Y} = -\left(\frac{3}{s^3} + \frac{5}{s^2} \right) \tag{B.3}$$

Using $N = 6$, the values for i range from 1 to 6. In the first calculation, we determine the values of V_i. For this procedure, it is useful to construct a table to determine the limits for Equation B.2. Remembering that k is calculated with integer arithmetic, we can construct Table B.2.

Based on Table B.2, V_1, V_5, and V_6 have one term each, while V_2, V_3, and V_4 consist of two terms each when evaluating the summation. Table B.3 gives a sample spreadsheet result for $t = 1$. The reader is encouraged to duplicate these results to verify his or her understanding of how Stehfest's formula is used. Calculating the results

Table B.1 Important Properties of Parameters and Limits Used in Stehfest's Formula

Parameter/Limit	Propery
t	Time or other dependent variable tansformed
i	An integer with values between 1 and N
N	N *must* be an even integer; usually a value of 10 is sufficient
k	k is calculated using *integer arithmetic;* a value of $k = 1.5$ becomes $k = 1$ when the fractional portion is deleted
V_i	The sum of V_i *must* equal 0 for a given value of N; V_i depends only on N and is calculated once
Smaller $(i, N/2)$	The upper limit of the summation in Equation B.2 dictates whether i or $N/2$ is used as the upper limit based on which is the smaller number
$\overline{Y}\left(\dfrac{i\ln(2)}{t}\right)$	Calculated by substituting $i\ln(2)/t$ for s in the Laplace transform

Table B.2 Determination of Limits for V_i When $N = 6$

i	$k = (i + 1)/2$	Smaller $(i, N/2)$
1	1	1
2	1	2
3	2	3
4	2	3
5	3	3
6	3	3

Table B.3 Determination of the Numerical Inverse of $\overline{Y} = -\left(\dfrac{3}{s^3} + \dfrac{5}{s^2}\right)$ Through Stehfest's Formula Using $N = 6$ for $t = 1$

t	N	i	k	$N/2$	V_i	P_{ai}	V_i*P_{ai}	F_a
1	6	1	1	3	1	−19.41518703	−19.41518703	−6.577231435
1	6	2	1	3	−49	−3.727753991	182.6599456	
1	6	3	2	3	366	−1.489958401	−545.3247749	
1	6	4	2	3	−858	−0.791183152	678.8351446	
1	6	5	3	3	810	−0.488340533	−395.5558319	
1	6	6	3	3	−270	−0.330784313	89.31176444	

for $t = 1$ through $t = 4$ and comparing the results with the exact inverse in Table B.4 illustrate that we can approximate these results within a small percentage error. The error can be decreased by increasing the value of N. N values greater than 18 or so are not recommended, however—after a certain point, rounding errors can lead to increasingly worse approximations.

VAN DER LAAN'S THEOREM

Van der Laan developed a useful method for determining moments of an equation in Laplacian space without inverting the equation (4). The moments are easily calculated with differential calculus.

The residence time or first moment is given by the following equation:

Table B.4 Comparison of Stehfest's Formula Calculation of the Inverse Laplace Transform for $N = 6$

t	$y = -\dfrac{3}{2}t^2 - 5t$	$F(t) = \dfrac{\ln(2)}{t}\displaystyle\sum_{i=1}^{4} V_i \left[-\left(\dfrac{3}{\left(\dfrac{i\ln(2)}{t}\right)^3} + \dfrac{5}{\left(\dfrac{i\ln(2)}{t}\right)^2} \right) \right]$	%Error
1	−6.5	−6.58	1.2%
2	−16	−16.33	2.1%
3	−28.5	−29.26	2.7%
4	−44	−45.36	3.1%

$$\mu = -\left(\frac{1}{C_0}\right)\lim_{s\to 0}\frac{\partial \overline{C}}{\partial s} \tag{B.4}$$

The variance, or second moment, is calculated by using the first moment and obtaining the second derivative with respect to s:

$$\sigma^2 = \left(\frac{1}{C_0}\right)\lim_{s\to 0}\frac{\partial^2 \overline{C}}{\partial s^2} - \mu^2 \tag{B.5}$$

When the solution to the inverted equation is symmetric, we can determine the moments and write a Gaussian equation as the inverted solution.

HUCKEL AND HELMHOLTZ-SMOLUCHOWSKI EQUATIONS

The Huckel equation for mobility is given by the following equation:

$$u = \frac{\varepsilon\zeta}{6\pi\eta} \tag{B.6}$$

The Helmholtz-Smoluchowski solution is defined as follows:

$$u = \frac{\varepsilon\zeta}{4\pi\eta} \tag{B.7}$$

The Huckel equation is applicable when the product $\kappa R < 0.1$, while the Helmholtz-Smoluchowski solution is valid for $\kappa R > 100$. These equations are derived from very different assumptions of the radius of curvature of the particles as compared to the thickness of the electrical double layer. For an in-depth derivation of these equations, see reference (5).

REFERENCES

1. Gaver GP. Observing stochastic processes and approximate transform inversion. Operations Res 1966;14:444–459.
2. Stehfest H. Algorithm 368: numerical inversion of Laplace transforms. Comm ACM 1970;13:47–49.
3. Moench AF, Ogata AA. Numerical inversion of the Laplace transform solution to radial dispersion in a porous medium. Water Resources Res 1981;17:250–252.
4. Ruthven DM. Principles of adsorption and adsorption processes. New York: John Wiley and Sons, 1984.
5. Hiemenz PC. Principles of colloid and surface chemistry. New York: Marcel Dekker, 1977.

INDEX

Note: Page numbers followed by *t* indicate tables; those followed by *f* indicate figures.

Abegg-Stevens-Larson
    equation, 131, 144
Absorbance detector, 180t
Absorption, rapid, 373
Acacia, 374
Acceptor numbers (AN), 86
Acetic acid
    adsorption data of, 371t
    dissociation constant for, 75t
    extraction of, 273
    ion-exclusion separation of,
        212–213
    recovery of, 370
    retention time of, 91t
    solvent extraction data of,
        371t
Acetone, partition coefficients
    for, 232t
Acid dissociation constant,
    74–75
Acid-base interactions, 67
    in organic-aqueous
        extraction, 234, 236
Acid-base scales, 84–88
    comparison and correlation
        of, 88
    in extractant selection, 236
Acids. *See also* Amino acids;
        Lewis acids; *specific
        acids*
    complexation of, 15
    hard, 85–88, 110
    organic
        extraction of, 273
        in ion-exchange, 15
    polarizability of, 87
    soft, 85, 86–88, 110
    weak, 74
Acrylamide, polymerized, 280
Additives, 367
    to maintain enzyme activity,
        368
Adenine, metal ion binding in,
    89
Adenine-thymine pairs,
    117–118
5′-Adenosine monophosphate,
    186t
Adsorbents, 53–54
Adsorption, 15, 23
Adsorption chromatography.
        See Chromatography,
        adsorption
Affinity chromatography. *See*
        Chromatography,
        affinity

Affinity ligands, 306
    in affinity chromatography,
        310
Agar, 374
Agarose, 184t
    molecular structures of, 278f
Agarose gel electrophoresis, 45
Agarose/dextran, 184t
Alamine 336, 370
Alanine
    dissociation constant of, 77t
    divalent metal ion stability
        constants with, 90t
    electrodialysis of, 34
    hydrophobicity of, 84t
    hydrophobicity scale for, 85t
    isoelectric point of, 83t
    temperature and solubility of,
        72f
β-Alanine, isoelectric point of,
    83t
Albumin
    isoelectric point of, 83t
    in virus vaccines, 18
Alcohols, partition coefficients
    for, 232t
Alumina, 53
Amines, temperature and
    extraction of, 237
Aminex-14, 41
Amino acids
    abbreviations for, 30t
    analysis techniques for, 29–35
    clusters of, 209
    compounds used to label, 31f
    depronation ratios of, 76–78,
        81–82
    dissociation constants of, 77t
    distribution coefficients for,
        322t
    divalent metal ion stability
        constants with, 90t
    grouping by hydrophobicity,
        84t
    groups according to side-
        chain or R-group, 76
    HFB derivatives of, 35
    high-performance liquid
        chromatograph of, 60f
    hydrophobicity scales for, 85t
    isoelectric points and charge
        of, 76–82, 83t
    labeling of, 32
    metal ion binding to, 89–90

negative charge of, 79
net charge of, 77–78, 79–80,
    81f, 82f
solubility of, 70–72
trifluoroacetyl alkyl esters of,
    35
UV absorbance spectra of,
    30f
Amino-bonded silica columns,
    47
Ammonium carboxylate
    perfluoropolymer, 246
Ampholine, 46
    removal of, 38
Amyl acetate, solubility of, 275
Analysis tools, 1
Analytic techniques, 57–61. *See
    also specific techniques*
    for amino acids, 29–35
    for biochemicals and
        biopolymers, 29–61
    for carbohydrates, 46–51
    for lipids, 51–56
    for nucleic and polynucleic
        acids, 41–46
    for peptides and proteins,
        35–41
    for steroids and antibiotics,
        56–57
    for vitamins, 56–57
Anion-exchange
        chromatography. *See*
        Chromatography,
        anion-exchange
Anion-exchange resins, 41
Ankyrin, 338
Antibiotics
    analysis techniques for, 56–57
    distribution coefficients for,
        232t, 322t
    extraction of, 316
    recovery of, 271
Antibody
    detection of, 40–41
    ELISA of, 40–41
    heavy and light chain
        regions, 116
    radioimmunoassay of, 41
    structure of, 116–117
Antibody-antigen complex, 40
Antibody-antigen interactions,
    116–117
    affinity, 114–115
        dissociation constant of,
            113t

409

Antifoaming agents, 366
Antigens
    binding studies of, 116–117
    ELISA of, 40–41
    epitopes of, 367
    radioimmunoassay of, 41
Antioxidants, 374
Aqueous colloids, 95
Aqueous extraction, two-phase,
    238–243
Aqueous layer, formation of,
    96
Aqueous solution
    chemical equilibria in, 67–69
    phenomena of in biological
        systems, 67
Aqueous two-phase magnetic
    separation, 311–312
Aqueous-aqueous extraction,
    two-phase, 316
Aqueous-phase reaction, 238
Arginine
    dissociation constant of, 77t
    divalent metal ion stability
        constants with, 90t
    hydrophobicity of, 84t
    hydrophobicity scale for, 85t
    isoelectric point of, 83t
    net charge of, 82f
Ascorbic acid
    dissociation constant for, 75t
    retention time of, 91t
Asparagine
    dissociation constant of, 77t
    divalent metal ion stability
        constants with, 90t
    hydrophobicity of, 84t
    hydrophobicity scale for, 85t
    isoelectric point of, 83t
    temperature and solubility of,
        72f
Aspartic acid
    dissociation constant of, 77t
    divalent metal ion stability
        constants with, 90t
    hydrophobicity of, 84t
    hydrophobicity scale for, 85t
    isoelectric point of, 83t
    net charge of, 81f
    production process for,
        360–361
    temperature and solubility of,
        72f
Aspergillus niger, in citric acid
    production, 22
Autochromic methods, 39
Avidin
    interactions of, 121–122
    use and structure of, 186t
Azeotrope, liquid-vapor

compositions for, 320f
Azeotropic mixtures, 319

Back-extraction processes, 238,
    275, 369
    in organic solvent extraction,
        237
Bacteria
    cell composition of, 352t
    cell walls of, 338–339
    fermentation of, 370
Baffle towers, 248
Band dispersion, 288–290
Base case, 386, 394
Base complexation, 15
Base pairing, 117–118
    specificity of, 118
Bases. *See also* Lewis bases
    characteristics of, 84
    hard and soft, 85, 86–88
    ion-exchange
        chromatographic
        separation techniques
        for, 41–42
    polarizability of, 87
    separation of, 43
Basic fibroblast growth factor,
    formulation of,
        382–383
Batch distillation. *See*
    Distillation, batch
Batch filtration. *See* Filtration,
    batch
Batch processing, staggered
    scheduling of, 394–395
Batch-basket centrifuge,
    172–173
Bead mill, 340, 346–347
Bench-scale laboratory
    crystallization, 127
Bentonite, 374
Benzene, in lipid dissolution,
    350
Benzene-toluene system,
    temperature-
    composition plot for,
        320f
β-ribbon domain, 118, 119
BHT, 374
Binders, 373–374
Binding cycles, 200–203
Binding energy, interactions
    contributing to, 108
Binding-elution cycles, 200–201,
    204, 217
    in affinity chromatography,
        213–214
Bioactivity, 372
    analysis of, 59
    in concentration

determination, 29–61
Bioaffinity, 108–124. *See also*
    Chromatography,
    bioaffinity
    definition of, 108
    interactions of, 108–123
        dissociation constants of,
            113t
    selectivity of, 123–124
BioAspen, 385
Bioassay, 180
Biochemicals
    analysis techniques for, 29–61
    removing from organic
        phase, 237–238
    uncharged low-molecular-
        weight, diffusivity of,
            73
Biocolloids, 95, 105–106
    forces of, 92, 104–105
    long-range interactions of,
        96–104
    short-range interactions of,
        95–96
Biocompatibility, polymer, 184
Biological component sizes, 15f
Biological compounds,
    distribution
    coefficients for, 322t
Biological interactions, 104
Biological properties, 14–16
Biomolecules
    factors affecting partitioning
        behavior of, 241–243
    thermodynamic and
        transport properties
        of, 67–93
Biopolymer analysis
    techniques, 29–61
BioPro Designer, 390t
Bioprocess industry, common
    analyses in, 389t
Bioprocessing
    clean, 368
    continuous, cost analysis of,
        399
    economics of, 385–399
    flow diagrams for, 12
    operating cost estimation of,
        388–389
Bioproducts, environmental
    impact of, 367–368
Bioreactor fluids, 357
    properties of, 14–16
    scale, concentration and price
        of, 13–14
    sizes of, 14–15
    spiked, 363–364
Bioreactor process
    concurrent with

bioseparation, 362–364
integration with
    bioseparation, 364–365
minimizing inhibition
    mechanisms in, 362
physical conditions in, 14
Bioseparation processes. *See
    also* Industrial
    bioseparation
    processes; Separation
    processes; *specific
    processes*
analysis tools for, 1
bioreactor step in, 366–367
concurrent with bioreactor
    process, 362–364
design and economics
    software for, 385
expert systems in, 364
final product formulation and
    environmental impact
    of, 367–368
heuristics of, 357–362
industrial, 11–28
integrated, 357–371
    with bioreaction steps,
        364–365
for lactic acid, 25f
magnetic, 299–314
mathematics in, 1
minimizing inhibition
    bioreactor
    mechanisms in, 362
popular chromatographic
    methods for, 181t
principles of, 11
product concentrations in,
    24t
product properties of, 14–16
reducing volume early in,
    358
resolving components early
    in, 361
selection of, 11–16, 23
Biostream separator, 297f
Biotin
    analysis of, 59f
    separation of, 57–58
Biotin-avidin/streptavidin
    interactions, 121–122
Biotin-binding domains, 367
Bond energies, 95, 96t
Bottle centrifugation, 168
Bovine serum albumin, 83t
Breakthrough curve, 202f, 204,
    215f
    width of, 217
Brownian motion, 96, 97
Buchner funnel filtration, 168
Buffer salts, mobile-phase, 211

n-Butanol, 232t
n-Butyric acid, 75t

Calcium precipitation, 397–398
Calmodulin, human, 83t
Canavalia ensiformis, 51
Capillary electrophoresis,
    281–282, 294
    in amino acid analysis, 32–33
    of oligosaccharides, 49
    system of, 50f
Capillary isotachophoresis, 45
Capillary zone electrophoresis,
    281–282
    in amino acid analysis, 33
    of fatty acids, 55–56
    for nucleic and polynucleic
        acids, 45–46
Capital cost estimation,
    386–388
Carbapenem, 57t
Carbohydrate-binding domains,
    367
Carbohydrates
    analysis techniques for, 46–51
    binding of, 122
Carbon, activated, 53
Carbon dioxide
    phase diagram for, 245f
    supercritical, 275
    supercritical fluid extraction
        with, 245
Carbon-13 nuclear magnetic
    resonance,
    monosaccharide,
    46–47
Carboxylic acids, 370–371
    adsorption of, 53–54
    capillary zone electrophoresis
        of, 282
    dissociation constants for, 75t
    low-molecular weight,
        separation of, 55–56
    partition coefficients for, 232t
    retention times of, 91t
    separation of, 51–52
Cash flow
    calculation of, 389, 390t
    cumulative net, 391
    graphing, 398
Catalase recovery steps, 27f
Cation exchange, 209–212
    in monosaccharide analysis,
        47
    in sodium form, 42f
Cation-exchange product
    isolation, 361, 362f
Cavitation
    devices for, 345
    in shock wave disruption, 345

ultrasonic, 343–345
Cavitation unloading, 343–344
Cebacron blue, 186t
Cell disruption techniques, 23,
    337, 339–352
    alternative, 366
    chemical, 348–350, 351
    debris released in, 367
    mechanical, 340–348, 351
    products requiring, 338t
Cell matrix permeabilization,
    366, 367
Cell membranes, 337–339
    disruption of, 345
    fragments of, 367
Cell paste, freeze-thawing of,
    340–341
Cell receptor-ligand
    interactions, 119–120
Cell walls, 338
    bacterial, 338–339
    fungal, 339
    yeast, 339
Cells
    bioproducts of, 337
    envelopes of, 339f
    immobilization of, 365, 366f
    magnetic particles in
        separation of, 312–313
    plant, 338
    recycling of, 365, 366f
    structures of, 337–339
    surfaces of in ion-exchange,
        15
Cell-surface receptors, classes
    of, 119, 120f
Cellulose, 184t, 338
Cellulose acetate membranes,
    51
Centrifugal contactors, 250–251
Centrifugal extractor, 273
Centrifugation, 23, 164, 176–177
    advantages and
        disadvantages of, 166
    centrifuge selection in,
        166–168
    centrifuge types for, 168–173
    conditions in, 175t
    governing principles of,
        164–166
    industrial-scale, 173–176
    performance data for, 176t
    techniques of, 168–169
    varying suspensions in, 174t
    volumetric flow rate in,
        173–174
Centrifuge separation, 14
Centrifuges
    batch-basket, 172–173
    bench-scale, 175

disc-type, 170–172
performance considerations
of, 166–167
sedimenting, 173t
selection of, 166–168
test-tube, 176
tubular bowl, 169–170,
176–177
types of, 168–169
Cephalosporin, 57t
Chaotropic buffers, 211
Charcoal absorbents, 53–54
Chemical equilibria, 67–69
Chemical potential, 230
Chemical precipitation
techniques, 70
Chemical stability, 372–373
Chemical thermodynamics of
partitioning, 230–231
Chemiluminescence detector,
180t
Chitin, 339
Chloroform-acetone system,
319
Chorismic acid, dissociation
constant for, 75t
Chromatofocusing, 210
Chromatography, 23, 223–228
adsorption, 217
in bioseparation, 181t
of fatty acids, 53–54
affinity, 179–180, 182,
213–215, 369–370
adsorption columns in, 200
binding-elution cycles in,
204
in bioseparation, 181t
examples of, 214–215
immobilized soft-metal,
87–88
in magnetic particle
separation, 310–311
of oligosaccharides, 49, 51
pH manipulation in, 218
for amino acid analysis, 29
analytic methods for, 187–206
anion-exchange
high-performance ion-
exchange, of
oligosaccharides, 49
in human insulin
production, 16
in monosaccharide
analysis, 47
bioaffinity, 15–16
detection methods in,
178–181
displacement, 218
elution
of fatty acids, 52–53

solid phases used for, 53t
gas
of amino acids, 35
of fatty acids, 51–52
gel permeation, 182, 206
in bioseparation, 181t
examples of, 206–208
in human insulin
production, 16
of oligosaccharides, 49
of peptides and proteins,
35–36
scale-up, 222–223
gradient, 217–218
high-performance
liquid, 59
in amino acid analysis,
31
of fatty acids, 52
of lipids, 56
thin-layer, 56
hydrophobic interaction, 15,
182, 216
in bioseparation, 181t
of peptides and proteins,
35–37
hydroxyapatite, 183
immobilized metal ion
affinity, 89–90
ion-exchange, 59, 182,
208–209, 208–213,
227–228, 369–370
adsorption columns in, 200
binding-elution cycles in,
204
in bioseparation, 181t
charged components in, 15
examples of, 209–213
of fatty acids, 54
for nucleic and polynucleic
acids, 41–42
of oligosaccharides, 49
of peptides and proteins,
35–37
ion-exclusion, 92, 182
in bioseparation, 181t
of fatty acids, 54–55
ion-pair, for nucleic and
polynucleic acids, 43
key variables for, 205–206
mass spectrometry-gas liquid,
51
membrane, 218
method decision tree for,
223f
micellar electrokinetic, 33–34
molecular-sieve, 376
moving bed, 225
of peptides and proteins,
35–37

perfusion, 181t, 182–183,
216–217
radial-flow, 218, 219f
response peaks in, 226f
reverse-phase, 15, 59, 182,
216
in bioseparation, 181t
of fatty acids, 55
of steroids and antibiotics,
56–57
reverse-phase high-
performance liquid
in amino acid analysis, 32
in amino acid labeling, 32
in human insulin
production, 16
for nucleic and polynucleic
acids, 42–43
of peptides and proteins,
35–37
scale-up strategies and
considerations in,
219–223
size exclusion, 14–15, 16
slalom, 43–44
stationary phases in, 183–186
system for, 179f
thin-layer
of lipids, 56
of oligosaccharides, 49
two-dimensional, 227
types of, 181–183
uses of, 178
of vitamins, 57–58
Chromogenic substrates, 39
Chymotrisinogen, 83t
Citric acid
commercialized production
of, 369
downstream processing of,
23f
ion-exclusion separation of,
212–213
partition coefficients for, 232t
production of, 22–23
recovery of, cost analysis for,
397–398
retention time of, 91t
Clean-room air handlers, 368
Clean-water production
facilities, 368
Coacervation, 376
Cocurrent contacting system,
255–263
Collapse temperature, 379
Colloidal forces, 104
Colloidal particles
dispersion force equations
for various geometries
of, 99t

velocity and electric field of, 284–285

Colloidal sphere size, 239–240

Colloid-colloid surface energy, 100

Colloids. *See also* Biocolloids
  definition of, 95
  smart responses of, 95

Complementarity-determining regions (CDRs), 116–117

Complexation chemistries, 234–236

Complexation extraction, 272–273, 370

Concentration polarization, 153–154, 162

Concentration-driven membrane processes, 147, 158–160

Concentrations, methods of measuring, 29–61

Conductometric detectors, 53

Constant pattern equilibrium, 200–204
  in analysis of chromatographic processes, 200–204

Constant patterns
  analysis of, 217
  development of, 202f

Contacting modes
  comparison of, 255–263
  concurrent, 252–253
  countercurrent, 254–255
  crosscurrent, 253–254

Continuous countercurrent extraction, 269–270

Continuous-contact extractors, 247–250

Continuous-flow electrophoresis, two-dimensional, 292, 293f

Continuum model, 197f

Controlled release drugs, 375
  encapsulated, 376

Convection
  accounting for, 5–7
  in electrophoresis, 286–287
  flux due to, 6
  in solvent evaporation and drying, 315

Copper ions, amino acid affinity of, 89–90

Cortisone, 58f

Cost analysis software, 398

Cost estimation
  of operations, 388–389
  resources for, 385–388

Coulombic attractive/repulsive

forces, 108–109

Countercurrent contacting, 270
  comparison with other contacting modes, 255–263
  double, 267–269
  systems of, 254–255

Countercurrent extraction, continuous, 269–270

Craig extraction, 268f

Craig extractor, 265–266, 267f
  double, 272f
  three-phase system in, 272

Crosscurrent contacting systems, 253–254
  comparison with other contacting modes, 255–263

Cross-flow contractor, 275

Cross-linked polysaccharides, 184t

Crystallization, 13, 15, 23, 143–145
  batch, 131
  crystal size distribution in, 134–136
  growth rate dispersion in, 137–139
  organic solvent and salt precipitation in, 136–137
  solid-phase balance in, 132–134
  solution balance in, 132
  characteristic zones exhibiting, 128f
  continuous, 140, 144, 145
  cooling-induced, 132
  definition of, 127
  in freeze drying, 382
  industrial, 129–130
  lab-scale, 129–130
  in liquid-solid systems, 323
  of l-lysine, 143–144
  nucleation phenomena and, 128–130
  in organic solvent extraction, 231–232, 237
  removal of solvent and diluent in, 141–142
  in resolving components, 361
  in solvent extraction, 275, 315–316
  supersaturation and, 127–128
  yield of, 141–142

Crystallizer
  continuous, 140
  hydrodynamic environment in, 130

liquid volume in, 132
  seeding of, 128–129, 134

Crystals
  growth of, 130–131
  growth rates of, 133, 142
    distribution of, 137–139
  size distribution of, 134–136
  size of, 132–133

Cycles, number of, 395t

Cysteine
  dissociation constant of, 77t
  divalent metal ion stability constants with, 90t
  electrodialysis of, 34
  hydrophobicity of, 84t
  hydrophobicity scale for, 85t
  isoelectric point of, 83t
  negative charge of, 80
  net charge of, 81f
  temperature and solubility of, 72f

Cytochrome C, 83t

Cytoplasm, 337

Cytosine, separation of, 43

Cytoskeleton, 338

CZE. *See* Capillary zone electrophoresis

Darcy's law, 147–148

de Laval, Gustaf, 164

Debye-Hückel approximation, 100–101

Debye-Hückel theory, 242

Dehydroshikimic acid
  Craig extraction of, 268f
  double-countercurrent extraction of, 269

Denaturing gel systems, 44

Density, 167

Deoxycytidine-terminated DNA fragments, 46

Deoxyribonucleotides, 117

Desorption, 381–382

Desthiobiotin, 57–58, 59f

Detection methods
  in chromatography, 178–181
  objectives of, 178

Detergent enzymes
  solubilization with, 349
  in surfactant solution, 367

Dextran, 184t
  in magnetic two-phase separation, 311–312

Dextran/acrylamide, 184t

Dialysis, 161. *See also* Electrodialysis
  in liposome entrapment, 376

Dialysis membranes, 146

Diamagnetic materials, 301–302
  magnetization curves of, 303f

Diamagnetism, 301–302
Diamond-Hsu expression,
    241–242
Dicarboxylic acids, species
    concentrations in,
    75–76
Differential pelleting, 168
Differential-scanning
    calorimetry, 35
Diffusion
    accounting for, 5–7
    rate of, 6
Diffusivity, 73–74
    Stokes-Einstein equation for,
    73
Diluent
    addition of in crystallization,
        141–142
    in organic-aqueous
        extraction, 233–236
    removal of in crystallization,
        141
Dipole-dipole interactions,
    95–96
Disc-stack centrifuges, 387t
Disc-type centrifuge, 169f,
    170–172, 173t
Dispersion and linear rate
    model (DLRM), 197,
    198–200, 203, 222, 224,
    228
    in analysis of
        chromatographic
        processes, 198–200
Dispersion forces, 96–100
    medium effects on, 99–100
Displacement, aqueous-phase
    reaction and, 238
Displacement chromatography,
    218
Dissociation constants, 124f
    of amino acids, 77t
    of receptor-ligand affinity
        interactions, 112–113
    temperature and, 76
    of typical bioaffinity
        interactions, 113t
Dissolution, lipid, 349–350
Distillation, 315
    batch, 325–327
    Rayleigh, 325–327
    temperatures in, 327
DLVO theory, 100–102
DNA
    binding of, 183
    double helix forms of, 118
    fragments of
        pulsed-field gel
            electrophoresis
            separation of, 45

single-stranded, 44
size-dependent separation
    of, 43–44
nucleotides of, 117
rapid recovery of, 92–93
separation of
    centrifugation in, 164
    by gel electrophoresis,
        278–279
    in two-phase aqueous
        system, 273
DNA domains, in binding
    proteins, 118
DNA-protein interactions,
    117–119
Donor numbers (DN), 84–85
Donor-acceptor theory, 84–86
Dosage forms, 375
Double helix forms, 118
Double-countercurrent
    extractor, 267–269
Double-layer forces, 105
Double-layer repulsion, 102
Drago E&C equation, 86
Dryers
    drum, 333, 334f, 335
    flash, 332
    fluidized bed, 331, 332f
    freeze, 333–334
    horizontal fluidized bed,
        334–335
    spray, 333
Drying, 23, 315. See also
    Evaporation; Freeze
    drying; Vaporization
    batch, 324
    continuous, 324
    equipment for, 331–334, 381
    in freeze drying process, 316,
        333–334, 381
    in liquid-solid systems,
        323–325
    spray, 376
Dyno mill, 347, 351

ECD. See Electron capture
    detection
Economic decision-making
    models, 389–394, 399
Economics, bioprocess, 385–399
Egg albumin, 83t
Egg lysozyme, 350
Ejector disc centrifuge, 173t
Electrical fields, 284–285
Electrically driven membrane
    processes, 147,
    158–159
Electrochemical detector, 180t
Electrode systems,
    discontinuous, 45

Electrodialysis, 14, 158–159. See
    also Dialysis
    of amino acids, 34
Electrolytic conductivity
    detector, 180t
Electromigration process,
    158–159
Electron capture detection,
    amino acid, 35
Electron correlation theory, 87
Electron impact mass
    spectrometry, 48–49
Electron spin motion, 299
Electrons
    spin-paired, 300–301
    susceptibility to alignment of,
        301
    unpaired, 300–301
Electro-osmosis, 286
    in electrophoresis, 287
Electrophoresis, 179, 277,
    294–298. See also Gel
    electrophoresis
    barrier, 287–288
    capillary, 32–33, 45–46, 49–50,
        55–56, 281–282
    charged components in, 15
    concepts of, 283–287
    free-flow, 277, 287–288
    gel, 278–281
    high-performance capillary,
        46
    isoelectric focusing, 282,
        290–291
    isotachophoresis, 283,
        291–292
    moving boundary, 283, 284f
    of peptides and proteins,
        37–39
    popular methods of, 277–283
    scaling up, 296–297
    two-dimensional, 292–294
    zone, 280–281, 287–290
Electrophoresis gel. See Gels,
    electrophoresis
Electrophoretic mobility, 104,
    284–285
Electrostatic interactions, 105
    DLVO theory and, 100–102
Electrostatic potential, 101f
ELISA. See Enzyme-linked
    immunosorbent assays
Elution chromatography. See
    Chromatography,
    elution
Emulsification, 159
Emulsion liquid membranes,
    159–160
Encapsulation, 375–377
Enrichment process,

mathematical theory of, 316–317
Enrichment techniques, 315–316
Enzyme-linked immunosorbent assays
competitive and non-competitive, 40
variations of, 40–41
Enzyme-linked receptors, 119–120
Enzymes
activity of, 39
detergent solubilization of, 349
digestion by, 350
disruption of, 347
as marker, 367
osmotic release of, 349
proteolytic, 367
release of during cell disruption, 351
Enzyme-substrate interactions, 120–121
Equilibrium
in bioseparation processes, 270
linear, 267–268
nonlinear, 267, 270
in receptor-ligand complex formation, 112–115
solute equations, 191–192
Equilibrium staging
configurations in, 252–270
graphical solution of, 263–265
Equilibrium-stage modeling, 252
Equipment
critical specifications for, 387t
size and cost of, 386–387
Ergun equation, modified, 219–220
Erythromycin
analysis of, 56
partition coefficients for, 232t
Escherichia coli,
homogenization of, 340
Ethanol
in lipid dissolution, 349–350
partition coefficients for, 232t
Ethyl acetate, 350
Evaporation, 23, 315, 316. *See also* Drying; Vaporization
calculations for, 334–336
in distillation, 326–327
equipment for, 327–331
in liquid-solid systems, 323–325

Evaporative light scattering detector, 180t
Evaporators, 327–328
agitated-film, 330–331, 335–336
classification of, 328
falling-film, 329–330
flash, 328–329
long-tube vertical, 329–330
rising-film, 329–330
short-tube vertical, 329
Excipients, freeze dried, 382
Expert systems, in process synthesis, 364, 365f
Extracellular receptors, 119–120
Extractant/diluent systems
extractant concentration in, 235
in organic-aqueous extraction, 233–236
Extraction, 15, 23. *See also* Fractional extraction
large-scale vessels for, 246–251
methods of, 230–275
organic-aqueous, 231–238
reverse micelle, 243–244
solvent, 316
stage-wise contacting configurations in, 252–270
supercritical fluid, 244–246
Extraction columns, 247–250
Extraction factor effects
on three-phase cross-counter contactors, 262f
on two-dimensional cross-flow cascade, 259f
Extraction towers, 247–249

Fats, chromatographic analysis of, 56
Fatty acids, 51–56
Ferguson plots, 295f
Fermentation
bacterial, 370
centrifugation in, 164
yeast recovery from, 176
Fermentation industry, vapor-liquid systems in, 317
Ferrimagnetic materials, 302–303
characterization of, 303–304
magnetization curves of, 303f
Ferrofluids, 312–313
Ferromagnets, 301
FIA. *See* Flow injection analysis
Fiber module filter, 150
Fick's law, 6

Filter cake, 148, 150
compressible, 149
incompressible, 148–149
Filtration, 14, 23
batch, 148
membrane, 146–163
Filtration tests, 161
Flame ionization detector, 48
Flash evaporation, 315
Flocculation, 15, 23
Fluorescamine, 39
Fluorescence detector, 180t
Formic acid
dissociation constant for, 75t
ion-exclusion separation of, 212–213
retention time of, 91t
Formulations, 12–13, 373–383
basic fibroblast growth factor in, 383
characteristics of, 372–373
definition of, 372
dosage forms of, 375
encapsulated, 375–377
freeze drying of, 377–383
Fourier transform infrared detector, 180t
Fractional extraction, 262–263, 265
with Craig extractor, 265–266, 267f
with double-countercurrent extractor, 267–269
Freeze drying, 316, 333–334
combined with packaging, 374
continuous, 382
example of, 382–383
in food and pharmaceutical industries, 377–378
technique of, 381–382
theory of, 378–381
Freeze-thawing, 337, 340–341
French press, 347–348
FSOT. *See* Fused-silica, open tubular capillary columns
Fumaric acid
capillary zone electrophoresis of, 282
dissociation constant for, 75t
ion-exclusion separation of, 212–213
retention time of, 91t, 213f
Fungal cell walls, 339
Fused-silica, open tubular capillary columns, 35
Fusion technology, 366, 367

G protein. *See* GTP-binding
    protein
β-Galactosidase
    cost analysis of producing,
        398
    net present worth for, 392t
    process design cash flow, 389,
        390t
Gas chromatography. *See*
        Chromatography, gas
Gaulin-Manton homogenizer,
        341
    arrangements of, 342f
Gaussian distribution
    equation for, 192–193
    in zone electrophoresis,
        288–290
Gaussian peak, 288–290
Gaussian probability
        distribution equation,
        187–188
Gaussian solution, 187–190
Gel electrophoresis, 278
    capillary, 294–295
    disc, 280–281, 283
    example of, 39
    media for, 294
    for nucleic and polynucleic
        acids, 44
    polyacrylamide, 37–39,
        280–281
    pulsed-field, 278–280
        for nucleic and polynucleic
            acids, 45
    slab versus capillary, 296f
    stacking, 280f, 283
Gel partitioning model,
        205–206
    in analysis of
        chromatographic
        processes, 205–206
Gel permeation
        chromatography. *See*
        Chromatography, gel
        permeation
Gelatin, 374
Gels
    electrophoresis, 37–38, 278f
    polyacrylamide, 46, 51, 146
    protease detection in, 38–39
    structures of, 184–185, 278f
Glass formation, amorphous,
        382
Glass transition temperature,
        378–379
gamma-Globulin, isoelectric
        point of, 83t
Glucan, 339, 367
Glucanase, 367
Glucose-containing

oligosaccharides, 51
Glutamic acid
    dissociation constant of, 77t
    divalent metal ion stability
        constants with, 90t
    hydrophobicity of, 84t
    hydrophobicity scale for, 85t
    isoelectric point of, 83t
    net charge of, 81f
    temperature and solubility of,
        71f
Glutamine
    dissociation constant of, 77t
    hydrophobicity of, 84t
    hydrophobicity scale for, 85t
    isoelectric point of, 83t
Glutathione, 186t
Glycine
    dissociation constant of, 77t
    divalent metal ion stability
        constants with, 90t
    hydrophobicity of, 84t
    hydrophobicity scale for, 85t
    isoelectric point of, 83t
    temperature and solubility of,
        72f
Glycoprotein analysis, 51
Glyoxylic acid, 75t
Gradient chromatography,
        217–218
Gram-positive cells, 344
Graphical solutions, 263–265
Gravitational sedimentation
        process modeling, 164
Gravity sedimentation test, 168
Gravity-operated extractors,
        247–249
Grinding, 337, 345–347
Growth media
    compositions of, 366
    costs of, 366–367
GTP-binding protein, 120
GTP-binding protein-linked
        receptors, 119–120
Guanidine, metal ion binding
        in, 89
Guanine, 43
Guanine-cytosine pairs, 118
Guoy-Chapman theory, 101
Gutmann donor-acceptor
        theory, 84–86
Gyro tester, 175

Hamaker constant, 99, 100
Hard-soft acid-base (HSAB)
        theory, 86–88,236, 84
Hartounian-Sandler theory,
        242–243
Helmholtz-Smoluchowski
        equation, 286, 407

Hemoglobin, isoelectric point
        of, 83t
Henderson-Hasselbalch
        equation, 75
Heterogeneous nucleation, 128
H.E.T.P., 205
Hexosamine-containing
        oligosaccharides, 50
High-performance
        chromatography. *See*
        Chromatography,
        high-performance
Histidine
    dissociation constant of, 77t
    divalent metal ion stability
        constants with, 90t
    hydrophobicity of, 84t
    hydrophobicity scale for, 85t
    isoelectric point of, 83t
    net charge of, 82f
    stability constant of, 89–90
    sulfur group of, 88
    temperature and solubility of,
        72f
Hofmeister series, 70, 71t
Homogeneous nucleation, 128,
        129
Homogenization, 337, 341–343
    of *Escherichia coli* cells, 340
Homogenizers, 352
    versus bead mills, 346
    Gualin-type, 340
Human growth hormone
    freeze drying of, 382
    recovery and purification for,
        26f
Hybrid myeloma cells, 16
Hydrodynamic cavitation
        device, 345
Hydrogen bonding, 15, 67
    in organic-aqueous
        extraction, 234–235
Hydrogen bonds
    energy of, 108
    formation of, 110–111
    purine-to-pyrimidine base
        pairs, 117–118
    between water molecules, 96
Hydrogenated vegetable oils,
        374
Hydrolytic enzymes, 349
Hydrophobic effects, 67, 96,
        102, 103f, 104
Hydrophobic interactions, 111.
    See also
        Chromatography,
        hydrophobic
        interaction
    binding energy of, 108
Hydrophobicity, 15

grouping amino acids by, 84t
magnetic material and, 306
Hydrophobicity scales, 85t
Hydrophobicity-hydrophilicity
    scales, 84

IMAC. *See* Chromatography,
    immobilized metal ion
    affinity
Imminodiacetate-Cu(II), 186t
Immobilines, 296
Immobilization, 365, 366f
Immunoassay, 40–41
    for blood hormone, 123
    magnetic particles in, 313
    methods of, 59
    of peptides and proteins,
        40–41
Immunoglobulin domains, 116f
Immunoglobulin G
    affinity chromatography of,
        215t
    characterization of, 116
    diffusivity of, 74
    ion-exchange
        chromatography of,
        228
    isolation in monoclonal
        antibody production,
        16
Inclusion bodies, 367
    human insulin in, 16
Industrial bioseparation
        processes, 11, 23–24
    for citric acid, 22–23
    environmental impact of,
        368
    for human insulin, 17
    for L-lysine, 21–22
    for monoclonal antibodies,
        16–17
    for penicillin, 19–20
    problems in, 24–28
    for protease, 20–21
    for rabies vaccine, 17–19
    selection of, 11–16
    steps in, 12
Infrared spectrometry, 46–47
Infrared spectroscopy, 49
Insulin
    production of, 16
    recovery and purification of,
        18f
    economics of, 395–397
Interferon, recovery of, 359–360
Internal rate of return, 391
Intracellular component
        separation, 164
Ion exchangers, 387t
Ion pairing, 234, 235–236

Ion spacer, 283
Ion-exchange chromatography.
    See Chromatography,
    ion-exchange
Ion-exchange column capacity,
    395–397
Ion-exchange membranes, 15
Ion-exchange packing, 224
Ion-exchange residues, 306
Ion-exclusion chromatography.
    See Chromatography,
    ion-exclusion
Isoelectric focusing, 38, 282,
    290–291
    solute trajectories in, 293f
    trajectories for, 294
Isoelectric points, 74–84
Isoleucine
    dissociation constant of, 77t
    hydrophobicity of, 84t
    hydrophobicity scale for, 85t
    isoelectric point of, 83t
    temperature and solubility of,
        71f
Isomerase, 83t
Isopropyl alcohol, 349–350
Isopycnic centrifugation, 169
Isotachophoresis, 283, 291–292,
    296
    in disc gel electrophoresis,
        280–281
    solute trajectories in, 293f
    trajectories for, 294
Itaconic acid, 54

Joule resistance heating,
    286–287

Kaolin clay, purified, 309
Ketones, 232t
King-Haynes expression, 242
Knudsen equations, 380

β-Lactam, 57t
Lactic acid
    bioseparation process for, 25f
    dissociation constant for, 75t
    retention time of, 91t
β-Lactoglobulin, 83t
Langmuir isotherm, 203, 214
Laplace transforms, 1, 137–138,
    400–403
    inverse, 400
    numerical inversion of,
        405–407
Lectin-carbohydrate
        interactions, 122
Lectins, 51
Leucine
    dissociation constant of, 77t

divalent metal ion stability
        constants with, 90t
    hydrophobicity scale for, 85t
    isoelectric point of, 83t
    temperature and solubility of,
        72f
Leukocyte purification process,
    360f
Lewis acid-base complexation,
    234, 236
Lewis acids
    characteristics of, 84
    classification of, 87t
Lewis bases, 67
    classification of, 87t
LFERs. *See* Linear free-energy
    relationships
L'Hopital's rule, 195, 401
Ligand-receptor affinity
    equilibrium approach to,
        112–115
    theoretical aspects of, 111
    thermodynamic theory of,
        112
Ligand-receptor interactions
    hydrogen bond, 110–111
    hydrophobic, 111
    ionic bond, 110
    specific types of, 116–122
    types of, 110t
    van der Waals, 111
Ligands
    affinity, 306, 310
    binding of, 122–123
    charge redistribution and
        conformational
        rearrangement in,
        109–110
Light absorbance, 179
Linear equilibrium, 196–200
    in analysis of
        chromatographic
        processes, 196–200
Linear free-energy
        relationships, 68–69,
        236
Lipids
    analysis techniques for, 51–56
    cellular, 338
    detergent solubilization of,
        349
    dissolution of, 349–350
Lipophilic substances, 316
Lipopolysaccharides, 350
Liposome entrapment, 376
Liquid chromatography
        detectors, 180t
Liquid-liquid extraction
    process of, 321–323
    two-phase, 321

Liquid-liquid extraction vessels, 246–251
Liquid-solid systems, 323–325
Liquid-vapor systems, binary, 318
L-Lysine
    production of, 21–22
    recovery and purification of, 22f
L-Lysine crystallization, 143–144
Loan, interest on, 391t
Local equilibrium theory (LET), 197–198
    in analysis of chromatographic processes, 197–198
    versus dispersion and linear rate model, 199
London dispersion forces, 87.
    See also van der Waals forces
Long-range forces, 104
Lyophilization. *See* Drying, freeze
Lysine
    bioseparation with cation-exchange product isolation, 361, 362f
    dissociation constant of, 77t
    divalent metal ion stability constants with, 90t
    hydrophobicity of, 84t
    hydrophobicity scale for, 85t
    isoelectric point of, 83t
    net charge of, 82f
    use and structure of, 186t
Lysozyme, 83t
Lytic enzymes, 350

Macrolides, 57t
Macromolecules
    centrifuge separation of, 164
    diffusivity of, 73–74
    partitioning of, molecular-level theories of, 241–243
Macroporous non-ionic polymers, 53
Magnetic bioseparations, 308–309, 313–314
    affinity chromatography in, 310–311
    applications of, 312–313
    aqueous two-phase, 311–312
    defined, 299
    high-gradient, 309
    magnetic particle classification in, 305–306

materials in, 299–305
    theoretical considerations in, 306–308
Magnetic dipoles, 299
Magnetic fields, 299, 300f
    around wire, 310f
    intensity of, 300
Magnetic particle separations, 308–312
Magnetic properties, 299–305
Magnetically stabilized fluidized bed, 310–311
Magnetism, 299
Magnetization, 301
Magnetization curves, 303f
Magnetophoretic mobility, 104
Malic acid
    dissociation constant for, 75t
    ion-exclusion separation of, 212–213
    retention time of, 91t
Mammalian cell separation, 164
Manton-Gaulin homogenizer, 352
Margules equation, 69
Mass conservation, 1–3
    differential manifestation of, 5–6
    diffusion, convection, and reaction in, 5–7
Mass spectrometry
    electrospray, 180t
    fast atom bombardment, 180t
    particle beam, 180t
    of single monosaccharides, 46–47
Materials
    magnetic properties of, 299–305
    selection of, 388
McCabe-Thiele technique, 263–264, 274
Mechanically agitated extractors, 249–250
Medium, 99–100
MEKC. *See* Micellar electrokinetic chromatography
Membrane. *See also* Cell membranes
    emulsion liquid, 159–160
    materials of, 146–147
    microfiltration, 147
    permeability of, 149–150
    porosity of, 146–147
    solvent flow across, 158
    ultrafiltration, 147
Membrane batch modules, 162f
Membrane filtration, 146, 160–163

driving forces in, 147
electrodialysis, 158–159
emulsion liquid membranes in, 159–160
flux equations in, 158
general theory of, 147–149
membrane materials in, 146–147
micro, 149–152
reverse osmosis, 156–158
ultra, 152–156
Membrane separation processes
    classification of, 147t
    concentration-driven, 147, 158–160
    driving forces in, 147
    electrically driven, 147, 158–159
    pressure-driven, 147–158
Metal ions
    binding constants of, 88–90
    binding of, 110
    chelation of, 350
    nitrogen affinity of, 89
    oxygen affinity of, 89
Methionine
    dissociation constant of, 77t
    divalent metal ion stability constants with, 90t
    hydrophobicity of, 84t
    hydrophobicity scale for, 85t
    isoelectric point of, 83t
    temperature and solubility of, 71f
Methyl cellulose, 374
Methyl ethyl ketone, 232t
5-Methylcytosine, 43
Microbial cell disruption, 339–340
Microbial glucanases, 350
Microcapsules, 377
Microfiltration, 14, 147, 149–150
    batch, 161
    general theory of, 147–149
    staging in, 151–152
Microgravity conditions, 287
Microspheres, 377
    basic fibroblast growth factor in, 383
Milbemycin
    chemical backbone structure of, 358, 359f
    extraction sequence for, 358, 359f
Mixer-settler extractors, 246–247, 251
    critical specifications for, 387t
Molasses fermentation liquor, 54

Molecular chain-cylinder
  interaction, 106f
Molecular recognition
  processes, 108–110
Monocarboxylic acid, 233
Monoclonal antibodies,
  production of, 16–17
Monosaccharides
  binding of, 122
  gas-liquid chromatography
    of, 48
  gas-liquid chromatography-
    mass spectrometry of,
    48–49
  high-performance ion-
    exchange
    chromatography of,
    47–48
  quantitative analysis of,
    46–47
  structure of, 46–47
M.S.E. Mullard Ultrasonic
  disintegrator, 343–344
Multiple bowl centrifuge, 173t
Multiple emulsions, 377
Myoglobin, 83t

Net charge, 284–285
Net present value, 391–392
Newtonian continuum
  mechanics, 196–200
  in analysis of
    chromatographic
    processes, 196–200
Nozzle disc-type centrifuge,
  173t
Nucleation
  phenomena of, 128–130
  rate of, 129–130
Nucleic acids
  analysis of, 41–46
  cellular, 337
  electrophoresis of, 44
  interactions of, 183
  in ion-exchange, 15
  metal ion binding to, 89
Nucleosides, 41–42
Nucleotides
  coenzymes of, 120–121
  ion-exchange
    chromatographic
    separation techniques
    for, 41–42
  ion-pair chromatography of,
    43

Oils, analysis of, 56
Oleic acid, 239f
Oligonucleotides,
  analysis of, 41–46

release of, 14
Oligosaccharides
  analysis of, 49
  capillary electrophoresis of,
    50
  fractionation of, 49
  high-performance ion-
    exchange
    chromatography of, 49
  high-performance liquid
    affinity
    chromatography of,
    50–51
On-off cycling, 222–223
Operating cost estimation,
  388–389
Osmosis, reverse, 14, 147,
  156–158, 160–161
Osmotic pressure, 156–157
Osmotic shock, 348–349
Overlapping resolution
  mapping, 33–34
Oxalic acid
  dissociation constant for, 75t
  ion-exclusion separation of,
    212–213
  retention time of, 91t
Oxaloacetic acid, 75t

Packed towers, 248–249
PAGE. *See* Polyacrylamide gel
  electrophoresis
Paramagnetic materials, 302,
  304
  magnetization curves of, 303f
Paramagnets, 301
Particles
  density of, 165
  dispersion forces and
    geometry of, 98–99
  rotational velocity of,
    165–166
  settling velocity of, 164–165
  shape of, 284
  size of
    in centrifugation, 167
    in electrophoresis, 284
Partition coefficients, 230
  for biochemicals using pure
    solvents, 232t
  colloidal sphere size and,
    239–240
  effect on three-phase cross-
    counter contactors,
    263f
  in liquid-liquid extraction,
    321–322
  measurement of, 275
  for monocarboxylic acid, 233
  pH and, 235–236

protein charge and, 240
  on two-dimensional cross-
    flow cascade, 260f
Partitioning
  chemical thermodynamics of,
    230–231
  due to size, 239–240
  factors affecting, 241–243
  protein charge effect on, 240
Payback period, 391
Peclet number, 222
Pelleting centrifugation, 168
Penicillin
  production of, 19–20
  recovery and purification of,
    20f
  reverse-phase liquid
    chromatography of,
    57t
Penicillin F, partition
  coefficients for, 232t
Penicillin K, extraction of,
  273–274
Pentoses, 47
Pepsin, 83t
Peptides, 35–41
Peptidoglycans, 338–339
Perforated bowl-basket
  centrifuge, 173t
Perforated-plate extractor, 250f
Perforated-plate towers, 249,
  250f
Perfusion chromatography,
  182–183, 183f, 216–217
  in bioseparation, 181t
Perrin factors, 73–74
PFGE. *See* Gel electrophoresis,
  pulsed-field
pH gradient
  focusing effect of, 290–291
  in isoelectric focusing, 282
Pharmaceuticals, controlled
  release of, 373
Phenyl, 186t
Phenylalanine
  dissociation constant of, 77t
  divalent metal ion stability
    constants with, 90t
  hydrophobicity of, 84t
  hydrophobicity scale for, 85t
  isoelectric point of, 83t
  temperature and solubility of,
    71f
o-Phthaldehyde, 39
Plate-and-frame filter, 150
Plug flow crystallizer, 145
Podbielniak extractor, 19, 273
Poisson-Boltzman equation, 101
Polyacrylamide, 184t
  molecular structures of, 278f

Polyacrylamide gel
    electrophoresis, 37–38,
    280–281
  in alkaline protease
    detection, 39
Polyacrylamide gels, 46
  in glycoprotein analysis, 51
Polyacrylonitrile materials, 146
Polyamino acids, 367
Polycytidines, 45–46
Polyethylene glycol
  as additive, 367
  in magnetic two-phase
    separation, 311–312
Polymer matrix, 377
  biocompatibility of, 184
  commercially available, 184t
  composition of, 184–185
Polymethacrylate, 184t
Porosity, stationary-phase,
    184–185
Potentiometric detectors, 53
Precipitation, 13, 15, 23
  definition of, 127
  of organic solvent and salt in
    batch crystallization,
    136–137
  in resolving components, 361
  supersaturation and, 127–128
Pressure shearing, 347–348
Pressure-driven membrane
    processes, 147–158
Process design software, 385,
    387
Process time, 395t
Product formulation, 12, 13,
    372, 373–383
  characteristics of, 372–373
Profitability, calculation of,
    394–395
Proline
  dissociation constant of, 77t
  divalent metal ion stability
    constants with, 90t
  electrodialysis of, 34
  hydrophobicity of, 84t
  hydrophobicity scale for, 85t
  isoelectric point of, 83t
Protease
  detection in gels, 38–39
  downstream processing of,
    21f
  in enzyme digestion, 350
  production of, 20–21
  SDS-PAGE detection of, 39
Protein
  analysis techniques for, 35–41
  binding of, 118
  cellular, 337

charge of, 82–84, 240
charge structure of, 284, 285f
conformational changes in,
    95
diffusivity of, 73–74
distribution coefficients for,
    322t
generic separation processing
    of, 370f
interactions of, 183
in ion-exchange, 15
ion-exchange
    chromatography of,
    208–216
isoelectric focusing of, 282
isoelectric points of, 82–84, 92
net charge of, 284
precipitation equations of, 70
processing of, 369–370
recovery of, 359–360
reverse micelles extraction
    of, 274–275
secretion of, 366
solubility of, 70–72
titration curves for, 295f
Protein A
  affinity chromatography of
    IgG using, 215t
  in IgG isolation, 16
  use and structure of, 186t
Protein G
  in IgG isolation, 16
  use and structure of, 186t
Pulsed amperometric detection,
    48
Pulsed extractors, 250
Purines bases, 43
Purine-to-pyrimidine base
    pairs, 117–118
Purity, binary system, 261
Pyrimidine bases, 43
Pyruvic acid, 75t

Quaternary amines, 186t
Quinic acid
  back-extraction of, 239f
  dissociation constant for, 75t
  retention time of, 91t

Rabies vaccine
  downstream processes for,
    19f
  production of, 16–19
Radial-flow chromatography,
    218, 219f, 225–226
Radioimmunoassays, 40, 41
Raoult's law, 379
Rate-zonal sedimentation, 169
Rayleigh distillation, 325–327

Rayleigh evaporator, 326
Rayleigh number, 287
Receptor-ligand complexes
  formation of, 115
  interactions of, 116–122
  selective affinities of, 123–124
  steps in formation of, 112
Receptor-ligand interactions,
    109. *See* Ligand-
    receptor affinity;
    Ligand-receptor
    interactions
Refractive index detector, 180t
Relaxation effect, 286
Relaxin, 373
Resolution
  in chromatography scale-up,
    219
  in DLRM model, 222
Retardation forces, 286
Return on investment (ROI),
    392–394
Reverse micelle extraction,
    protein, 274–275
Reverse micelles
  formation of, 243
  size of, 243
  in supercritical fluids, 246
Reverse osmosis. *See* Osmosis,
    reverse
Reverse-osmosis membrane,
    361
Reverse-phase
    chromatography. *See*
    Chromatography,
    reverse-phase
RIA. *See* Radioimmunoassays
Rotating disc extractors,
    249–250, 251f
Roxithromycin, 57t

Salt
  concentration in protein
    partition behavior,
    242–243
  effect of type and
    concentration on
    reverse micelles,
    243–244
  minimizing addition of, 366
  precipitation of, 136–137
Saturation, 127–128
Saturation equilibrium, 200–204
  in analysis of
    chromatographic
    processes, 200–204
Scale-up strategies, 219–223
Scatchard-Hildebrand equation,
    70

Scatchard plots, 123f
SDS, removal of, 38
SDS-PAGE, 281, 294
    in alkaline protease
        detection, 39
Secondary nucleation, 128–129
Sedimentation centrifuges,
        173t
    volumetric flow rate in,
        173–174
Seed crystals, 128–129
Sensitivity analyses, 394–398
Separation factors, 316–317
    in liquid-liquid extraction,
        321
    in liquid-vapor system, 318,
        319
Separation processes. *See also*
        Bioseparation
        processes; *specific*
        processes
    integrating steps of, 357–371
    size-based, 14–15
    techniques of, 14–16
Separation technologies,
        164–177
Serine
    dissociation constant of, 77t
    divalent metal ion stability
        constants with, 90t
    hydrophobicity of, 84t
    hydrophobicity scale for, 85t
    isoelectric point of, 83t
    temperature and solubility of,
        72f
Serum albumin, 83t
Settling velocity, 164–165
Sharp zones, 283
Shearing devices, 347–348
Shikimic acid
    analysis of, 55
    Craig extraction of, 268f
    dissociation constant for, 75t
    double-countercurrent
        extraction of, 269
    partition coefficients for, 232t
    retention time of, 91t
    temperature and extraction
        of, 237
Short-range forces, 104
Silanol groups, 55
Silica, 184t, 374
    absorbents, 53
    fused, 55
Single bowl centrifuge, 173t
Sitaraman's equation, 73
Size-exclusion chromatography.
        See Chromatography,
        gel permeation;

Chromatography, size
        exclusion
Sodium alignate, 374
Sodium dodecyl sulfate, 281
Soft ice technique, 382
Solid-liquid interfacial tension,
        144
Solid-phase balance, 132–134
Solubility, 69–70
    as function of pH, 143f
    as function of temperature,
        143f
    of protein and amino acid,
        70–72
Solvatochromic comparison
        method, 86
Solvent flux equations, 158
Solvents
    displacement of molecules of,
        109
    extraction of, 316
        alternative, 370f
    no loss of in crystallization,
        141–142
    organic
        in extraction procedures,
            231–238
        precipitation of, 136–137
    partial loss of in
        crystallization, 142
    removal of, 334–336
        in crystallization, 141
    equipment for, 327–334
    methods for, 315–316
    by Rayleigh distillation,
        325–327
    theory of, 316–325
Sonication, 337, 343–345
Spiral-wound membrane filter,
        150, 151f
Spray drying, 376
Spray towers, 248
Stability, 372–373
    preservatives for, 374
Stability constants
    for amino acids with divalent
        metal ions, 90t
    metal ion complex, 88–89
    parameters in, 89
Stabilizers, 383
Staged models
    in analysis of
        chromatographic
        processes, 190–196
    results of, 194t
Stage-wise contacting
        configurations,
        252–270
Stationary phases, 183

chemical functional groups
        on, 185, 186t
    components of, 184
    particle size in, 220–223
    porosity, 184–185
Steam
    condensation of, 324–325
    sterilization, 167–168
Stearic acid, 374
Stehfest's formula, 405–406
    compared to inverse Laplace
        transform, 407t
Steroids
    analysis techniques for, 56–
        57
    chemical structures of, 58f
Stokes-Einstein equation, 73,
        207
Streptavidin
    interactions of, 121–122
    use and structure of, 186t
Streptomyces avidinii, 121
Sublimation, 381
Succinic acid
    dissociation constant for,
        75t
    ion-exclusion separation of,
        212–213
    partition coefficients for,
        232t
Sulfur group-soft metal
        interactions, 87–88
Supercritical fluids, 244–246
Superparamagnetic materials,
        304–305
    in cell separation, 312
    dispersal of, 305–306
Superparamagnets, 301
SuperPro Designer, 385
Supersaturation, 127–128
    in batch crystallization, 136,
        137
    degree of, 128
Surface chemistry, 306
Surface energy change, 239–
        240
Surface-active agents, 374
Surfactants
    in emulsion liquid membrane
        process, 159
    reverse micelles and
        concentration of, 244
    in reverse-micelle extraction,
        246

Talc, 374
Tartaric acid
    dissociation constant for, 75t
    retention time of, 91t

TATA box binding domain, 118, 119
Temperature
    in bioseparation processes, 327
    in electrophoresis, 286–287
    in organic-aqueous extraction, 232–233
    swing of in organic-aqueous extraction, 237–238
    unaccomplished, 136
Testosterone, 58f
Tetracycline, 57t
Therapeutic proteins, recovery of, 359–360
Thermodynamic properties, 67–93
Thermodynamics
    of partitioning, 230–231
    of receptor-ligand complex formation, 112
Thermogravimetry, 35
Thiamine triphosphate binding, 121f
Thickeners, 373–374
Thiourea, 88
Threonine
    dissociation constant of, 77t
    hydrophobicity of, 84t
    hydrophobicity scale for, 85t
    isoelectric point of, 83t
Thymine
    metal ion binding in, 89
    separation of, 43
Thymol, 374
Tiselius, Arne, 283
Tocopherols, 374
Toluene, 349–350
Transport properties, 67
Tributyl phosphate, 236
Tricarboxylic acids, 75–76
Tridodecylamine, 237
Triethyl phosphine oxide, 86
Trioctylphosphine oxide, 236
Tryptophan
    dissociation constant of, 77t
    divalent metal ion stability constants with, 90t
    hydrophobicity of, 84t
    hydrophobicity scale for, 85t
    isoelectric point of, 83t
    temperature and solubility of, 72f
Tryptophanyl residues, 121–122
Tubular bowl centrifuge, 169–170, 173t, 176–177
Turbidity, 372
Turnaround time, 395t
Two-dimensional contacting system, 257–260

Two-dimensional cross-flow cascade, 258f
    extraction factor effect on, 259f
    partition coefficient effect on, 260f
Two-dimensional electrophoresis. *See* Electrophoresis, two-dimensional
Two-phase aqueous extraction, 238–239
    effects on partitioning in, 241–243
    partitioning due to size in, 239–240
    protein charge effect on partitioning in, 240
Tyrosine
    dissociation constant of, 77t
    divalent metal ion stability constants with, 90t
    electrodialysis of, 34
    hydrophobicity of, 84t
    hydrophobicity scale for, 85t
    isoelectric point of, 83t
    macro forms and microspecies of, 80t
    negative charge of, 80
    net charge of, 81f
    temperature and solubility of, 72f

Ultracentrifugation, 376
Ultrafiltration, 14, 147, 152–154
    application of, 154–155
    batch, 154–155
    modification of, 156
Ultrafiltration membrane, 162, 163, 387t
Ultrasonic cavitation, 343–345
Ultraviolet absorbance, 29
Uracil
    metal ion binding in, 89
    separation of, 43
Urease, 83t
UV photometric detectors, 52–53

Valence electron redistribution, 109–110
Valine
    dissociation constant of, 77t
    divalent metal ion stability constants with, 90t
    hydrophobicity of, 84t
    hydrophobicity scale for, 85t
    isoelectric point of, 83t
    temperature and solubility of, 71f

van Deemter equation, 205, 216–217
    in analysis of chromatographic processes, 205
van der Laan's theorem, 137–138, 193, 195, 406–407
van der Waals attraction, 102
van der Waals forces, 96–99, 111
    versus hydrophobic effects, 102
    medium effects on, 99–100
van der Waals interactions, 104, 105
van Laar equation, 69
Van't Hoff's law, 157–158
Vapor
    enthalpies of, 325t
    pressures of, 317–318
Vaporization, 315. *See also* Drying; Evaporation
    energy-intensive, 358
    in liquid-vapor system, 318–319
Vaporizer, flash, 328–329
Vapor-liquid systems, 317–320
Virus vaccines, 16–19
Vitamin H. *See* Biotin
Vitamins
    analysis techniques for, 57–58
    fat-soluble, 57–58
    water-soluble, 57–58

Waste
    minimization of, 368
    streams, 368
Water
    in bioseparation, 67
    destabilizing effect of, 374
    energy-intensive removal of, 358
    as hard acid, 236
    in organic solvent extraction, 232
    solubility in extract phase, 274
Water molecules
    hydrogen bonds between, 96
    properties of, 67
Water-protein mixtures, 378
Wetted-wall towers, 247–248
Wilke-Change equation, Sitaraman's modification of, 73
Williams-Landel-Ferry equation, 380–381

X-press shearing device, 347

Yeast cells
    disruption of, 340
    lipid dissolution of, 349–350
    ultrasonic cavitation of, 344
    walls of, 339
    composition of, 367

Zeta potential, 286
Zinc finger protein domain,
        118, 119
Zone electrophoresis, 287–290,
        296
    in disc gel electrophoresis,
        280–281

solute trajectories in, 293f
    in two-dimensional
        electrophoresis,
        292
Zwitterionic character, 182
Zwitterions, 208–209